SCIENCE, BEING, & BECOMING

THE SPIRITUAL LIVES OF SCIENTISTS

SCIENCE, BEING, & BECOMING

THE SPIRITUAL LIVES OF SCIENTISTS

PAUL J. MILLS, Ph.D.

Science, Being, & Becoming: The Spiritual Lives of Scientists
Paul J. Mills, Ph.D.

Tradepaper ISBN: 978-1-958921-05-0
Electronic ISBN: 978-1-958921-03-6
Library of Congress Control Number: 2022942596

Published by Light on Light Press
An imprint of Sacred Stories Publishing, Fort Lauderdale, FL

Printed in the United States of America

Advance Praise for

SCIENCE, BEING, & BECOMING

Professor Mills is a shining example of a rigorous scientist and classic mystic, whose mind and heart are dedicated to the pursuit of both objective knowledge and subjective wisdom. This book tells the stories of scientists and scholars who walk the line between science and spirit, weaving them together into a beautiful tapestry that should validate the personal experiences of many more and bring us just a little bit closer to a glimpse of the ultimate and infinitely faceted nature of reality.

~ Cassandra Vieten, Ph.D., Director of Research,
Arthur C. Clarke Center for Human Imagination,
The University of California, San Diego, La Jolla, CA

I found this book riveting. I couldn't put it down until I had devoured the whole thing. This book is a rare and engaging account of the spiritual transformation of the scientific mind, writ large on the one hand but also very much from the inside as well, from the "horse's mouths" of the author and his interviewees and commentators as well. The many individual transformations accounted here span a vast variety and tremendous depth of the spiritual-nonrational realms while maintaining scientific solidity. Together, these point the way to a tremendously expanded and bright potential future for science as it begins to embrace spirituality and consciousness.

~ Thomas Brophy, Ph.D., President,
California Institute for Human Science, Encinitas, CA

Over the course of his distinguished career, Professor Paul Mills has inspired many scientists through his ability to navigate the competitive and rigorous landscape of academic science while at the same time nurturing a deeply spiritual life. One only has to spend a few moments with Paul to sense and benefit from his compassion and wisdom. In this much needed work, Paul provides an intimate and personal view into the spiritual lives of scientists. In the midst of these rapidly changing times, their stories of inspiration, dedication, and struggle can encourage and steady each of us along the path.

~ Thomas Liu, Ph.D.,
Neuroscience and Director of the Center for Functional MRI,
The University of California, San Diego, La Jolla, CA

Professor Mills reminds us in *Science, Being, & Becoming* that some scientists, perhaps many, have had the same kinds of profound psychic, mystical, and spiritual experiences that people have reported throughout history. Long-held taboos in science once prevented serious exploration or even discussion of these realms, but books like this are slowly but surely dissolving those constraints, and in the process whole new realms of reality are poised to emerge. A delightful and inspiring book.

~ Dean Radin Ph.D., Chief Scientist, Institute of Noetic Sciences, Petaluma, CA, and author, *Real Magic*

In the stories of people's lives recorded in this book, virtually all combinations and permutations of human spiritual experience and aspiration are represented. The reader will very likely find many aspects of themselves described in some way in Professor Mills' exploration of the spiritual lives of scientists. In my case, it was very helpful to see how others experienced transcendent phenomena and spiritual teachings of all kinds. The way some scientists quoted in this book understood certain teachings, for example the Advaita Vedanta teachings from Paramahansa Yogananda, helped me understand that philosophy and teaching better and has enriched my own spiritual journey. Ultimately Professor Mills' book is a spiritual guide, a road map depicting many pathways to self-realization (and becoming anchored in the divine) which will inspire one's further spiritual exploration in directions one may not have thought of or encountered yet. This is a page turner. Experience the joy of seeing one's own experiences, shared in some way by others, become even more illuminating and meaningful in the moments of reading a given section and reflecting.

~ Hollis H. King, D.O., Ph.D., Center for Integrative Medicine, The University of California, San Diego, La Jolla, CA

Scientific method, inspiration, consciousness -- all of them work together to bring scientists to the point of discovery and sharing their work with responsible and ethical methods. There are a lot of books about science and scientific inspiration, but this is the only one that delves into how mysticism and consciousness within the scientific community influence the gifts that community can bring to the world. Bravo to Dr. Mills, consummate scientist and mystic, for this mighty contribution!

~ Julia Mossbridge, Ph.D.,
Co-Founder, The Institute for Love and Time

Dr. Paul Mills' collection of life stories of scientists passionate about spirituality is a gem. If scientists are weird, spiritual scientists are even more bizarre, and we get to understand this rare breed better. Science and spirituality are opposite, right? In this collection of stories, Paul shows us that it is not the case. Spirituality and science can and do co-exist harmoniously, and as surprising as it is, science, as a method, may be used to pursue spiritual goals.

~ Arnaud Delorme, Ph.D.,
Swartz Center for Computational Neuroscience,
The University of California, San Diego, La Jolla, CA;
Institute of Noetic Sciences, Petaluma, CA

Professor Mills is a pioneer in the world of science and research. He's courageous and committed to reducing the chasm between science and spirituality. I have the utmost respect and admiration for Dr. Mills as a scientist; his credentials are impeccable. More than that, I'm honored to know him as a colleague and mentor and look forward to his next trail-blazing accomplishments in bridging the science-spirituality gap.

~ Tamara Goldsby, Ph.D.,

University of Louisville, Louisville, Kentucky

Science, Being, & Becoming is about a subject that is very seldom talked about and yet very important: the spiritual journey of scientists. Most first-class scientists such as Albert Einstein, Neils Bohr, and Erwin Schrodinger wrote about the spiritual side of their life and even about the fundamental spiritual nature of the universe. Yet, these writings are mostly forgotten. This book is well written, entertaining, and very informative. I hope you enjoy it as I did!

~ Gaetan Chevalier, Ph.D.,

California Institute of Human Science, Encinitas, CA

Professor Mills brings us back home to the questions that drive our curiosity as human beings and our authentic endeavor to know and understand ourselves, one another, and this grand dance of integrally interdependent life in which we are partaking and co-creating. In only the way one of our eminent scientists could tell the story of the profoundly spiritual journey that has shaped his path and led him on an intrepid journey of the human spirit, the consciousness of humanity, and his scientific achievements, Paul inspires us to envision the boundless possibility of our collective being, this remarkable planet, and our capacity as an integrative whole. This book honors the masters of both science and spirit that have come before and that are with us at present. It also welcomes us into the lineage of life inquiry to enact their tools and methods to cultivate an all-encompassing manifestation of love and realization that infuses innumerable future generations to come.

~ Tawni Tidwell, T.M.D., Ph.D.,
Center for Healthy Minds,
University of Wisconsin, Madison, Madison, WI

Dr. Mills has given scientists back their right to be spiritual and at the same time be capable scientists. The world is richer for his work and his wisdom.

~ Melinda Connor, D.D., Ph.D.,
Akamai University, Hilo, HI; Chair,
National Alliance of Energy Practitioners, Redondo Beach, CA

Listening to spiritual guidance is a critical resource for us to tackle the significant issues facing each of us individually and collectively. In *Science, Being, & Becoming*, Dr. Mills speaks about how many scientists have found access to their inner voice. This voice lays the foundation for inspired research and living that points the way to healthy outcomes for all, not only a select few. Find and access your inner voice that can come from daily spiritual practices.

~ Mary Jo Bulbrook, R.N., Ed.D.,
President, Akamai University, Hilo, HI

For many years, science and medicine have focused on the individual reductionist approach that excludes more whole-person factors such as spirituality. This book is a refreshing counterpoint to that approach. Dr. Paul Mills is a visionary scientist who takes a more holistic perspective of human life and the factors that truly drive health and well-being. His many talks on this perspective have inspired me to apply new methods for understanding the human condition. I'm proud to have worked with him.

~ Michael Goldsby, Ph.D.,
University of Louisville, Louisville, Kentucky

In this pioneering work, Dr. Mills takes us on a journey through the interconnected worlds of science and spirituality as seen through the previously unseen lives of scientists who have personal spiritual practices. As the reader realizes the elegant weaving of spiritual experience with the development of scientific thought through each scientist's story, the notion that these topics are separate and insignificant in the development of ideas is reconciled. Dr. Mills' work will surely increase the dialogue across previously disparate disciplines such as these and encourage the use of spirituality as a catalyst and tool for creative thought.

~ Dr. Christine Tara Peterson,
Research Scientist, The University of California, San Diego, La Jolla, CA

CONTENTS

RECOGNIZING AND EVOLVING OUR SPIRITUAL AND HUMAN NATURES

Ken Wilber

Human beings have at least two quite different but equally important types of spiritual development. One has been understood for millennia and is found, in some form, in most of the world's great religions and can be referred to as "Waking Up." Recognition of the other is much more recent and not found in the major world religions and can be referred to as "Growing Up."

Waking Up is marked by a direct experience of an Awakening to, or Realization of, what is said to be an ultimate Reality (Spirit, Godhead, Ground of All Being, Tao, Brahman, Great Perfection), an experience known around the world as Enlightenment, Metamorphosis, Satori, Moksha, the Great Liberation, the Supreme Identity. Although there are certainly differences in the ways that realization is experienced by different cultures, different historical periods, and so on, there are some general resemblances that make them all generically recognizable. There is almost always an expansion of self-identify from an isolated, separate self sense to a oneness with this ultimate Reality, and, therefore, also a sense of being one with the entire universe. This is often described as an "awakening" or "enlightenment" because, by comparison, a person's typical pre-awakening consciousness was felt as being lost in a "dream world," a world that is, by comparison, fallen, alienated, and fragmented, marked by inherent suffering (e.g., dukkha) or original sin. The realization of this Supreme Identity is usually said to be the *summum bonum*, the ultimate good, of humankind.

Many of the great traditions have specific systems of meditation, contemplation, deep prayer, yoga, or voyaging that, through a specific set of steps or injunctions, are designed to help move a person from their limited and narrow identity with the separate self sense (often called the "ego") to a deeper and wider self (generally described as one's "True Self" or "Real Self") that is made to be one with Spirit or ultimate Reality itself. For example, both Vedanta Hinduism and Tibetan Buddhism describe five major states of consciousness possessed by all humans. They are the waking state (which is said to be an example of the "gross state" of consciousness); the dream state (an example of the "subtle state"); deep dreamless sleep (an example of the "very subtle" or "causal" state); then *turiya* (which is simply the Sanskrit term for "fourth," and is given that name simply because it is the fourth major state—after the gross, subtle, and causal—and it indicates a pure content-free Awareness, also called the Witness or True Self); finally, *turiyatita* ("beyond turiya," which is beyond pure Awareness because the Witness itself becomes one with, or nondual with, everything that it witnesses—in other words, a type of "ultimate Unity consciousness" or "nondual Oneness"). Similar states to these five major states can be found in many of the world's great mystical and spiritual traditions. The first three states are said to be relative and the home of the ego; the second two states are ultimate and the home of the True Self and its nondual Suchness or ultimate Oneness.

For these traditions, when a person takes up an effective meditative or contemplative practice, they are usually only fully conscious of the first major state—the waking state, marked by the separate self or ego. As meditation progresses, they will move through each of the five major states (not like rungs in a rigid ladder but as flowing states of increasing expansion), until they have a fully conscious realization of the highest or ultimate state itself (*turiyatita* or nondual Oneness), which is then said to be the Enlightened or Awakened state. In other words, the major states of consciousness become major stages of development in the overall process of Awakening, or a general process that is called simply "Waking Up." Waking Up refers to an overall process that is said to result in the ultimate transcendence of the relative, conventional self (in a selfless True Self and nondual Ground of Being).

Growing Up, on the other hand, refers to the actual development of the relative self itself—the specific stages that it goes through as it grows and develops and evolves in the conventional world. And why is this important especially for spirituality? One of the most significant features of the Growing Up process is that no matter what type of immediate awareness or experience a person has (sensory, mental, or spiritual experience), the stage of Growing Up they are at will be the primary interpreter of that experience. This means that no matter what state of Waking Up a person is at, they will interpret that state using the stage of Growing Up they are in.

One way to explain this is that Waking Up consists of major *states* of consciousness while Growing Up consists of major *structures* of consciousness. Both of them can unfold in stages (which are often called state stages and structure stages), but otherwise, they are quite different. States are immediate, first-person, direct experiences or direct awareness;

when you have a state experience, you are directly and fully aware of it. But structures are not like that at all; they are more like grammar—they are third-person rules and patterns that cannot themselves be experienced but govern how experience is interpreted, edited, parsed—and, most significantly, they cannot be seen by introspecting or looking within or meditating.

Most of the world's great religions were founded by individuals who had experienced some of the higher states, and they began teaching others how to attain or Wake Up to these enlightened states. The stages of Growing Up turn out to be incredibly important and are profound factors in determining just what a person's spiritual worldview will be. Waking Up is interpreted by Growing Up. A very high spiritual experience, for example, may indeed be thought-free or involve an awareness that rises above any thought or any mental forms, a purely formless experience of unmanifest reality. But as soon as one comes out of that direct formless experience, one begins to interpret it, to explain it, to try and make sense of it—and in doing that, one definitely makes use of the mind, even if the experience itself was supramental; the person begins to think about it, even if the experience was thoughtless.

Humans possess what's referred to as "multiple intelligences." However important cognitive intelligence certainly is, there is also emotional intelligence, moral intelligence, aesthetic intelligence, mathematical intelligence, interpersonal intelligence, and so on—and this definitely includes a *spiritual intelligence* as well. The discovery of this spiritual intelligence is the second major form of spiritual engagement that human beings have. Notice that this specific spiritual intelligence, *as one of several multiple intelligences*, is not the same as a direct Waking Up or personal experience of anything claimed to be an ultimate reality; these multiple intelligences are all aspects of relative mental growth and development.

They are lines of development in the Growing Up process; they all consist of structures, not states. What we're talking about here are stages in spiritual Growing Up, not states of spiritual Waking Up. In my book *Integral Psychology*, I include charts of over 100 different models of the Growing Up process taken from around the world.[1] A central point about the relation between Waking Up and Growing Up —and the crucial importance of Growing Up—is simply that a person can have a profound Waking Up experience, but they will interpret that experience very differently depending on the stage of Growing Up that they are at.

For any truly integral spirituality, both the Waking Up and Growing Up paths need to be fully considered and consciously included. There is another major developmental dimension that we can call "Cleaning Up." Cleaning Up refers to the general processes whose discovery is most often associated with the name of Sigmund Freud, but actually includes an entire panoply of folks such as Janet, Jung, Adler, Rank, Binswanger, Sullivan, Perls, Kernberg, and others. It deals with the fact that human beings can, under extremely stressful and traumatic conditions, deny or repress or split off or alienate significant facets of their own being, pushing them from consciousness into an unconscious state. Denying or repressing these aspects—often known as "shadow"—does not actually make them go away, but instead forces this shadow material to create warped, disguised, and painful symptoms (of anxiety, depression, obsession, etc.), a dysfunction usually known as neurosis or a more serious psychosis. The cure—according to C. G. Jung and others—is to contact the alienated, unconscious, shadow material and make it conscious again, rejoining and integrating it with the rest of the conscious mind.

It's important to note that Cleaning Up is a process that in itself is profoundly different from both Waking Up and Growing Up. It has different causes; it produces very different results; it is addressed by very different practices and techniques. And they are all relatively independent—Growing Up and Waking Up and Cleaning Up—you can progress incredibly well in any one of them and still be totally lagging in the others, and this can happen in any combination. But it seems undeniable that many, perhaps most, individuals on a spiritual path seem on occasion to be infected with significant amounts of shadow material, and the drive to (unconsciously) deal with that shadow material can become disguised as the drive to find Spirit—with the added disaster that most spiritual paths teach very little, if anything, about Cleaning Up and ways to both recognize and treat shadow material. If severe, this shadow material can derange the spiritual journey in profound ways, including completely derailing it.

Any spiritual path in today's world needs to take all three of those processes—Growing Up and Waking Up and Cleaning Up—into account and actively address all of them if it wishes to provide anything like a comprehensive, competent, and legitimate spirituality.

The final domain I want to note is called "Showing Up" which refers to being fully open to, or showing up for, the major fundamental perspectives available in a person's life. There are many different and equally legitimate ways to refer to these basic perspectives. The Greeks called them "the Good, the True, and the Beautiful." Virtually all languages around the world contain pronouns that refer to 1st person, 2nd person, and 3rd person perspectives (with 1st person being "the person who is speaking," an "I" or "me"—and this includes aesthetics or the Beauty that "is in the eye— the 'I'—of the beholder"; 2nd person is "the person being spoken to"—or a "you"—with the Good being the ethical ways that "I" should treat "you"; 3rd person is "the person or thing being spoken about"—which includes the objectively True of the natural sciences). These are Karl Popper's "Three Worlds" (subjective, sociocultural, objective). They are Jürgen Habermas's "three validity claims" (aesthetics/truthfulness, goodness, objective truth).

They are Immanuel Kant's three major critiques of Pure Reason, Practical Reason, and Judgement.

These perspectives directly apply to spirituality as well because Spirit itself can legitimately be viewed through all three of them. Notice, as but one example, that Christ spoke ABOUT God (that's 3rd person), he spoke TO God (that's 2nd person), and he spoke AS God (that's 1st person). As usual, it's fairly rare that an integral approach is taken and all three perspectives are realized to be equally valid so that all of them are fully embraced—usually just one perspective is chosen and proclaimed the only correct view. The Spanish Inquisition was fine if you discussed God in 2nd or 3rd person, but if you started talking in 1st person with reference to God, they would have a nice long chat with you. To entertain a 1st person view of Spirit was literally heresy; only Jesus could do that. This meant that the great Christian mystical saints had to spend their time tiptoeing around language to make sure they didn't step over that line. Giordano Bruno went too far and was burned at the stake; St. John of the Cross and St. Teresa of Avila walked the line very carefully and got away with it (Teresa, for example, would simply say things like, "Beyond this point cannot be put into words"); Meister Eckhart slipped a bit and, although he himself was not condemned, his theses were (so apparently while Eckhart is in heaven, all of his ideas are now burning in hell). An Integral approach—which we also call the "1-2-3 of Spirit" (because it honors and embraces all three perspectives) is indeed relatively rare—but there is no reason whatsoever that such a fractured approach needs to continue.

Whether you recognize the reality of Spirit or not, these three perspectives, in themselves, are very real, which is why every mature language in the world contains them. Many of the greatest philosophers, from the early Greeks to today's Jürgen Habermas, have recognized these perspectives and have realized that they all contain different (but equally real) kinds of truth, are accessed by different methodologies, and give entrance to very important (if different) dimensions of awareness and being. The Integral approach agrees and suggests that Showing Up for all of these truly fundamental perspectives is a crucial aspect of any approach, in any discipline, that genuinely wishes to be comprehensive, inclusive, competent, and integral. Needless to say, this applies to spirituality as well.

An *Integral Spirituality* includes Waking Up, Growing Up, Cleaning Up, and Showing Up, each being equally important dimensions/perspectives that any spirituality would want to fully incorporate to be authentic.

In the pages of this book, we read about the Waking Up experiences of many scientists and how those experiences transformed their personal lives and the direction of their scientific work. Additionally, we learn about other consequences of their Waking Up experiences, including how these scientists subsequently went through stages of maturation, of Growing Up, and eventually began to Show Up, aligning their lives to their most fundamental truths and sharing those truths through teaching and service.

INTRODUCTION

THE GALACTIC CENTER

A few years ago, I was with a group of friends discussing what was then a current astronomical event called the Galactic Alignment. This is when the earth and our Sun's positions change in relation to the center of the Milky Way Galaxy when we pass through what is called the "Galactic Plane." This event occurs every 26,000 years. It was all news to me, as other than the cycle of our own moon, I typically didn't track such things. In addition to the astronomical features of what had occurred, our conversation also included somewhat esoteric astrological aspects, and we wondered how our new position in relation to the Galactic Center might affect us here on earth, if at all.

After the conversation ended, I decided to sit and meditate on what we had been discussing. This was a contemplative kind of meditation, something I like to do to have more insight into a topic that I've just learned about. I closed my eyes with the simple intention of wanting to better understand what we had been discussing, to possibly glean some new insights. Within a moment's time, I found myself encountering a vast consciousness, a "consciousness mind" if I may put it that way. This consciousness was utterly foreign to anything I had ever come across before. It began to communicate with me telepathically, i.e. I heard no words, but I started receiving images in my mind.

The first thing it communicated was that it had no interest in individual human beings but was keenly interested in humanity itself. I immediately thought this was odd in the sense that my study of religions indicates that beings greater than ourselves, such as the great devas and godheads, are indeed interested in helping human beings. People routinely pray to such beings for intercessions where possible and appropriate. This consciousness indicated it had been watching the development of humanity for eons and that one day in the future, humanity would come into its own in terms of fulfilling the reason it was created. It was that future event that this consciousness was interested in and waiting to occur.

As I wondered what this future event might be, the consciousness communicated to me that humanity is unique in the universe and that, after further developing our consciousness, one day, humanity will manifest something that has never before been manifested in our physical dimension. As it communicated this into my mind, I saw an image of the earth from space, and while it "spoke" there emerged spreading across the globe a vast white, goldish-colored light that will one day be produced by a more evolved collective consciousness we call humanity. It was a beautiful and inspiring sight. What is it that will one day be manifested? Acknowledging what I consider my limited capacity to fully understand the vastness of that consciousness I encountered, what I understand will be manifested into physical creation is love itself. Love, that force of and behind creation, is to come into physical existence itself.

This was a momentous experience for me. With a firmness in my mind, I thought that this was indeed a future destiny of humanity. I started looking for books that might help me understand what I had experienced. I eventually found a few that speak about vast consciousness minds, including Jude Currivan's *The Cosmic Hologram: In-formation at the Center of Creation*, Carl Johan Calleman's *The Nine Waves of Creation: Quantum Physics, Holographic Evolution, and the Destiny of Humanity*, and work by the great Indian Sage Sri Aurobindo, who in *The Life Divine* describes vast levels of mind consciousness that are beyond typical human mind consciousness but of which we are destined to one day embrace. I'll share more on this later, but after reading *The Life Divine*, I wondered if the vast consciousness I encountered was what Aurobindo calls the "Overmind," or perhaps it was the "Supermind," which is a level of mind beyond the Overmind. I wondered, too, whether that transformational vision I had of the earth's future corresponded to what Aurobindo describes as the eventual "descent of the divine into the world."

I also began to ponder whether this future event is definitely going to occur—or is it a "probable", a "maybe" depending on a host of variables unknown to me. If the latter, I wondered if it is a responsibility of each of us as individual human beings to foster humanity's eventual spiritual awakening to help that eventual manifestation of love. I found myself wondering, "What have I as a biomedical scientist been doing to help or hinder this eventual development of humanity?"

More broadly, I wondered, "What are the biomedical sciences and the scientists working in them doing in terms of helping or hindering this future goal of humanity? Are we in service to humanity's future advancement or not?" I was aware that the vast majority of the scientific endeavor is highly materialistic, with metaphysical and mystical considerations outright rejected as nonexistent; love itself is considered only an emotion and even then, not routinely studied. Yet, I also knew there were biomedical scientists, including some of my own colleagues, working within the mainstream sciences who have deeper spiritual understandings of who we are as people and of the conscious universe at large. I also knew that many of them were working in their own way to bridge the so-called "gap between science and spirituality." They were working to overcome the materialism and its choking effects on the progress of the sciences that could ultimately help support a future spiritual life of humanity.

This is the primary reason for my writing this book. I wanted to understand what such scientists experience in terms of insights into their own spiritual nature and how those insights and experiences influence their scientific work. I thus set about reaching out to scientists and asking if I could interview them about their spiritual life. The book is a compilation of transpersonal, metaphysical, and mystical experiences and explorations of leading senior national and international scientists and clinician scientists, as well as more junior scientists who are at the beginning of their scientific careers.

TRANSCENDENTAL MEDITATION AND THE OAK TREE

My own spiritual journey, if I may call it that, started in 1972 while I was in high school when I learned to practice transcendental meditation (TM). TM was introduced to the West by Maharishi Mahesh Yogi, and back then, it was sweeping the country, much in the way mindfulness meditation has done in more recent times. I was drawn to meditation because I heard it could help me unfold my potential. I was also drawn to it for another less concrete reason: it simply seemed like a good thing to do.

I had attended an introductory lecture on TM, and the next step was to individually meet with the teacher and learn the actual meditation technique. The TM teachers lived in an old stone farmhouse just down the road from where my family lived in the small town of Pineville, in Bucks

County, Pennsylvania. In the room where I was to learn to meditate was a painting of the Maharishi's guru, Swami Brahmananda Saraswati. The Swami had been a highly significant spiritual figure in India, serving as the Shankaracharya of the Jyotir Math monastery in Northern India. After a brief ceremony, my TM instructor, Robert Wallace, gave me my personal mantra (or sound) and told me to repeat it quietly in my mind. I followed his instructions and very quickly found myself in a deeply relaxed, peaceful, and expansive place. After some time, he told me to stop repeating the mantra and open my eyes. When I opened my eyes, I looked at the painting of the Swami, and then at Robert, and asked him if the Swami had given him the mantra to give to me. He smiled and said no.

My first few months of practice, while pleasant and deeply relaxing, were otherwise uneventful. That changed one day when I went outside to meditate under a large oak tree near our home in the countryside. The tree was situated in the middle of a nearby field of wild grasses. It was midafternoon on a warm summer day. A few minutes into the meditation, my awareness suddenly shifted, and I found myself hovering above my body, looking down upon myself seated under the tree. I could clearly see the top of my head, my shoulders, and my back, as well as the ground around me, which was mostly covered in dried leaves. It was initially shocking, but I quickly settled into it.

Although it was a completely new experience for me, it at the same time felt familiar. While looking down upon my body, I found myself in an expansive sea of silence, peace, and tranquility. It was both me, Paul, having the experience and a much bigger "me," the one to whom it all felt so familiar and okay. After a few minutes, I was "back in my body." I sat there in awe; I had never heard of such a phenomenon. In the TM introductory lecture I had attended just prior to learning the actual meditation technique, my teacher Robert hadn't mentioned this as a potential experience. Rather, the lecture focused on things like quieting the mind and experiencing relaxation.

Over the following weeks, as I pondered the experience, a seed was formed that would eventually direct me to a scientific career. I decided I wanted to learn the methodologies of science to study how meditation worked and to better understand our human potential. I particularly wanted to know about the capacity and limits of further developing our consciousness, what lies beyond the normal everyday perception of our sense of self and mindbody. Prior to that meditation under the tree, I had not known we could experience consciousness without boundaries. I also wanted to understand how is it that I was both the "Paul" I had always experienced myself to be and yet also a much bigger "me" to whom the expansive and peaceful experience was all so familiar. Who was the "real" me, or were both me? In addition to helping me understand who I was, I wondered if conducting scientific studies on meditation would help encourage others to learn it too.

STATES OF CONSCIOUSNESS

That experience under the oak tree galvanized me to become devoted to my meditation practice. I also made a commitment to myself to learn all I could about the nature of such consciousness experiences. As part of that commitment, I devoured books on the topic, including books by the Maharishi. In his books *The Science of Being and Art of Living* [2] and *Commentary on The Bhagavad Gita*, he describes seven different states of consciousness that are available to human beings. These included the ordinary states of consciousness we are all familiar with, i.e., waking, dreaming, and deep sleep, but also an additional four "more elevated states." Each of these elevated states is characterized by the experience of "pure unbounded awareness," yet the context of what that awareness experiences is markedly different in each of the four states:

- Transcendental consciousness. The state of pure unbounded inner self-awareness, the true Self, without perception of the other three ordinary states of consciousness.

- Cosmic consciousness. The state of pure unbounded inner self-consciousness found in transcendental consciousness, yet simultaneously maintained while experiencing any of the other three states of ordinary consciousness of waking, dreaming, and deep sleep. This state is often called "witnessing consciousness" because of the perceived disparity of the unbounded inner self-consciousness while at the same time perceiving the apparent bounded limited objects in everyday life.

- Divine consciousness. This state of consciousness is considered a partial resolution of the disparity of cosmic consciousness in the sense that the separation and witnessing of cosmic consciousness begin to change such that perceptions that had appeared to be spatially bounded objects in everyday life now begin to be seen for their true nature as unbounded consciousness.

- Unity consciousness. Unity represents a resolution of the illusion that the infinite unbounded Self is any different from all that is being perceived. The infinite is seen in each and every object of perception and is known to be what those objects are. Despite also seeing the unique appearance and form of every object of perception, it is simultaneously seen as consciousness itself. There is no longer a sense of witnessing, no longer any duality; all is seen as one infinite totality, as Self. Prior experiences of separation are now known to have been illusion. The Maharishi also spoke of a further refinement of unity consciousness called Brahman consciousness. The Sanskrit word Brahman refers to the eternal absolute reality, the totality of all existence. Brahman consciousness then is the experience of the totality of all existence, its essential oneness. Interestingly, Maharishi said that the transition from the state of unity to Brahman consciousness takes place on the level of the mind. It is some kind of realization on the level of the intellect that enables one to come into this state of experience.

What Maharishi taught was not new. In fact, many meditation traditions provide insight into what are often called non-ordinary, unitive states of consciousness. In his Special Commentary for this book, Ken Wilber shares such insights from other traditions, including the fourth state called turiya and the fifth state beyond turiya called turiyatita. I found reading Maharishi's books particularly relevant to my experience under the oak tree that day. For the first time, I could put a name to my experience, and I knew someone else had experienced it too. The teachings provided me with an important roadmap as the years unfolded and I continued to experience further transformations of my experience of consciousness.

MAHARISHI INTERNATIONAL UNIVERSITY AND MEDITATION RESEARCH

In September 1974, I arrived in the small town of Fairfield, Iowa, where I entered the newly established Maharishi International University (MIU). MIU was and remains an unconventional university. It was a risk of sorts for me to choose it for my university education, but since my meditation practice continued to unfold, I felt compelled to do whatever I could to deepen my understanding and further develop my consciousness while simultaneously becoming a scientist. MIU was billed as a place to get a traditional education in terms of the academic curriculum, but also to receive deep learning on the nature of consciousness itself and its further development.

I had learned of MIU some months earlier in Livingston, NY, while I was attending a one-month Science of Creative Intelligence (SCI) course taught by the Maharishi. The course provided an understanding of the principles of consciousness and intelligence that underly creation, essentially showing how the universality of consciousness and its inherent creative intelligence are the very foundation of creation and govern its activity.

When I returned home from the SCI course, I told my father about MIU and that I would like to attend the university and see how it would go. I have such a clear memory of that conversation. It was late afternoon, and we were seated on a bench facing west, overlooking the fields where he kept his sheep. The sheep were a hobby of his that brought him joy and satisfaction. I have many fond memories of watching him care for them, or simply roaming around the fields with them. Fortuitously, he and my mother had learned TM while I was away on the SCI course, and they both were enjoying the practice. They thought it was a good thing and saw no conflict between it and the Catholicism they actively practiced. So, my dad said he would send me there for one or two quarters, and then we'd decide if it would be worth staying longer. Three weeks later, I was on a flight to Fairfield.

Back then, the MIU faculty consisted primarily of young, highly enthusiastic professors from institutions such as MIT, UCLA, and Princeton, all of whom were TM practitioners. Like me, these academics had learned TM, had some kind of transformative experience, and were intent on studying consciousness within their respective discipline, be it the arts, physics, physiology, chemistry, or psychology. Their personal experiences with meditation drew them to MIU.

At that time, the unique vision of MIU was depicted in its logo, which was a large tree with four main branches symbolizing the branches of knowledge from the ancient Indian philosophical texts known as the *Vedas*. Placed in a circle around the tree were the words, "Knowledge is Structured in Consciousness." The tree's deep roots merged into the word "consciousness," symbolizing it as the root of all experience. It was an exciting time for me and the many hundreds of young people like myself who came to MIU from around the US and beyond to participate in this particular approach to education, which was not yet available in other universities or colleges. Over the decades since then, I've remained close friends with many of my MIU classmates.

At MIU, I was fortunate to meet my first mentor, Robert Keith Wallace, Ph.D. (I know, a remarkable coincidence that my first two major TM-related teachers had the same name!). Dr. Wallace had taken his own unusual meditation and scientific path by conducting some of the very first studies on the physiology of meditation, publishing his initial work in the journal *Science* in 1970.[3] From there, he went to Harvard University in Boston to work with Dr. Herbert Benson who later developed *The Relaxation Response*. It's fair to say that Keith's early research helped set the stage for the subsequent surge of meditation research, one that continues to this day, with currently over 8,500 scientific publications on meditation.[4] From there, Keith continued his non-conventional journey and became the first President of MIU. Keith was easygoing and friendly.

Fortunately for me, when I approached Keith and said I wanted to work in his lab to support research on meditation, he enthusiastically said yes. I was, of course, thrilled. In retrospect, at this stage in my life, now having run a large research lab at a university where student help with research is always welcome—actually often vital—I'm sure he was happy that I showed up.

From 1875 to 1973, the MIU campus had been the home of Parsons College, which had unfortunately gotten into accreditation problems during the Vietnam War, forcing its closure. Keith's lab was in the science building of Parsons, which, at that time, was still filled with old scientific instruments and supplies, some of which we were able to use, but much of it was discarded. The truth be told, I was extremely excited to be working with Keith and to have the opportunity to conduct meditation research with him. I was fortunate that early on at MIU, I found myself in a position where I could immerse myself in my original desire—that is, to conduct research on how meditation works to advance our human potential. Over the next couple of years, we worked together designing and conducting studies focusing on the neurophysiology of meditation. I think we made a good team. As a result, I was able to publish papers in scientific journals from research conducted while I was still an undergraduate student.[5,6]

Several years into my time at MIU, I attended the annual Congress of the International Society of Electrophysiological Processes in Boston, MA. I was there as a student to present findings from one of our physiology studies.[7] We had examined the effects of meditation practice on what is called the Paired-Hoffman Reflex. The reflex is elicited by administering a series of mild electric shocks to the tibial nerve in the popliteal fossa (back of the knee) and then recording on an oscilloscope the subsequent nerve action potential. Despite the pain involved, many of my fellow MIU students and several staff and faculty members lined up to be research subjects. Their willingness was despite the fact that it often took me some time to find the precise location of the tibial nerve. That is, I'd often have to deliver multiple shocks before I found the nerve and could secure the electrode over it to then begin the actual experimental protocol. As far as the results, we found that the amplitude of the reflex response was significantly facilitated in the more experienced meditators, suggesting distinct physiologic effects in response to a more prolonged meditation practice. It was "good data," meaning that the findings were statistically significant and made sense in terms of prior findings from EEG studies on TM meditators.[8]

My conference presentation was in the form of a poster. In contrast to giving a talk, posters are a much more informal tool that researchers use to present their findings at scientific meetings. Typically, poster sessions take place in a large hall, with dozens and sometimes hundreds of posters lined up end-to-end. Conference attendees walk up and down the rows of posters, and each presenter vigilantly stands by their poster ready to explain their work and answer any questions. I was relieved that the conference organizers had scheduled my presentation as a poster and not a talk, as at that time early in my career, I didn't yet have the confidence to speak in front of an audience.

While I was standing at the poster, a conference attendee walked up to me and said, "That is really interesting looking data, but what is meditation?" I was so surprised to hear this question as I'd been living in my own bubble of meditation for the past few years, and it all seemed rather normal to me. But how far we've come, given that meditation is now a household word. During my subsequent years at MIU, I helped conduct more research on meditation and presented at other scientific conferences, including the Iowa Academy of Science, the Society for Neuroscience, and the American Psychosomatic Medicine Society.

TIME FOR A CHANGE

A few years later as I was getting close to graduating with my Ph.D. in Neuroscience, I was wrestling with where I was personally heading on my spiritual journey. For much of my adult life, I had been very deep into TM. I had even spent six months in Europe in an intensive meditation program to learn how to teach TM, which the Maharishi instructed us how to do.

Approximately a year before my graduation, I had an experience that showed me it was time for a change to the journey I had been on. While I didn't understand the full implications of the experience at the time, it was immediately clear to me that the life I had been living was now over.

Before sharing the experience, I must say that, while writing this book, I deeply wrestled with which and how much of my more metaphysical and/or mystical experiences to share. In addition to sharing my own such experiences, this book is a compilation of interviews with dozens of scientists who shared with me their transpersonal, metaphysical, and mystical experiences. Many of them too shared with me their worry about what to divulge and what to keep to themselves.

While seeking my own answer to this question, many times I felt a deep dread and would ask myself, "Why do it?" I wondered how these sharing's would be received. Would it be something like, "Too bad, he's so delusional," or "Yes, he was once a respected and productive scientist, but now it's such a shame he's gone over the edge," or "We've seen this before. He couldn't handle all the academic stress and he's lost touch with reality."

Over the decades, centuries really, there are many examples of scientists who choose to share such experiences and the career-adverse effects that resulted. In the end, I and my interviewees chose to proceed with the belief that sharing such personal experiences will, at the end of the day, have beneficial outcomes. Each of us is interested in seeing an end to the metaphysical and/or mystical being so off-limits to Western scientific inquiry. Another reason for the book is to take the opportunity to educate about the nature of human consciousness development, a path that is open to us all. With that said, I'll proceed.

One afternoon, I went to meditate in one of the large meditation halls on the MIU campus, which was my normal routine. In this particular hall, there was a separate side room where I often sat with some of my fellow Neuroscience graduate students. The room had been set up so we could conduct research on meditators while they were participating in the larger group meditation. That day, while the hall was full of many hundreds of students, staff, and faculty members, I was the only student in the research room.

I closed my eyes to meditate and soon heard a distinct inner voice clearly and very unexpectedly state, "You can no longer serve two masters." While I didn't immediately know what the "two masters" meant in terms of who is "the other master," I did immediately know on a deep inner level that my relationship with the Maharishi as my teacher was about to end. What then unfolded was something I witnessed as a spectator and in which I participated. No sooner were those words spoken to me than the Maharishi appeared in front of me and looked at me, and my inner voice said, "I release you," He then promptly vanished.

It was fortunate I was alone in the separate research room because I began to cry. It was painful for me to let go of the relationship I had with the Maharishi. He was my teacher. I had learned so much through him, and my life was much better because of it. Internally, I wondered what was going on and if what was happening was correct or some kind of terrible mistake. Yet at that same time, I felt there was a larger part of me orchestrating the event and that it was indeed time for a change. That sense provided me some solace.

What happened next, though, solidified my confidence in that what was occurring was correct for me. The Maharishi's guru, Swami Brahmananda Saraswati, who had passed away in 1953, then appeared before me. He was sitting in full lotus, and his face was serene. When he appeared, I immediately felt that everything was okay, to trust the process of what was occurring. As I looked at him, he began to expand out into the universe, and his form began to dissolve into the infinite consciousness. My own consciousness followed and expanded to that state of no boundaries. A new life had started.

Approximately two weeks later, with that prior experience lingering in my awareness, I had a subsequent experience that continued the process that had started in the research room. The experience involved encountering a spiritual consciousness of such beauty and magnitude that there is no language I can possibly use to adequately describe it. During the encounter, my consciousness was further transformed. The expanded consciousness I had experienced following the Swami now became self-illuminated.

For weeks afterward, it was as if I was simultaneously existing on several planes of existence—the one I was used to living in and another plane more vast and cosmological in nature. I recall thinking to myself then that prior to that encounter, my experience of life had been like living in a small closet illuminated by a single dim lightbulb hanging from a ceiling wire. Now, my awareness was not bound spatially, and that prior dim illumination was replaced with a bright inner illumination, one that filled my being with a sense of peace I had not known before.

This part is hard to describe, but while I "Paul" still existed as I had always known myself, a consequence of this new state was that everything I perceived now existed within my own consciousness. That is, I found that nothing existed separately outside of my perception. Nothing was experienced as separate from it—not people, trees, hills, the sun or the moon in the sky, nor the stars themselves.

It took me time to adapt to this new perception, particularly my sense of distance, which no longer existed as it used to. Previously, I experienced a sense of separation and distance between myself and whatever I was seeing. It was where I was, and whatever or whoever I was seeing was over there somewhere. Now there was no distance or separation of anything; somehow, it all resided in my own consciousness, within my own perceptual awareness. With time, my perception recalibrated, so that despite everything existing within and not outside of myself, I managed to regain a perceptual perspective.

As I acclimated to this new perception and sense of self, I realized I needed to focus as best as I could on completing my dissertation research and graduating from MIU. This happened approximately one year later. Curiously, while MIU had at its core a strong central focus on the development of consciousness, and, as I described earlier, the Maharishi elegantly described the transformations of consciousness available for human beings, I did not feel comfortable sharing any of what had occurred. Other than my then wife Kim, as well as two close friends, I didn't share these experiences with anyone.

THE UNIVERSITY OF CALIFORNIA SAN DIEGO AND INTEGRATIVE MEDICINE

Three years after my graduation from MIU, I was interviewed for a faculty position in the Department of Psychiatry at the University of California (UC) San Diego in La Jolla, California. I had spent the prior three years at UC San Diego successfully completing a post-doctoral Fellowship in Behavioral Medicine and had secured independent grant funding from the National Institutes of Health, scoring in the first percentile. I was now eligible to be considered for a faculty position.

For scientific research impact, since its founding in the 1960s, UC San Diego has had a meteoric rise, having been ranked by *Nature Reports* second in the US among all public universities (with UC Berkeley ranking first) and fourth in the world. While I had published quite a few papers since arriving to UC San Diego and met the faculty eligibility requirements, during one of the interviews, the two senior faculty members interviewing me skipped through the more recent post-doc publications in my CV and went to my earlier publications on meditation from when I was at MIU. These senior faculty wondered if I wanted to start a meditation research program in the psychiatry department. One of them sternly said to me, "This is not a department that conducts research on meditation, and it won't be accepted!" The truth is, I was fine with that as, in reality, I didn't have an interest then in resuming meditation research. I was very much enjoying the behavioral medicine research I was conducting during my fellowship with my mentor, Professor Joel E. Dimsdale, M.D., and wanted that to be my continued focus.

Years earlier, while still at MIU, I had attended a psychosomatic medicine society conference in Philadelphia, and there I met Professor Dimsdale. Years later, I would become actively involved in the governance of that society, including becoming president. Joel was at UC San Diego, having recently arrived there from Harvard. He was hired to be the Director of the Consultation Liaison Psychiatry service at UCSD Medical Center. I was attending the conference to give a talk on data from my Ph.D. dissertation research, namely, on the effects of meditation on adrenergic physiology. Despite a few nervous stumbles during my talk, I think it went fairly well, overall. Joel also gave a talk on the topic of his latest research, which was on the effects of stress on the sympathetic (adrenergic) nervous system and cardiovascular health. While listening to his talk, I realized that we shared a mutual interest in adrenergic systems. I also felt a personal affinity with Joel; I liked his presence.

After Joel's talk, I approached him and expressed an interest to work with him, asking if he had any openings in his lab group. To my good fortune, an opening did emerge a few months later, and I soon found myself in San Diego. Joel was a remarkable research mentor and also a wise guide through the complex and competitive world of academia. I remain very grateful to him. I'll add that I did get that psychiatry faculty position. I'll also note that in 2002, approximately eleven years after that interview with the senior faculty stating, "no meditation research in this department!", under the direction of mindfulness educator Dr. Steven Hickman, the Department of Psychiatry established a Center for Mindfulness Meditation, which is now one of the largest and most successful in the US.

Over my subsequent years at UC San Diego, as Complementary and Alternative Medicine became more accepted throughout the US, the UC San Diego medical system became more and more open to including meditation and yoga practices within the clinical operations. In 2011, under the leadership of Dr. Gene (Rusty) Kallenberg, M.D., a Professor of Family Medicine and Public Health and Director of Family Medicine services, UC San Diego opened a Center for Integrative Medicine.[9] To this day, the Center offers "evidence-based" modalities such as meditation, acupuncture, and yoga to outpatients and inpatients across the large UC San Diego Health system. There are currently over 80 such Centers for Integrative Medicine at major academic medical institutions in the US and Canada.[10]

By 2015, the Center had seen such growth within the medical system that the university's leadership recognized the need for a more centralized location for conducting research on integrative modalities, in order to strengthen the evidence base for integrative medicine. Dr. Bess Marcus, then Chair of the Department of Family and Preventive Medicine, led the effort to create such a new research center to centralize these efforts. After a national search, I was hired to lead this effort and became director of what would be called The Center of Excellence for Research and Training in Integrative Health.[11]

It was an exciting and rewarding time for me. I felt like Bilbo from the *Lord of the Rings* trilogy, in his journey "there and back again." In fact, when I gave my job talk in front of the Chair and faculty of the department, I included a slide in my presentation of an image from *Lord of the Rings*, and I explained why I included it, that I had started my scientific career studying meditation and was now returning to it. I would now take the lead on a major institutional effort to conduct research on meditation and other complementary practices.

Since the Center of Excellence's opening, we've published many novel studies that have further expanded the evidence base for integrative health. At present, I've published approximately 400 scientific papers, reviews, and book chapters and have presented at hundreds of scientific meetings around the world, including presenting at the United Nations on the scientific training our Center of Excellence provides to students.

THE SCIENTISTS AND THEIR INTERVIEWS

This book is a compilation of stories from primarily biomedical scientists I interviewed about their own transpersonal and metaphysical explorations. I envision a second book wherein the scope of interviews will be expanded to additional scientific fields such as cosmology, physics, and the life sciences, including evolution, the environment, and philosophy.

During the interviews for this book, I was humbled to be the recipient of such personal sharings, and for that, I am deeply grateful. I must add the sense of regret I felt at times while interviewing scientists I knew. The reason for the regret was that during the interviews, I learned about the depth of their spiritual experiences and their journeys, and all the while knowing them over the years, we had never spoken of these things. I understood why, as they, like me, kept that part of them under a different hat than their scientist's hat, which was the circumstance under which we typically met and conversed. Personally, I never spoke of such things with colleagues, assuming none of them had an interest or perhaps had nothing in particular to share. Since then, when speaking with colleagues, under the appropriate circumstance, I do find a way to open the door to such a conversation if there might be an interest or perhaps even a need to speak about it.

When I embarked on the interview process, I had a list of sixteen prepared questions to guide the course of the interviews. By the fifth interview, a definite pattern had emerged, with each interviewee resonating with only a subset of those sixteen questions and not so much the others.

As I looked closely at those questions, a theme emerged in my mind, one of which I couldn't help but connect with Joseph Campbell's book *The Hero with a Thousand Faces*.[20] In the book, Campbell summarizes a narrative pattern found across countless cultures of what is commonly called the "monomyth," or what Campbell called "the hero's journey." Across cultures, the monomyth journey has common themes and stages, with anywhere from eight[21,22] to seventeen[23] different stages. Features common across all of the different models include what is called the initial "call to adventure," where the hero is called to leave behind the ordinary world they've known and felt comfortable in and venture into the unknown, into what the respective culture they were born into doesn't readily accept. Other common stages among the models include the hero meeting a mentor who helps them on their journey and the hero facing tests and adversities along the way, including encountering enemies and trials that discourage the hero from proceeding.

If the hero fulfills the impulses for the call, the journey's fruition includes a type of resurrection and rebirth with a new self, with the hero then returning to share their newfound gifts with others. The hero/heroine's spiritual journey is not an easy one, as you will see through the stories in this book. While "heeding the call" and undertaking a journey is a way to achieve a meaningful and foundational life, there are losses along the way, some imagined, some real, including the loss of the self that was previously known and, in some ways, cherished. Demons must be faced and learned from; that is part of the rebirth process.

One never fully knows what we will find when we open ourselves to our subconscious mind and what has been suppressed or repressed there. On his hero's journey, Jedi Knight Luke Skywalker had to come to terms with the fact that he was the son of Darth Vader, an evil person in the universe. Upon coming to terms with this, he was able to accept all the parts of himself and be reborn as a Jedi.

Based on my knowledge of the monomyth, and what domains I found interviewees resonating with during my initial set of interviews, I restructured and simplified the questions to the following:

1. Tell me about your spiritual journey in terms of specific transpersonal, metaphysical, or mystical experiences that might have changed you and your perceptions of the world.

2. To what degree did these experiences lead you to become a scientist? If you were already a scientist when you had the experiences, how

did they influence your approach to and/or understanding of science?

3. What mentors, either academic or spiritual teachers, were instrumental in shaping and supporting your spiritual unfoldment?

4. What challenges have you faced, personally or academically, in the form of circumstances or people, if any?

5. How are you sharing your insights and giving back?

These are what I used for the remainder of the interviews. The questions cover common themes of the monomyth. Chapters 2-5 are structured with the content from these questions, following the classical monomyth. Chapter 2, titled "Heeding the Call: The Makings of a 'Spiritual Scientist,'" parallels the "Call to adventure" and covers material from when the scientist's share stories of their initial inspirations and experiences that took them on their journey of self-exploration and, for some, eventually science. The other chapters, too, follow major themes of the monomyth, including Chapter 3 on Mentors and Other Allies, Chapter 4 on From Trials to Commitment and Transformation, and Chapter 5 on Giving Back and the Next Generation, where the hero/heroine returns and shares their gifts with others.

In the introduction to this book, I shared highlights from my own journey, which shares many of these monomyth features. The truth is each of our lives follows the monomyth, with an overall grand cycle through our lifetime and with many shorter monomyth cycles occurring, particularly across our major maturation stages. In his Spotlight feature for this book, Dr. Robert Atkinson describes common monomyth patterns along the evolutionary trajectory of consciousness towards wholeness, showing that "science and spirituality are not only representations of this wholeness but also equal partners in it."

Throughout this book, I use the terms "spiritual," "metaphysical," and "mystical." By spiritual, I'm not referring to religion but rather our human experience that is beyond what our physical senses present to us. While religion can be spiritual, it is not necessarily so. The word metaphysical refers to what is beyond (meta) the physical and is not commonly used in the context of religion. I use the word mystical to describe experiences of nonphysical—sometimes called supersensible—entities or beings not accessible by way of our normal sense perception. I also use it to refer to the deeply intimate experience of awe that can accompany spiritual and metaphysical experiences.

The psychologist Abraham Maslow became famous for articulating what he called the "hierarchy of needs," a description of the universal needs of an individual within a society.[24] At the time of its development, the theory culminated in what Maslow called "self-actualization," the highest level of psychological development a person could attain, where a person had fully realized their individual potential. Later in his career, he added another level to this hierarchy, what he called "self-transcendence." This is when a person has moved beyond self-actualization. Self-transcendence is the movement that occurs at the "very highest and most inclusive or holistic levels of human consciousness."[25] I believe that the many scientists I interviewed for this book have been achieving this level of understanding of their true nature as the holistic and unitive levels of human consciousness, of awareness itself, of the true Self.

I want to add that some of the scientists I approached for an interview acknowledged they had experiences and trajectories that would fit into the topic of the book but were unwilling to be interviewed. One of the reasons was that their particular meditation tradition actively discouraged such sharing. Other scientists were concerned about the potential negative impacts of sharing on their professional careers. For those who had already attained tenure at their respective academic institution, this was less of a concern. In other cases, concerns were simply the pros and cons of publicizing intimate spiritual experiences, which, as I previously noted, I fully understand. I continue to have some haunted feelings about doing so.

One of my interviewees said, "Science is a mystical path." That is, in its essence to explore and help us understand ourselves and the world around us, it has the potential to unlock many metaphysical and mystical aspects of our existence. May all scientists awaken to that perspective. The spiritual is the final frontier for science.

PERSPECTIVES ON TRANSPERSONAL, METAPHYSICAL, AND MYSTICAL EXPERIENCES

Kyriacos C. Markides, Ph.D.

W hen I crossed the Atlantic and began my university education in America in 1960, one of the greatest challenges I faced was the conflict between my traditional Greek Orthodox upbringing and the prevailing secularism that dominated academia. I tried to resolve that conflict by switching my educational focus from business to sociology. By the time I received my doctoral degree, I thought of myself as an agnostic, albeit a reluctant one. I felt that I had no other choice but to internalize the prevailing consensus among my professors and peers that the only reality is that of physical matter. God is a projection of society, and spiritual or mystical experiences are, at best, forms of illusion, or at worse, mental aberrations. Furthermore, I accepted the notion that the only legitimate form of knowledge is that of empirical science grounded on rationalism.

It was later in my academic career that I came full circle, realizing that the materialist worldview is itself a grand and destructive illusion.

I understood that the seed for the increasing eclipse of materialism as a viable worldview was, in fact, planted during the troubled 60s and sprouted in full force during the 70s and beyond. Those two decades were not only a revolutionary period in American social and political life but they were also the beginning of a radical transformation in our scientific and collective consciousness.

In this essay, I will consider some key factors that have birthed a new understanding of reality, a new emerging consensus that may accommodate mystical and religious experiences as legitimate forms of awareness and knowledge about reality. This development may bring an end to the conflict between science and religion that emerged during the Western Enlightenment. Specifically, I will comment on the following interdependent factors which are leading to this potentially monumental transformation: the Big Bang theory of the universe and quantum physics, the rise of transpersonal psychology and studies of psychic and mystical phenomena, as well as the rise of studies of near-death experiences.

BIG BANG AND QUANTUM PHYSICS

Until the early 60s, the dominant view among cosmologists concerning the origin and nature of the world was the so-called Steady State theory. This means that the physical universe must be seen as a brute fact in no need of an explanation and having no beginning or end. Consequently, there is no role for a creator and no apparent meaning for human existence.

Based on that underlying presupposition, leading thinkers of the last two centuries assumed that God is a transient illusion, and that mystical experiences and transcendent visions are nothing other than wishful thinking or regression to infantile stages of human development.

By the middle of the 60s, pioneering astronomers came to the radical conclusion that the universe had, in fact, a beginning and an end, having been created about 13.7 billion years ago as a result of a stupendous explosion that set it on its evolutionary trajectory. Whereas our galaxy, the Milky Way, was previously seen as the totality of the universe, astronomers now concluded that it is only one of at least 200 billion such constellations, each one housing billions of solar systems like ours.

This new understanding among astrophysicists unavoidably led contemporary philosophers and thinkers to revisit the old question as to who or what was behind this original explosion that virtually *ex nihilo* gave birth to these billions and billions of galaxies. Anyone who seriously contemplates the Big Bang, mused philosopher Ken Wilber, cannot but become an idealist.

Parallel to the questions raised by the Big Bang theory of the universe, another scientific breakthrough, in regard to the micro-level of reality, emerged at the beginning of the 20th century and became prominent during the following decades. Quantum physics revealed a universe "deep down" that is as unfathomable as the Big Bang of the outer universe. It is also compatible with what the mystical traditions of all the great religions have been teaching throughout the ages—namely, that consciousness plays a decisive role in how subatomic particles behave.

Mind, we are told, is not the byproduct of matter but the other way around. It is the foundation of everything that exists. Quantum physicists today write books on *The Self-Aware Universe* or *The Immortal Mind* suggesting that the materialist, reductionist worldview that we have inherited from the 18th and 19th centuries is no longer tenable.

I believe these two revolutions—the Big Bang theory of the macrocosm and quantum physics of the microcosm—are setting the foundation for further developments toward the emergence of a more integral vision of reality. Transcendent and non-ordinary or mystical experiences can no longer be easily dismissed nor marginalized by the gatekeepers of mechanistic, scientific orthodoxies.

To sum up, recent developments in physics have paved the way in bringing Spirit back into the picture through the emergence and appearance of a self-aware universe inhabited by self-conscious beings. As paradoxical as this may sound, a spiritual understanding of the foundation of the universe is beginning to make its first steps inside the corridors of science and academia.

TRANSPERSONAL PSYCHOLOGY, PSYCHIC AND NEAR-DEATH EXPERIENCES

Parallel to the above developments in the hard sciences is the emergence during the late 60s of Transpersonal Psychology. A group of pioneers like Abraham Maslow and Stanislav Grof established this sub-field within psychology based on the conviction that there are stages of human cognition and experience lying beyond the rational ego. This branch of psychology focuses on the study of phenomena that are neglected or shunned by the mainstream. There is increasing recognition of widespread religious, mystical, and peak experiences, including clairvoyance, precognition, telepathy, and psychokinesis (known as PSI). These phenomena are compatible theoretically with the changes taking place in physics, as discussed above.

To be sure, transpersonal perspectives existed before the 60s but remained unrecognized as such because of the dominance of the materialist worldview. Pitirim A. Sorokin, for example, the Russian American sociologist who set up the Harvard sociology department during the 30s, taught that we can know the world in three ways: via our senses (empirical science), our mind (reason, mathematics, logic), and through intuition (extraordinary mystical experiences, psychic phenomena). The first two are acknowledged and cultivated in our Western educational systems, but intuition—which, according to him, is a superior and *sui generis* form of knowledge—has been ignored.

Sorokin advocated the development of what he called "integral truth" that would honor all three strands of knowledge. He was a sociological precursor to Transpersonal Psychology insisting that human beings have not only a conscious and an unconscious mind but also a supra-conscious mind that can account for phenomena like extrasensory perception and mystical and psychic experiences. Sorokin's position was similar to what William James, the celebrated father of American Psychology demonstrated in his *Varieties of Religious Experiences* and what psychoanalyst Carl Jung termed the "collective unconscious."

Our understanding of the self, based on the perspectives of such pioneers, is continuously being validated by contemporary empirical research. The beginning of the 70s offered a fertile ground for spirituality as it coincided, among other things, with the creation of *The Journal of Transpersonal Psychology*. Through that journal and other scientific venues like the Esalen Institute in California and the Institute of Noetic Sciences— created by astronaut Edgar Mitchell right after his own mystical experience during his return voyage from the moon—pioneering research has been offering ongoing support to the reality of psychic and mystical experiences. Such experiences are real and deserve serious study by the academic community as shown by the work of psychologists such as Dean Radin and Charles Tart, as well as physicists like Russell Targ and Harold E. Puthof.

I have personally witnessed such phenomena in my own field research, exploring the world of unconventional psychic healers, mystics, and Christian monks and hermits. For myself and social anthropologists like the late Michael Harner of the New School of Social Research—*The Way of the Shaman*—there is no doubt that such phenomena of transcendence can be real and authentic rather than illusory or forms of psychopathology.

We are becoming increasingly aware that such experiences seem to be widespread around the world among people of all classes and professional backgrounds. I am amazed to be contacted by international readers who are eager to let me know that they live within the realities of my research subjects. Yet, they remain silent out of fear of being labeled as mentally ill. Among the categories of people who approach me, there is a noticeable number of scientists and fellow academics who seem to live a double life. They keep their extraordinary experiences secret out of fear of being misunderstood and thus have their professional careers undermined.

It would be a great omission not to mention the impact of the revolutionary work in thanatology of pioneer Dr. Raymond Moody, dubbed the new Columbus. His seminal work on *Life after Life*, published in the 70s, launched a cottage industry of near-death studies showing that consciousness is not confined within the physical brain. The Near Death Experience (NDE) is by now a household word demonstrating that many people who have been declared by doctors as clinically dead, having no medical signs of life, return to consciousness with extraordinary tales of their near-death state. Moody's research supports the notion that the brain is not the same as the mind but rather that it is a vehicle of the mind. It also means that the mind of a human being seems to survive the physical death of the body. The ongoing and promising research related to NDE supports the notion that it is not the consciousness of human beings that is dying but rather the mechanistic worldview that appears increasingly outdated.

As a postscript, I would like to quote from my latest work *The Accidental Immigrant: A Quest for Spirit in a Skeptical Age*.[26] "Mystical, transcendent, ecstatic experience is an ever-present potentiality within human nature itself. There will always be human beings who have such experience, and under certain historical conditions, this experience may spawn movements that culminate in institutional religions. Mystical transcendence could never be eradicated by secularization, and religion could never be annihilated from human history… In the words of Aldous Huxley, 'The mystics are channels through which a little knowledge of reality filters down into our human universe of ignorance and illusion. A totally unmystical world would be a world totally blind and insane.'"

For all of the above reasons, the current volume of Professor Paul J. Mills could not be more timely. It promises a major empirical contribution toward the ongoing transformation of science through the spiritual and mystical lives of its practitioners. It is poised to contribute in a major way to the emergence of Sorokin's integral truth, a prerequisite for the long-term survival of our species.

CHAPTER 1

THE NEED TO TRANSFORM MATERIALIST SCIENCE

Contemporary science is based on the philosophy of materialism,
which claims that all reality is material or physical.

— Rupert Sheldrake

W hen Joseph Campbell visited India in 1954, he had already written *The Hero with a Thousand Faces,* which recognizes the universal "monomyth" motif of adventure and transformation found in nearly all of the world's cultures. He decided he wanted to meet a major guru, so he went to see a celebrated teacher named Sri Krishna Menon who lived in Trivandrum. As soon as Campbell sat down, the teacher asked him, "Do you have a question?"

Campbell learned later that he had the good fortune of asking the same question that Sri Krishna Menon had first asked his own guru: "Since all is the divine radiance, how can we say 'no' to ignorance, to brutality, to anything if the universe is a manifestation of divinity?" If that were the case, wouldn't it mean that "even the base elements of the world are manifesting divinity?" The teacher replied, "For you and me, the way is to say yes." It confirmed for Campbell his thoughts that anything that fosters separation is false.[27]

Unfortunately, the materialistic sciences do just that—foster false dualities of spirit and matter where none in fact exist. Modern science casts a judgment upon the nature of the world and discounts the realm of inquiry called the spiritual. One of my interviewees, Julia Mossbridge, beautifully stated that "Judgment is observation without love." The great Tibetan teacher Djwhal Khul considered judgement "humanity's greatest heresy." Our sciences foster duality and judgment - observations without love - resulting, as we will see, in numerous downstream negative personal, social, and environmental consequences.

Campbell left this teacher with new insight and a new way to meditate. He was instructed by the guru to focus his attention between his thoughts. From there, he could gain an intuitive flash of the source field out of which all thoughts and energies arise, including those that structure his own sense of egoic identity. The goal was to move between apparent opposites: stillness and thought, good and evil, up and down, the seemingly material and spiritual, into an experience of unity.

Campbell was on the journey on which all scientists embark to find answers to fundamental questions of self-identity, life, and the broader universal existence. Many scientists hear the call of an ephemeral experience that makes them wonder who they are and how the world works. They then thirst for knowledge and experience that will let them discover the answers for themselves. They choose the scientific path of knowledge and inquiry because it is considered a reliable way of discovering valid knowledge.

Since humanity's inception, we have been on a path of seeking and attaining increasing amounts of knowledge. Be it Adam and Eve approaching the Tree of Knowledge in the Garden of Eden, or Prometheus stealing knowledge of fire from the Greek gods, we are driven to learn. The so-called Scientific Revolution in 1593 marked the beginning of scientific methodologies as the dominant path to attaining new knowledge. While there have been clear benefits to the emergence of this way of gaining knowledge, there have also been significant downsides to how it has ultimately progressed.

THE CASE AGAINST MATERIALISTIC SCIENCE

Reductionism is a hallmark of the modern scientific approach to knowing, one of its foundational attributes and strengths. The idea is to reduce down to its constituent parts whatever is being studied, and once the parts are understood, reconstitute the parts into the original whole. For a variety of reasons—including the phenomenon of "emergence"—this approach does not work for complex organisms. The belief that one *can* take the reductionist approach to know complex structures and living organisms is called "naive reductionism."

Among scientists, Dr. Rupert Sheldrake stands out as having spoken against naive reductionism and materialism more broadly. In one of his pioneering books *Science Set Free,*[28] he lists the "ten core beliefs that most scientists take for granted." Included among these ten creeds are that there is no reality but material reality, that human consciousness is a byproduct of the brain, and that matter is unconscious.

Sheldrake writes, "These beliefs make up the philosophy or ideology of materialism whose central assumption is that everything is essentially material or physical, even minds. Science is being held back by centuries-old assumptions that have hardened into dogmas." He then offers ten solutions that can pave the way for much-needed new discovery. "By freeing the sciences from the ideology of materialism, new opportunities for debate and dialogue open up, and so do new possibilities for research," Sheldrake writes optimistically, adding, "Much remains to be discovered and rediscovered, including wisdom."

Concurring with this position, Ken Wilber describes the current sciences as "narrow science," only allowing evidence from the lowest realm of consciousness, the five senses and their extensions.[29] Other authors too address the "consciousness-matter relationship," including in the books *You are the Universe*, and *The Self-Aware Universe: How Consciousness Creates the Material World*. They argue that consciousness is not a byproduct of matter but the *source* of matter.

In 1926, J. C. Smuts published the book *Holism and Evolution* in which he argued against the growing tide of reductionism he observed in the sciences. He foresaw a time like now when the appreciation of the innate holism of all living things and a holistic and integrative approach to science and its related fields would eventually reemerge. Other more current books about this emergence characterized by the recognition of wholeness and interconnectedness include Duane Elgin's *Choosing Earth: Humanity's Great Transition to a Mature Planetary Civilization*, Dr. Robert Atkinson's *The Story of Our Time: From Duality to Interconnectedness to Oneness*, and *The Coming Interspiritual Age* by Kurt Johnson and David Robert Ord. The latter shares an evolutionary vision of consciousness development and an eventual synthesis of world religions and spirituality.[30]

Leon Kass, the former chairman of the President's Council on Bioethics, writes about the culture of materialistic science and its "soulless scientism" as being a great threat to our human nature. He defines scientism as "a quasi-religious faith" that "eliminates all mystery," "giving purely scientific explanations of human thought, love, creativity, moral judgment, and even why we believe in God."[31]

Science in the form of scientism is currently a kind of religion that has become an ultimate authority, the arbitrator of truth of the nature of existence, including how our own bodies and minds work and our relationship with the natural world around us.

I, too, have considered scientism a kind of religion among highly materialistic scientists, and in some ways, it's an understandable consequence of the roughly four centuries-long split between science and religion, the denying of the existence of the spiritual within the scientific endeavor. Scientism represents our overly mentalized minds, a state in which it's not easy to join our spiritual nature with the mental nature of our human heritage. With its overbearing emphasis on the material world, on everything that can be objectively parsed and measured, the scientific endeavor has strictly limited or outright avoided its explorations into human spirituality.

Rupert Sheldrake speaks of the "scientific priesthood."[32] This is highly unfortunate and—in some ways—mirrors, in reverse, what took place during the birth of Western science. The Roman Catholic Church deemed Galileo and Copernicus as heretics because their scientific discoveries challenged religious doctrine and authority. Today, science not only avoids studying the spiritual but actively denies its existence, thus implicitly ridiculing the very foundations of the church which spawned its current beliefs.

Science and biomedicine have made our bodies machines and don't consider the emergence of consciousness, the "ghost in the machine."[33] Richard Dawkins, author of *The God Delusion* and perhaps the world's most famous materialist, writes, "There is no spirit-driven life force," no soul nor consciousness, and we are simply the sum total of our genes. This deeply embedded closed-mindedness has limited the potential for important discoveries in a broad range of science. Can someone who is closed-minded to the possibility of what is outside their realm of experience really be considered a scientist?

Kass writes that the strictly scientific *outlook* on life and the world that characterizes scientism needs to be distinguished from science itself, which "is a methodical art for gaining knowledge and the accumulated knowledge itself." One might argue that scientific inquiry has its own inherent limitations and, as such, could never actually know the ineffable. Kass rightly points out that the currently considered limits of science derive not only from our current scientific knowledge but also more foundationally "from science's assumptions about what sorts of things are scientifically knowable. They stem from science's own self-proclaimed conceptual limitations—limitations to which neither religious nor philosophical thought is subject."

Owen Barfield, described as one of the most original thinkers of the 20th century, was part of an Oxford literary group known as "The Inklings," among whose members included C.S. Lewis and J.R.R. Tolkien. Barfield was keenly interested in the effects of materialism on our cultures and saw imagination as a remedy. For him, imagination was not something to be discounted but a deeply important faculty of the human being, something to be cultivated. It is through imagination, whether conscious of it or not, that we apprehend reality. He wrote, "Mere perception—perception without imagination—is the sword thrust between spirit and matter."[34] Barfield agreed with the English poet Samuel Taylor Coleridge's perspective that true imagination is creative, it being "the living power and prime agent of all human perception." Echoing Barfield was physicist and Nobel Prize recipient Frank Wilczek, who said that knowledge without imagination is barren. Albert Einstein said that imagination is more important than knowledge. Imagination is, in fact, one of our spiritual faculties.

For Barfield, without imagination, matter becomes a "thing." This "thing" is a hallmark of today's materialistic sciences and what Kass' "soulless scientism" has given us. If Barfield could look upon our materialistic sciences today, I suspect he would describe them as a form of idolatry, studying appearances with no appreciation of what is within them.[35] Science, as it is practiced today, doesn't support perception of an intimate relationship with matter, with the world, but maintains an artificial separation from it. "Mother Earth" became "Nature" which then became the "environment" and more recently is considered only a "resource." The scientific revolution created a material world no longer to be understood for its spiritual nature but rather to be quantified and controlled.

Over the past two centuries, science has provided us with deep knowledge of the complexities of life and the beauty and vastness of the universe. One would think this would instill in us such a great sense of awe and connection that we would quite naturally overcome the belief in materialism. It's odd because that hasn't been the case.

Joseph Campbell used the term "mythic dissociation" to describe this phenomenon in the broader culture. The term refers to the loss of meaning in a culture when a *culture* moves away from its founding mythologies and evolves into a *civilization*. For Campbell, mythologies keep the individual, as well as their culture at large, "transparent to the transcendent." Mythologies engage the imagination and thus avoid Barfield's "sword thrust between spirit and matter."

Campbell's mythic dissociation reminds me of the term "disenchantment," which Max Weber uses to describe the disappearance of the "mysterious and incalculable" from our natural world.[36] Today, our cultures have been "demythologized" as a result of our materialistic sciences.[37] As such, science is not "transparent to the transcendent" and doesn't provide that window into the transcendent which would ultimately benefit society from what new discoveries could be had.

In his book *Esotericism and the Academy*, Professor Wouter J. Hanegraaff tracks historically how all things esoteric and mystical were purged from our philosophical and ultimately scientific traditions.[38] All such topics became the "Other," to be scorned with contempt and avoided at all costs by academics.

In the 1968 movie *Planet of the Apes,* there was the "Forbidden Zone," which apes of the society were forbidden to enter. What resided in the forbidden zone was knowledge that would reveal their origins and true identity, which was contrary to what the apes were taught to believe. Any ape entering the zone, risking banishment and even death, would ultimately learn that the mythology of their origins was not true. The areas of spirituality, mysticism, and metaphysics are such forbidden zones in our modern world of science.

Scientists who do venture into these zones, while no longer persecuted as heretics, are persecuted in new ways, being labeled "pseudoscientists" or "woo woo scientists," and can become objects of campaigns to discredit them and their scientific research. The degree to which materialistic scientists truly believe in their worldview and that science as they understand it has the responsibility to protect society from scientific quackery is an understandable position. What, however, are the downsides of that approach?

During my interview with Dr. Dean Radin, he said he understands why many scientists are close-minded, sticking with materialism, and not wanting to go anywhere near metaphysical phenomena. "It would suggest there's something about materialism that is incorrect in some way." The materialist position is "well, metaphysical phenomena such as telepathy cannot possibly be true because if it is we'd have to throw away all of our textbooks and start over again." Of course, he said that's nonsense; we don't throw away what works. It would simply mean that all of our assumptions about materialism don't hold 100% of the time.

Within these considerations, healthy skepticism must be given its due. A degree of skepticism is important in science. In some ways, it plays a central role in guiding the progression of science. The theoretical physicist Richard P. Feynman, who in 1965 won the Nobel Prize in Physics for his work in quantum electrodynamics, said, "Religion is a culture of faith; science is a culture of doubt." Skepticism that has transformed into closed-mindedness, however, is a kind of poison—a poison of the scientific mind and a poison to the process of scientific inquiry.

There was much evidence of scientific closed-mindedness when the complementary and alternative medicine movement emerged in the late 1970s. Closed-minded scientists and clinicians were very averse to these efforts. As a result, the ability to conduct solid systematic science into a variety of complementary and alternative modalities was thwarted because few scientists were willing to study it. The result was that the practitioners of these modalities, in their frustration to see studies conducted, began to perform their own research as best they could. Given few of these practitioners had proper training in scientific methods, many of the early studies were deeply flawed and rightly ridiculed for being of such poor scientific quality. As a result, the field was further disparaged.

It was only after more than a decade that research started to open up with more scientists and clinicians being willing to study these modalities. Granting agencies also joined the endeavor with much-needed funding. Now, some forty years later, there is a robust field with tens of thousands of scientific publications on the benefits to physical health, psychological health, and spiritual health of individuals who adhere to a variety of complementary and alternative medicine practices, particularly for certain yoga asanas, meditation, and pranayama (breathing) techniques. The evidence is so significant that the field of Integrative Health emerged, helping to provide more validity to these practices and their eventual integration within many clinical settings.

During our interview, Dr. Wayne Jonas asked the question, "What is the biggest obstacle to discovery?" He went on to share one of his favorite phrases by Daniel J. Boorstin, who said, "The greatest obstacle to discovery is not ignorance—it is the illusion of knowledge." Wayne went on to say that his mother is one of the wisest people he knows, and she has a simpler way of saying that "when you're green you're growing, when ripe you're rotting." Science misses so much by thinking it has all the answers without actually having the data.

Personally, I don't understand this phenomenon of rigidity in science. By the very definition of the scientific method, there will never be a definitive "truth" of anything. Every scientific discipline rests on theories, and theories are assumptions. Oftentimes, these assumptions turn out to be either partially or completely wrong. A brief study of the history of science will reveal this to be the case. Pick up a medical or physics textbook from twenty or even ten years ago, and you can find many examples of what was previously believed about something is now discredited. With each new discovery, an older theory and its assumptions are modified or completely discarded.

In school, theories are not often taught as being provisional but, by definition, that is the case. The Library of Medicine hosts the largest biomedical database in the world, called PubMed. As of today, PubMed houses over 34 million scientific publications. It is currently estimated that anywhere over fifty percent of the scientific reports in that database are false, i.e., they are incorrect observations. Such a statistic makes one question the idea of science delivering ultimate truths.

It seems reasonable to expect that, as an endeavor, science would be open to all realms of inquiry. That it hasn't been is an odd and unfortunate situation. Odd because if we think of the purpose of science—to explore the unknown and make discoveries that could benefit humanity— science has consciously chosen to ignore fundamental characteristics of human experience. It is unfortunate because mainstream science is a vast and wealthy enterprise, with worldwide spending of over $2 trillion per year. To quarantine an entire area of inquiry as being unsuitable, out of bounds, pseudoscience, and quackery means those vast resources are not being used to provide much-needed discovery and application of those discoveries for the betterment of humanity.

What have been the costs to scientifically driven progress? I think the most significant adverse consequence is that the materialistic sciences have supported a false picture of duality, of separation of the individual from the world, and a denial of our fundamental nature.

In contrast to Western scientific traditions, there are long-standing traditions in the East of scientific inquiry into the metaphysical, including the nature of human consciousness. While the West focuses on the physical world of the senses, some Eastern traditions focus on what is sometimes called the "supersensible," that is, studying what lies in the spiritual world beyond perception with our external physical senses but available to study by using our inner senses, including imagination.

It would surprise some Western scientists to learn that the methodological approaches taken by these traditions have a great deal in common with how they themselves approach their science. That is, subjective approaches to spiritual inquiry can fulfill the two primary observation criteria of validity and reliability, hallmarks of Western scientific methodology. These models show that it's possible to bridge the perceived gap between the scientific and spiritual worlds.

OPENING THE GATES FOR A RENEWED SCIENCE

Fortunately, at least for some scientists and clinician scientists, this approach to scientific inquiry is being actively explored. Some have actively moved forward with exploration of their own spiritual nature and brought their insights and realizations to bear on their research. How might this be affecting the quality of their science? I would think it leads to more inspiration and certainly greater creativity. Albert Einstein said, "The most beautiful and profound emotion we can experience is the sensation of the mystical. It is the source of all true science. He to whom this emotion is a stranger, who can no longer wonder and stand rapt in awe, is good as dead."

The scientists interviewed in the pages of this book share their own personal stories about how they are pursuing the spiritual and mystical and bringing new insights to bear on their research. I've observed that more spiritually oriented scientists don't fall into the trap of materialism. They can follow the reductionistic model of science, looking at the parts, but they have an easier time keeping sight of the whole because they are in active touch with their own inner holistic nature.

Science has achieved great feats that have changed the world and the course of humanity, but if it remains only materialistic, its potential for more meaningful contributions will remain vastly limited. Beyond any of the benefits to be seen in healthcare and other sciences, the lasting and most meaningful benefit will be a greater understanding of ourselves, for what could be a greater gift to humanity of the sciences than an appreciation and realization of our own nature? The mystics and so-called spiritual teachers have been trying to awaken us to such knowledge and experience for millennia. For the Western scientifically minded, it will be science itself that provides the context for this to finally find its landing.

As I see it, this will be a repudiation of the premise of John Horgan's book *The End of Science*, which argues that "the era of truly profound scientific revelations about the universe and our place in it is over."[39] In actuality, further profound scientific revelations will be had when we recognize that the spiritual is the final frontier for science. Support of this endeavor will be at hand when scientists can successfully investigate the nature of the subject known as "I."

An overarching goal of this book is to help heal the illusion of separation within and beyond the sciences and to educate about these domains of consciousness open to us all. By embracing their spiritual nature, not rejecting it, many scientists are changing the landscape of a variety of scientific disciplines. I consider this vital at this time because the Western sciences, including the biomedical sciences, need to transition from the current strongly materialist scientism worldview to a more inclusive and unitive worldview.

In his essays on the nature of consciousness,[40] Rupert Spira discusses the consequences of the "calamitous assumption of materialism" that lies at the root of our cultures. It's not just the scientist and science that are adversely affected, it filters down through all stratas of society. Spira writes that this assumption of materialism "informs almost every aspect of our lives and is the root cause of suffering within individuals, the conflicts between communities and nations, and the degradation of our environment."

Professor Mattias Desmet at Belgium's Ghent University describes another highly significant downside from our living in a "dead universe" materialist worldview without a foundation in Spirit - the ability of totalitarianism to take hold more readily in a culture.[41] In the book *Dark Persuasion*,[42] my mentor Professor Joel E. Dimsdale does a deep dive into the history of brainwashing and describes how totalitarian governments routinely use it to serve their goals. Brainwashing and its consequences are possible when individuals are living only from the mind and not their deeper spiritual nature.

Spira's calamitous assumption is perhaps akin to science's foundational "conceptual limitation" that Kass refers to. It is science's blind spot, the elephant in the room. While today, it is certainly proudly "self-proclaimed" by many scientists, historically, it wasn't self-proclaimed but more so imposed. It was the Roman Catholic Church after all that told nascent science it couldn't study the non-physical—the spiritual—because scientific ideas and discoveries contradicted church teachings.

Science was hamstrung right out of the gate, having an entire world of potential exploration closed off to it. For scientists, over the centuries we've gone from knowing we needed to avoid the spiritual to now having entire generations of scientists who don't believe there actually is a spiritual world. The church, too, lost out as it disavowed interest in the physical world, essentially divorcing itself from the earth and the beauty of human incarnation itself.

Recognizing the existence of the calamitous assumption of materialism is the first step to healing the sciences so they can better contribute to supporting the spiritual lives of scientists and their ability to meaningfully contribute to our understanding of what humanity is and to a healing and well-being of humanity.

The theoretical physicist Max Planck, who won the Nobel Prize in Physics in 1918 for his discovery of energy quanta, said, "It was not by accident that the greatest thinkers of all ages were deeply religious souls." I believe he's not speaking here of religion per se, but of a deeply soulful way of approaching existence. Toward this end, Planck also said, "All matter originates and exists only by virtue of a force. We must assume behind this force the existence of a conscious and intelligent Mind. This Mind is the matrix of all matter."

It's time to finally eliminate the false divide between scientific discovery and our spiritual nature in order to support the creation of societies and healthcare systems that actually support the health and well-being of ourselves and the natural world. It is surely time to liberate the materialistic scientific mind to embark on a new epoch of truly transformative discoveries for the greater good. Let us embrace and expand our appreciation of our own nature, which is that of the universe itself. Scientists like the ones interviewed in these pages are leading these efforts for all of us.

CHAPTER 2

HEEDING THE CALL: THE MAKING OF A "SPIRITUAL SCIENTIST"

We must have perseverance and above all confidence in ourselves. We must believe that we are gifted for something and that this thing must be attained.

— Marie Curie

W hat prompts an individual to become a scientist and take on this pursuit? Throughout the conversations I had with my fellow scientists, I found remarkably different reasons. Some were deeply sensitive to nature as a child or had innate intuitive or telepathic abilities. For others, it was trauma, a mental illness, or a death in the family. Still others first encountered the mystical through the use of psychedelics or, like me, pursuing a meditation practice. Still others accessed the mystical through a profoundly deep desire to know themselves and their place in the universe.

Varied as these reasons were, they were pivotal moments in each interviewee's life—moments that opened the door to their journey. It was the proverbial "heeding the call to adventure" stage of the monomyth. A consistent feature among them is that they were all deeply personal experiences. The memory of these experiences and the lasting imprints they left on their consciousness continue to be guiding forces through their lives.

THE CALL DURING CHILDHOOD

For Rudolph (Rudy) Tanzi, his opening into the metaphysical wasn't through trauma, drugs, or meditation, but through playing a game with his twin sister. When they were children, they played a game where they tried to convince each other that nothing existed. They would begin the game with a series of questions to each other, "What if there was no house?" "What if there was no neighborhood?" "What if there was no city?" "What if there was no earth?" The questions would go on and on until they arrived at the question, "What if there was no universe?"

If they were successful with the game, as they arrived at the place where nothing indeed existed, not even the universe, they would get a unique experience in the pit of their belly that they called the "flip." They would race each other to see who could get to the flip first, to experience the feeling that absolutely nothing existed.

Rudy told me that his experience of the flip was deeply visceral and energetic in how it affected his consciousness. It was something beyond anything he could experience in any other circumstance in his life. It wasn't until many years later that he was able to categorize the flip experience as a window into the metaphysical.

The more they played the game and experienced its rewards, the more Rudy's interests in the mystical grew. He said, "The game led me to wonder, 'What is consciousness, what is awareness, what is the universe?'" He eventually decided that since science allows us to measure what meets the eye, perhaps if it could also lead him to what lies beyond the senses, he should learn the scientific method in his quest to understand the metaphysical.

He said those earliest passionate interests and his willingness to explore the metaphysical made him a different type of scientist, helping him maintain an open mind and keep thinking about the big questions. He said it kept him thinking "outside of the conventional box of science." He also shared that he's "always been titillated by the idea of new discovery. I love when you discover something new and for that moment, or for that day or perhaps a few weeks, you're the only person that knows something really important about the universe that nobody knew before."

Science fiction, too, became a passion because, through it, he could learn about the paranormal, telepathy, and mediumship. He read voraciously on these topics. He recalled that his friends wondered, "What's wrong with Rudy?" because all he was interested in talking about were these strange things.

Rudy is an amazingly talented human being. In addition to being a top-tier scientist—having discovered three of the four genes linked to Alzheimer's disease—he's also a professional musician, having played with the rock band Aerosmith. Some years ago, he appeared in a *GQ Magazine* photo shoot for a campaign called "Rock Stars of Science." Today, he is the Joseph P. and Rose F. Kennedy Professor of Neurology at Harvard University and Director of the Genetics and Aging Research Unit. He's shared his explorations of consciousness at numerous meetings around the world, including the annual Science and Nonduality conference.

Other interviewees had a different kind of sensitivity as children, not in a form that manifested while playing a game, but a sensitivity to the natural world around them, prompting in them a deep yearning to understand the world and their relationship to it.

Neil Theise is a Professor of Pathology at NYU Grossman School of Medicine, a diagnostic pathologist who also does academic work in stem cells, liver disease, and the anatomy of the human interstitium, and "dabbles in complexity theory." Neil grew up in a household that was open to spiritual exploration. He said that religion "was a lovely thing as it was practiced by my family. It was very rich, warm, and non-punitive." His father was actually part of the *Kindertransport,* or Children's Transport, a rescue effort between 1938 and 1940 that brought thousands of refugee children, the majority of them Jewish, to Great Britain from Nazi Germany. "That's how he survived." His father's parents, however, did not survive. "A lot of our family was killed. I grew up in a world of survivors where there was an imperative to find a way to flourish."

As a young child, Neil experienced that world with "a deep-felt sense." He felt the world was highly personalized and rich with feeling. "If I kicked a rock, and it went into a sewer, I felt bad because the rock would be unhappy. If I found ants in our home, I would carry them outside because I didn't want my mother to notice, since she would then call the exterminator, and there'd be an ant Holocaust." That intimacy with the world turned into curiosity, which eventually turned toward medicine and the sciences.

His mother was from England, from a family with a long line of rabbis going back to the Middle Ages—a true lineage, I thought. She "operated in the world through magical realism, which included her being able to communicate with the deceased." He explained, "Magical realism wasn't a style; it was how my forebears experienced the world, and my mother very much experienced it that way. There was the emotional sense of being connected to God, which included ritual, holiday celebrations, and family relationships."

Magical realism is something that Neil, too, has direct experiences with. I found that linkage interesting because there is evidence that lineages can carry abilities such as clairvoyance and clairaudience through generations. After his mother passed away, Neil told one of his mother's friends that his mom had recently been communicating with the deceased. The friend replied, "Your mother was always seeing dead people!"

One of Neil's dearest friends, Richard, died during the AIDS epidemic. "Before he died, I said to him, 'Look, you will probably have more important things to do, but if, after you're dead, you find yourself aware of things and able to interact with this world, and don't have anything more important to do, I'd love to have you around for a year as a guardian angel.'" Richard promised he would stick around if he could. Neil shared that he did stick around. "He was a very vivid presence for me."

While Neil was in college, a cousin from Canada visited and gave him a copy of James Michener's *The Source*, which describes the rise of Jewish mysticism during the medieval period. He had never heard of Jewish mysticism; Michener's novel opened him to those new ideas.

As he learned about Jewish mysticism, he was moved by the idea that, as a human, we can experience God's perspective of the world, particularly as he contemplated the Jewish Holocaust. He then related a story of a walk on the beach with a friend shortly after gaining these newfound insights from Jewish mysticism. As he related the story to me, he choked up. While walking, he was sharing with his friend what he was learning about Jewish mysticism, and he had a sudden insight. He said to his friend, "The idea is that if a human can have an experience of God's perspective of the world, then everything can make sense." He was thinking about the Holocaust, in particular. "So that's when I started thinking about mystical stuff and wanting to learn about it."

A rub for Neil, however, was in the classic teachings in Judaism, you can't formally practice Jewish mysticism until you are forty years old, married, and have attained a full understanding of the Talmud. "I was gay and couldn't be married, so I couldn't go that route." During his senior year of college, someone gave him the book *The Three Pillars of Zen*, by Roshi Philip Kapleau. The book contained meditation instruction and descriptions of people's firsthand experiences of enlightenment. "I thought to myself, 'This is the kind of stuff I was reading about in Jewish mysticism, but more so, this is a how-to guide rather than a novel.'" Zen offered him a route to study mysticism that Judaism would not permit him to pursue.

A few years after that his husband Mark read an article about a Zen monastery in the Catskill mountains where one could do retreats. Mark said, "You've been talking about this Zen thing forever. Why don't you go and do it?" So he did and started a deep dive into Zen practices. His practices included the use of Zen Koans, which are "the sort of thing that, when you're working on it, there's no intellectual way through it. You can't answer it, but you can't let it go either. It sort of wears away, even when you're not thinking directly about it."

His Zen practices eventually led to a deep realization of the "emptiness of inherent existence." He was in the Zen center in New York City, sitting on his cushion. "I looked up across the room, and on the altar was a burning stick of incense turning into smoke. This was just one of those moments where you get it. I realized the emptiness of inherent existence. Something is a thing, and at the same time, from another perspective, it is a phenomenon, and one does not exclude the other. But you can only see one or the other, depending on your perspective." He said, "I got very excited; it was a blissful moment."

The next time he had the opportunity to meet with his Zen teacher, he described what happened and asked, "Is that what emptiness of inherent existence means?" She responded matter-of-factly, simply saying "Yeah," confirming his experience. He said, "That's it?" She said, "Well, yes, that it's. It's not easy, but it's simple," meaning the emptiness of inherent existence is simple, whereas getting to that realization is not. "Since then, for me, it's been just how those two things dance with each other."

He described it "as a gifted moment" because it helped solve something he had been living with since he was a young boy. He had his Jewish tradition and its theology, and he had what he was learning about in science. They were separate, and that was okay. He didn't think they had to be connected. "They were in two separate boxes in my brain." He wanted to do both and did not want to choose sides. Going forward, his Zen realization brought those two boxes together for good. The spiritual informs the science, and the science informs the spiritual; they're no longer separate things to me— they came together." Questions that he had thought were spiritual have scientific answers, and questions he thought were scientific have spiritual implications.

Christine Peterson's early spiritual experiences came through as a child with her deep connection with the natural world. Christine grew up on a 50-acre farm in West Virginia, where the nearest town was an hour away. Other than her parents, she wasn't in contact with many people. As a child, if she wasn't tending to the cows, she would climb up to the top of the mountain behind their home and stare at the sky. "I had some of my first nondual consciousness experiences on the top of that mountain," she said. Nonduality, or living in nondual awareness, refers to a state of consciousness where the dichotomy of "I and other" is transcended and awareness is experienced as oneness.

"I would stare into the expansiveness of the blue sky, with maybe a cloud floating by, and just feel my body start to melt away and merge into that expanse." She didn't have the words then to describe it but can now say that it was a dissolution of her sense of individual self into a much larger non-personalized Self.

Christine said that losing the typical sense of self boundary that most people carry around wasn't in any way upsetting or disconcerting. In fact, she preferred it. In that nondual awareness state, she would feel more love for and connectedness with the cows, the rocks, and the plants—all of the natural world. She added that those early experiences of the expansive nature of her own consciousness sparked her eventual movement into science.

While an undergraduate student at the University of Virginia she found an ashram, the Satchidananda Ashram called Yogaville. There she was exposed to "hatha yoga and, luckily, to the concept of living yoga alongside of the asana itself."

I asked her what she meant by "living yoga." She clarified that she learned all the limbs of yoga, not just hatha (the asana postures), but also bhakti, jnana, raja, karma, and japa yoga. She said people who go to yoga studios don't often learn the full context of what yoga actually is. After our interview, I went to the Yogaville website and learned that they teach what is called Integral Yoga, which "synthesizes the various branches of yoga into a comprehensive lifestyle system, the purpose of which is to support the harmonious development of every aspect of the individual."

While later getting her Ph.D. in Microbiology & Immunology from the Institute for Biomedical Sciences at the George Washington University School of Medicine, she started taking yoga teacher training programs. She also started taking formal meditation classes.

While learning meditation, she met someone who would later become her Tibetan philosophy teacher. His name was Llama Tsoknyi Rinpoche. While on a meditation retreat with the Llama, he taught her and the rest of the students a specific meditation technique called Sky Gazing, the purpose of which was to "help the practitioner find release from the narrow confines of the individual personality to connect to the expansive experience of their authentic nature as awareness."

As Christine was learning this, she thought to herself, "Wait a minute, I know this technique!" It was what she had been doing as a child on the mountains of West Virginia. She learned there was a name for what she had practiced and experienced. It was part of a formal meditation system in Tibet.

I asked her how learning this information might have affected her spiritual experiences. She said, "It allowed me to go deeper with the practices themselves and also, importantly, to have a context for what I was experiencing." She felt she grew more from the experiences because she now understood what was going on.

During graduate school, she went on a three-day retreat with the spiritual teacher Amma and there learned about Ayurveda. "When I heard what Ayurveda was, I had a recognition that I needed to study this ancient medical system. It lit a spark in me." It gave her goosebumps speaking about it during our interview.

Over the subsequent years, she's taken intensive training in Ayurveda, learning the different daily and seasonal practices which benefit her own health and well-being. In addition, she graduated from the California College of Ayurveda and, with her certification, helps other people benefit from Ayurveda.

Christine said that the "purpose of Ayurveda is to help us realize our connectedness to all of life, to balance and extend the life of the human being and support self-realization so the natural state of Samadhi can arise." As she said these words, I realized that Ayurveda was much more than I had previously thought it to be. A few years ago, I had the pleasure of traveling with Christine to Kerala in Southern India, the acknowledged birth pace of the Science of Ayurveda. It was a wonderful and adventurous trip.

Today, Christine is on the research faculty at UC San Diego, at the forefront of scientific research showing how traditional herbal medicines cultivate health for our gut microbiome and can have positive effects on diseases such as Parkinson's. She's published many groundbreaking papers on these topics.

I asked Christine if she thinks there is hope for the materialistic sciences to be transformed. She said, with the rise of systems biology, she sees hope because that approach to science is by nature more inclusive and expansive, taking the scientist further and further beyond what were previously thought to be the limits. She explained that we'll need a new system of science with new methodologies and tools of assessment. "I've heard people complain and 'call out' science as being too materialistic, but they never really offer a new system or framework to replace it."

She recalled the book *Flatland*, by Edin A. Abbott, where everyone can only see in two dimensions, so their view of reality is skewed and not completely relevant. They needed a new perspective. I hadn't read the book in ages but wondered if the people in that book who lived in 2-D were deeply prejudiced against concepts of 3-D and beyond, as we see in scientism today.

Menas C. Kafatos is the Fletcher Jones Endowed Professor of Computational Physics at Chapman University. Like Christine Peterson, Menas' realizations started by gazing at the sky—not the blue sky of the day, but the star-filled sky of the night. At age ten, Menas, who co-authored the book *You Are the Universe*, started watching the night skies of Crete. His fascination was with "the sense of infinity" he would personally experience in the Milky Way. It kept him fixated, and it accompanied a strong felt sense. "I had the sense that I came from the stars. It was strange because when you are that young, you don't rationalize such experiences away. You don't say to yourself, 'Oh, this is crazy.'"

Menas also had a great gift in terms of artistic ability, which his family recognized. When he was thirteen years old, his father sent him to Paris to live with his uncle so he could attend art school there. He stayed there for a while and then, one day, reached out to his dad and said, "I don't want to be an artist. It's too easy for me." His father asked him what he wanted to do. Manos said, "I want to be a scientist." That was hard on his father, as well as his mother because they both knew that for him to become a scientist, he'd have to go to America, which he would eventually do four years later. As a sign of support for Manos' decision to study science, his father bought him a three-inch Edmond Scientific telescope so he could start studying the planets and stars.

He ended up at Cornell University as an undergraduate, thinking he would become an astronomer. One of his professors, Thomas Gold, one of the three founders of steady-state theory, talked him out of it. He told Menas, "Don't study astronomy. Study physics."

Another of his Cornell professors, Phil Morrison, told him the same thing. Professor Morrison then went over to MIT, and upon graduation, Menas was admitted to MIT to become a physicist, so he looked up Professor Morrison to study with him. Morrison had been a student of Oppenheimer and worked on the Manhattan Project. After the atom bombs were dropped on Japan, he went there, saw the devastation, and said, "We have to stop this." He was then called a Communist.

In the 1980s, Menas became more interested in metaphysics and learned meditation. He did deep dives in Advaita Vedanta and Kashmir Shaivism, which he continues to this day. I asked him about those two paths, of which I have much more familiarity with the former than the latter as the Maharishi taught from the Advaita Vedanta tradition. He said, from his point of view, Advaita Vedanta is a lot like mathematics, the exercise called infinite limits, where you keep taking away until you have what you're looking for. In Advaita, the path is, "I'm not this thought, I'm not the body, etc., until you arrive at the realization of what you actually are, which is awareness itself."

Kashmir Shaivism has the same goal but is by nature more inclusive. It's a path for householders, not monks. He finds it "an easier path" and a better fit with his scientific views because it's all-inclusive. "It matches better the Western mind."

I responded that for some Westerners, Advaita, the process of rejecting aspects of things, was easier, including for me. I explained that I grew up Catholic and basically anything below the waist was rejected and to be avoided for its potential to lead you to sin. Encountering a meditation teaching as mental as TM was a good fit for me, for what I had been used to experientially.

I asked Menas if he wanted to add anything before we closed. He said, "Let me finish by recalling the great mystic and Western philosopher Teilhard de Chardin and his theory on the Noosphere." The Noosphere is a kind of group mind and evolving superconsciousness of humanity. Menas said de Chardin had it right in the sense that "we now have a global society" that is evolving into a unity through the internet.

I'll close by saying that I have a fond memory of sitting at a dinner table with Menas and other attendees at a consciousness conference when the topic of "complementarity" came up. The phenomenon of complementarity is important in quantum mechanics—it cites the coexistence of multiple properties in an object's behavior and that objects have pairs of complementary properties that cannot all be observed or measured simultaneously. It's fair to say that Menas loves this topic and can easily and humorously launch into citing endless examples of its relevance to every aspect of our lives.

A DESIRE TO KNOW THE TRUTH OF LIFE

Arnaud Delorme came to spiritual exploration through a headier desire to know the meaning of his life. He had the realization when he was just twelve years old in the Paris suburbs where he wanted to "Know why I'm here. Why am I thinking right now?" Years later, he completed his Ph.D. in computational neuroscience, then moved on to the Salk Institute for a post-doc fellowship in Terry Sejnowski and Francis Crick's laboratory, where he focused on statistical analysis of electro-encephalographic (EEG) signals recorded during cognitive tasks.

One of Arnaud's amazing and generous contributions is that, in collaboration with a UC San Diego researcher, he developed the free EEGLAB software for advanced analysis of EEG signals. This software is now among the most widely used in EEG research worldwide. He was awarded a Brettencourt-Schueller young investigator award and a ten-year anniversary ANT young investigator award for his contributions to the field of EEG research.

Early in his career, he had a kind of awakening about science: "I discovered that science can tell us *how* things work but can never tell you *why* things work." Prior to that realization, he said he was "more in the camp of the religion of science, that science has all the answers to all the questions." After that, he started to think of science as a method you can apply to questions, but it will never tell you the why of things.

Back then, the Salk Institute offered meditation classes, so he took these in part so he could better understand the psyche. Over time, he developed a keen interest in the scientific study of consciousness and spirituality. He is a long-term Zen meditator and has taught in India on the neural correlates of conscious experience. The more his experiences deepened, the more he came to appreciate the Eastern tradition's view of the relationship between consciousness and matter—that consciousness is not an emergent property of the brain but rather fundamental. He then decided to direct his efforts to incorporate this understanding into his research, as he sought "any way to show this connection."

I asked him how his meditation experiences informed his under-standing of his own self. "Definitely yes," he said. It became clear to him that he wasn't all of the thoughts he experienced. This insight tied into one of the research areas he embarked on, the phenomenon of "mind-wandering," the incessant movement of the mind from one thought to another. There are many research studies showing how meditation practice helps reduce this phenomenon and make us more of an observer, rather than being too closely creating an identity with each thought. Arnaud published some of the earliest papers on this phenomenon.

At one point, as described later in this chapter, Arnaud teamed up with Rael Cahn and traveled to Switzerland to study the psychedelic psilocybin. He said psilocybin was transformative for him. I asked how it affected him. He said, "The experience was extremely insightful." His great insight was that as psilocybin wore off and he started to return from it, his first impression was that he was exactly the same as he had been, but then he clearly experienced his rational thinking mind slowly coming back online, like "building blocks." As that happened, he realized he could once again "elaborate complex thoughts." This insight showed him that he was experiencing things before his thinking mind awoke from the experience. He knew then he was not his mind, but his mind is something he experiences. He learned about his relationship with his own mind from a new perspective.

I commented to Arnaud that his meditation practice gave him the insight that he wasn't his thoughts, while psilocybin gave him the insight that he wasn't his mind. He then commented that psilocybin also gave him the experience of "oneness and connection with all things—a level that was 'organic.'" Arnaud also traveled to India with Rael to set up EEG labs in ashrams to study the effects of different forms of meditation on the brain and thoughts.

There was a period when he started becoming disillusioned with science but then fortunately came across Dean Radin's book *The Conscious Universe: The Scientific Truth of Psychic Phenomena*. Soon thereafter, he met Dean at one of the annual Science of Consciousness conferences. Dean said, "It's great to meet you, I've been using the EEG software you developed!" This led Arnaud to the Institute of Noetic Sciences (IONS) in Petaluma, CA, in 2010 for an internship with Dean to learn more about the work and experiments he was engaged in.

It was around then I met Arnaud, and soon thereafter, we co-wrote a grant to the Bial Foundation in Portugal to study mediumship. The grant was funded, and I started traveling back and forth to IONS where the project was being conducted. Arnaud had devised a very sophisticated methodology with controls to test the mediums' EEG signals while they were supposedly communicating with the deceased. After the testing sessions, we would often have a meal with the mediums, during which time they would often "read" any deceased people they were seeing in our midst. They were quite accurate in the things they shared about the people they saw, often deceased parents of those of us running the project, sharing things they couldn't have known about. I recall one of the mediums sharing about my father that was very accurate, including the farm property where he had lived and the garden setup he loved to spend time in, including his specific tools. Our findings from the study, which we published in the mainstream scientific journal *Frontiers in Psychology* in 2013, suggested that the process of communicating with the deceased is a distinct mental state from ordinary thinking or imagination.

Gaetan Chevalier was an avid reader from a young age. At thirteen years old, he read about the travels of Julius Caesar. He wanted to read the Bible, but his father told him that he shouldn't do so "because it was too complicated; it was only for the priests to read." He went ahead and read it anyway.

When he read the New Testament, he realized those teachings were very different from what he had been learning in church. He said, "I closed the book and said to myself, 'This isn't what my religion is teaching me, that love is the central thing in life.'" He said that he was taught only the fire and brimstone, mortal sins, hell, and "all the rest of it." He felt a crisis brewing. He went into what could be construed as a deep depression but even that was not there. There was only a vacuum where even his identity was swept away. He did not know who he was and if his environment was real or just a dream. That period lasted approximately six months. "That was the start of my real spiritual journey." He decided he wanted to learn the truth about life.

At some point, he found René Descartes' book *Discourse on the Method for Conducting One's Reason Well and for Seeking Truth in the Sciences.* The "I think, therefore I am" felt true to him, so he went with that as a start. He soon realized, however, that he could be aware of his existence even without thinking, so he changed Descartes' statement to "I am aware that I am aware, therefore I am." That was helpful but left him with the question of the apparent existence of his environment. "How real is that?"

Gaetan then recalled that Jesus said in the New Testament that the knowledge of truth will set you free. He said to himself, "Yeah, that makes sense. I'm going with that." He also said, "I'm going to find the truth behind my environment. Even if it's a dream, a careful study of it should get me to the truth." But he needed something more specific. He wondered where to start his study of the environment and what would be the simplest thing to focus on. "Objects!" So at fourteen years of age, he decided to find truth by studying objects, which eventually led him to pursue the physical sciences.

During his undergraduate training, he worked as a nurse helper in a major university hospital in Montreal. In fact, it was the hospital where he was born. While speaking to a nurse during a night shift, he learned about Rosicrucianism. From her description, he got interested in it and started reading about it. Rosicrucianism is an esoteric Christianity-related group that is said to have started in the 17th century. The Rosicrucians themselves, though, say that it started with the Pharaoh Akhenaton in ancient Egypt. While reading about it, Gaetan learned that "Lo and behold, Rene Descartes was a Rosicrucian! How about that?" This helped confirm for him that he was on the right path. He then learned that Gottfried Leibnitz, who developed integral calculus, was a Rosicrucian, as was John Dalton, the "Father of Chemistry." So Gaetan joined the Rosicrucian order to learn more, which led him to study meditation and esoteric teachings. Later, he read *Autobiography of a Yogi*, by Paramahansa Yogananda, which influenced his spiritual journey and view on what science could potentially reveal about the nature of the world.

After graduating with his Ph.D., he worked in Montreal as a specialist analyzing light emitted from plasmas, working with spectroscopy and lasers. He told me, "I'm not really a theoretical guy, I need to do experiments— that is how I learn and have insights into things."

While working there, he read an article about Hiroshi Motoyama, a Japanese scientist and Shinto priest, who was doing very interesting experiments on meridians and chakras using novel equipment including a Faraday cage and a device of his invention called the Apparatus for Meridian Identification or AMI. There was an address at the bottom of the article. Gaetan wrote to that address saying that he was interested in helping with his experiments in some way, if possible. He was not really expecting a response but did receive a letter telling him that Dr. Motoyama would be giving a lecture at a yoga center in Santa Monica. He attended the lecture and learned that Dr. Mototyama just started a graduate school in Encinitas, CA. Dr. Motoyama asked him if he could teach basic principles of electricity to his students so that they could understand his devices. Gaetan responded with an enthusiastic "yes."

After teaching for a year at the California Institute for Human Science, Dr. Motoyama invited him to join the Institute full-time as faculty, an offer he accepted. One of the frequent lecturers at the Institute was Valerie Hunt, herself from UCLA. She was running bioenergy field experiments there. Gaetan became friends with her and offered to help with her experiments. In exchange, she offered to do past life regressions using a method she developed based on mental imagery. Through these regressions, it became more and more clear to Gaetan that "I am not the body. I am a soul, a center of consciousness moving through space and time." He added, "Oh yeah, many years earlier, I learned about astral travel and got pretty good at it and spent time traveling outside of my body, which helped me realize, too, that I am not the body." All these things he did were part of his search over years and years to answer that basic question of "Who am I?"—the search that he started as a young boy.

He spoke about the mathematician Cantor, who developed theories of infinities, that some infinities are bigger than others. "If I take the symbol of infinity, I can square it, prove mathematically that infinity squared is greater than infinity and it has something to do with dimensions." A line, even if it's finite, has an infinite number of points. If you expand that line and move that line into another dimension, then you can make that line become a plane. Every point in the line becomes itself a line. "To me, it's like God replicating itself, the line replicating itself an infinite number of times. Dimensions are what creates expansion. To me, that's what God is doing. There is an expansion that makes a reason for the universe." He added, "It's love—that's all there is to it. It's all much more than I could have imagined, and it took me years to get there, to realize that, but I needed all the details to figure it out."

This brought his mind back to his reading of the New Testament as a young boy, which is all about love, and here he was realizing this firsthand. "I recall a book about the life of Swami Muktananda entitled *The Play of Consciousness*, where Muktananda talks about the Vedic philosophy concept of the Divine Lila, the Divine Play, that formed and transforms the entire universe."

MANAGING PSYCHIC ABILITIES

For some of the interviewees, their childhood sensitives were not so much in response to the natural world as seen through the five senses, but through different forms of psychic abilities. While psychic abilities have historically not been part of mainstream Western scientific investigation, there are many scientists who indeed personally experience them.

Thomas Brophy had such psychic experiences as a child in a form of telepathic knowing. He felt constrained because he didn't know how to pursue learning about them. What's more, his father was an "extreme rationalist," severe in his dismissal of such things, and told Thomas, "Your experiences could not be real because science has disproved them." But his experiences were real.

Thomas shared that he would play cards with his father and sometimes have clear telepathic knowings. While a grade school student, one day he was seated at the kitchen table with his father playing with a deck of cards. An inner voice "told" him to spread all the cards out and ask his father to pick one randomly. "When my father viewed the card he had picked, it flashed into my mind, and I told him the correct card." There was a large picture window behind the chair his father was seated in at the table. His father was so astonished that he swung around to look behind him to examine whether Thomas could have possibly seen a reflection of the card in the window. Despite such repeated evidence, his father would still deny it as a possibility.

Thomas was astonished by that and similar experiences, as well. His fascination with how those experiences happened fueled his pursuit of fundamental science. It also sparked his interest in stage magic, and in junior high school, he took a "mini-course" in magic and mastered the performance of some sleight-of-hand card tricks. Throughout his career, he has used this knowledge and skill in stage magic to heighten his discernment between fraudulence and authenticity in investigations of psychic phenomena.

Despite the lack of encouragement, Thomas' experiences drove him to science. He decided to become a scientist to see if it was indeed true that science had disproved his psychic experiences. He would eventually obtain his Ph.D. while simultaneously reading books on mysticism and consciousness. As an undergraduate, his dual interests in science and the metaphysical took him to Japan to study Zen. While in Japan, he had psychic experiences that helped him connect more to their reality. Like many of the scientists in this book, it was highly important to Thomas to understand his experiences in scientific terms. The more he learned about physics, however, the more he came into a quandary. One of the rubs for him was if non-physical consciousness does interact with physical reality, it would violate the law of conservation of energy.

After receiving his Ph.D., he returned to Japan for further study, having been awarded a post-doctoral fellowship to study there. All the while, the nagging issue of psychic phenomena violating the laws of physics had remained in his consciousness, and he couldn't resolve it. It seemed therefore that his father had been correct that science had disproved psychic phenomena. He was fortunate though because, one day, he experienced a revelatory insight that solved his dilemma.

As part of his fellowship, he had an office at the University of Tokyo that overlooked an interior courtyard. He would often sit at his desk, peering through the window into the courtyard to take a break from his work. One day, he saw a small crow land in the courtyard, and as often happens in moments of scientific discovery, he had a deep insight. "Just seeing the crow land, I don't know, my consciousness shifted, and I had a clear realization that life itself is something very different than physics!"

"In the end, you can connect through quantum mechanics, the collapse of the wave function, a collapse with multiple different outcomes. So the crow or anything living can choose different outcomes." He further explained, "They all have the same total energy. That's part of quantum mechanics, and so it doesn't violate conservation of energy if it happens that way. It was a very visceral realization."

I told Thomas that I so appreciated hearing his experience. How beautiful it is when the natural world presents us gifts of insight that help us solve things with which we are grappling. The history of scientific discovery is full of such examples when, during a period of restfulness or quiet reflection, a long-sought insight reveals itself to the scientist, helping to further their theories and work.

Soon afterward, while back in the U.S. and writing his first book, *The Mechanism Demands a Mysticism: An Exploration of Spirit, Matter and Physics*, Thomas read an article about the work of Dr. Hiroshi Motoyama. As a Shinto priest, scientist, and master of yogic meditation practices, Dr. Motoyama shared many of his own interests, so Thomas wrote him a letter. Dr. Motoyama wrote him back and invited him to come to Tokyo and visit him, which he did—and stayed there for a month. "I was so fortunate. We spent much of our time speaking about how to prove scientifically that consciousness interacts with the mindbody."

I said that must have been so rewarding because that was why he got into science in the first place, to prove this stuff is real, to disprove his father to some extent, and here he is discussing ideas about how to do that with Dr. Motoyama, a mystic and clairvoyant scientist.

During his visit, Thomas learned that in the US some years earlier, Dr. Motoyama had founded the California Institute for Human Science in Encinitas, California. Fittingly, Thomas' academic journey eventually took him to visit the California Institute for Human Science. He currently serves as president there, where he recently oversaw the institute receiving its national accreditation. He is also currently serving as President of the Society for Consciousness Studies.

Sometimes psychic abilities are revealed in a setting where they are not discounted—as Thomas had experienced—but rather actively acknowledged. Such was the case for Hollis King who was spiritually inclined from a young age. For him, spirituality came first, and like several other of my interviewees, that led him to use the principles of science to understand the spiritual.

At around fifteen years of age, he started dabbling in spiritual phenomena like Ouija boards and automatic handwriting. I asked him about automatic writing and how he learned it while in high school. He said it was through his mother, who was studying to be a religious science practitioner. Her teacher did automatic writing. "So, it seemed to be an okay thing to try," which he did. It turned out he was pretty good at it.

Soon after, Hollis came to the attention of Edgar Casey's son Hugh Lynn, who asked him if he had a spiritual practice. Young Hollis said, "Yes, I'm doing automatic handwriting, and by the way, I've been in touch with your father" (Edgar Casey was deceased by that time). Hugh Lynn then invited Hollis to Cayce's Association for Research and Enlightenment (ARE) headquarters. At age sixteen, the summer between his junior and senior year in high school, Hollis hopped on a bus from San Diego to Virginia Beach, VA, to spend time at ARE.

Hollis said while at Virginia Beach, he picked up a meditation practice and has been a regular meditator since then. He also started a deep study of spiritual readings on topics including reincarnation, karma, and healing.

He started recording his dreams in journals, which he's done ever since. "I've got dream books from 1961 to now," he said. I asked him if his dreams have changed much over the decades. He said that in the last ten years, there's been more spiritual content and more guidance that has been supportive of his work and life's direction.

Hollis ended up at Duke University for college and, while there, participated in J. B. Rhine's parapsychology research. He said he "was really good at guessing the Zener cards." Zener cards were developed by psychologist Karl Zener to conduct extrasensory perception (ESP) experiments. The deck has twenty-five cards across a mix of five different symbols. After graduating from Duke, he went on to complete his Master's in Psychology on dream analysis at Trinity University in San Antonio, TX.

While getting his Master's degree, people there knew he had spent time at ARE and wondered if he could do medical readings like Edgar Casey had. They asked him to try, and he said sure. He was put into a mild trance, and while under trance, the person conducting the process asked Hollis, "What should Hollis King do with this life?" Out of Hollis' mouth came the answer: "He should either be a psychiatrist or an osteopathic physician." After coming out of the trance and learning what he had said, Hollis thought to himself, "I'm just about to finish my Master's which I want to do, and I'm therefore close to being able to get a Ph.D., so I'll continue with psychology for now."

He went on to Louisiana State University and received his Ph.D. in Clinical Psychology. Within two years, while practicing as a psychologist, he had a dream he was wearing a white lab coat and working on patients on a treatment table. He interpreted it as guidance that he should pursue osteopathic training, so he and his wife and two kids moved to Fort Worth, TX, to attend the Texas College of Osteopathic Medicine. After completing four years of medical school and doing an internship, he started his medical practice.

Hollis said that osteopathy brought him back to his spiritual roots because of its three basic tenants. The first is that the person is a dynamic unity of body, mind, and spirit. The second is that the body is self-regulatory, self-healing. The third is that the body has a structure-function relationship. During his training, there would be guest osteopaths who did cranial work involving meridians and energy vectors, things we would now call the human biofield. "With all that, I knew I was on the right track for my career."

I asked Hollis if he could share with me what spiritual insights he gained from his patients as a result of doing his osteopathic cranial treatment. He explained, "The cranial work I do is really a euphemism for doing energy work." He said he's learned about the energy field of the human body and working on that brings almost more healing than anything else he can do for a person. He also said that he's had instances when his treatments opened a person to their spiritual nature, which they were previously unaware of. That is, they had a spiritual experience during treatment; then, he would help guide them through it. I told Hollis I thought it was fascinating that his life's profession now fulfilled his life direction as prescribed by him when he was under trance those many decades earlier.

I want to add that during his career, Hollis became actively involved in scientific research, acquiring grants, and publishing scientific papers on osteopathy. Some of that work brought him to join the faculty at the School of Medicine at UC San Diego, which is where I met him. In addition to his clinical duties at the UC San Diego Center for Integrative Medicine, he's been conducting research on osteopathic cranial adjustments to demonstrate their beneficial effects. He and I worked together on a grant from the American Osteopathic Association, and we are currently in the process of analyzing the data in preparation for writing and publishing a paper.

Gita Vaid, a psychiatrist and Indian by heritage who grew up in Wales, had a strange experience as a child that put her on a trajectory to dive deeply into our capacities as human beings. While a young girl, she had a deep insight "where I knew that everything I was living in was, in fact, a story and not true at all." The insight merged into her consciousness one afternoon: "It was unbidden really."

While in that state of realization, she had the thought that she was "very interested in doing acrobatics, such as cartwheels, splits, and handstands." She wasn't, however, athletic and had never before actually been able to do acrobatics. "It was the strangest thing. Since I now knew I was living in a story, I realized my thoughts about limitations not being able to do acrobatics were part of the story, and, therefore, I could change the story." For the next twenty-four hours or so, she indeed stepped out of those limitations and could suddenly and easily do all the acrobatics she previously couldn't. She rejoiced in it while showing her younger sister. The next morning, she awoke and was no longer in that state of realization and thought to herself, "I bet I can't do them anymore," and, indeed, she couldn't.

Gita had somehow managed to find a way out of her normal perceptual and thinking limitations and enter a state where she could physically do things that she couldn't normally do. It wasn't until decades later she could even begin to conceptualize what had happened and what that might mean for how she lives her life. "However that happened, it involved the falling away of some of those structures or concepts that I had inhabited or thought. Their falling away was a really highly important moment for me."

Gita grew up "with a strange mixture of the myths of India and some of its spiritual teachings that my mother would impart to me while my father, on the other hand, was a cardiologist who would speak more about science and Darwin and natural selection." Growing up, she gravitated more toward being a kind of atheist or agnostic.

Gita's journey was also deeply influenced by the significant amount of anxiety she experienced in her young life, as she "thought people were just the strangest thing." She was constantly noticing that "what people say and what they do are very different things." This was not lost on her, even as a child. In order to be able to enter into people's worlds, to feel more comfortable around them, she was always looking at people, studying them, yet often still "just really confused about what's going on with people." She wondered how she could enter into experience with people and have a connection with them in order to survive and keep herself safe.

She took to reading literature as a way to better understand people, to better understand the human condition. The more she immersed herself in literature, she would try to imagine the life story of a particular person she would encounter in her real life, "so she could have a scene with them, like in the books."

Those early efforts to try to figure people out to alleviate her anxiety eventually led her to become a psychiatrist. During our interview, after telling me all this, she laughed and said, "I think nothing has really changed since then. I still find people just as confusing and just as interesting." The more she pays attention and thinks "I've got this figured out," the more she realizes she doesn't.

She said the same holds for herself. She laughed and said, "I feel I don't know what informs my actions most of the time, and this is after a lot of self-therapy, psychoanalysis, and self-exploration. I feel the more I become knowledgeable, the less I know, which actually is kind of in some ways liberating and spiritual simultaneously."

Years later, while working with patients, the moments she experienced as a child would emerge again. Gita described when she is in session with someone, when all of the safety structures and ways in which we as people have locked ourselves in, when we can escape those cages of projections and protections we routinely live in, then there is healing—a movement out of ordinary consciousness. "The phenomenon is unmistakable, and it's when a deep emotional contact occurs. It's the experience of true connection with another human being. That is the real thing, the magical moments, the spiritual moments when true healing occurs."

Gita explained that, for her, it often leads to a state of ego dissolution, a "getting out of separation and into a state of oneness or unity." She also realized that the experience hearkens back to her earliest crises of separation and disconnection as a child. "My own reflection on this is that oneness, unity, is indeed our natural state, and thus living in perceived separation is unnatural and leads to anxiety in our human psyche." Another of her insights is that "to have the deeper connections we all seek, we need to disconnect from the false constructs of ourselves we make. It takes courage to do so."

THE CHALLENGE OF THE CLAIRES - CLAIRVOYANCE, CLAIRAUDIENCE, CLAIRSENTIENCE, CLAIRCOGNIZANCE

So far, I've used the word psychic to denote abilities such as telepathy or getting into an altered state of consciousness. Other psychic abilities include clairvoyance, the ability to see things beyond the physical sphere, like auras and the deceased; clairaudience, the ability to hear beyond normal human auditory perception; clairsentience, the ability to know things without any previous knowledge of the subject; mediumship, the ability to communicate with the deceased. What happens when a young child has these types of psychic abilities yet no understanding of them? Imagine a child being inundated with such levels of perception of which she/he has no real ability to manage.

Tiffany Barsotti, as a child, had all the claires but no point of reference for the experiences she was having, nor any real support from her family. During our interview, Tiffany called it "claire overwhelming," because she had so much information continually coming in from multiple psychic levels of perception that she had great difficulty managing and understanding it all. "Every part of my existence was on overwhelm. It wasn't an easy on-ramp of developing intuition or psychic abilities as an adult. It was hard, and it was filled with trauma." She said, "Even my dream state was not safe from these perceptions."

I asked her about her sleep and dreams. She said she had prophetic dreams about all kinds of catastrophes such as car accidents or train wrecks, and then they would be in the news the next day. As a child, she thought she had caused them. "I don't know how, but I thought I was a cause of these things." She then mentioned Erik Erikson's psychological model, that as children, for some reason, we take everything as our responsibility, whether our parents are fighting or not—everything becomes self-referenced for a child.

Fortunately, her mom started to provide some solace around her experiences and even gave Tiffany books to read about psychic things, including books on Rosicrucianism and ancient Egypt. "I felt safe when I read those types of books." Tiffany also shared that she never felt afraid of the psychic things she was hearing and seeing. It was just the amount of it and trying to figure out what it was that was so overwhelming. She gained quite a reputation with her parents' friends because of the things she would say that she could not possibly have known otherwise.

Tiffany told me that her mom and her friends liked to play with the Ouija Board, and this was something that Tiffany was intrigued about. The Ouija Board was popular in the 60s and 70s, and people played with the intention of communicating with the dead. One evening, her mom and friends were using it, and one of the women asked a question about herself. No answer was given because, immediately, the planchette of the board spelled out, "Don't let Tiffany use this game." Years later, Tiffany came to understand that message as one of protection for her. At that age, given all that was going on psychically, she needed protection. The last thing she needed was to open herself psychically in the context of such a game.

When Tiffany was in elementary school, she had the thought that she needed to turn it all off. She found it impeded her learning and her ability to build relationships with other children. "For instance," she said, "One day in class, I was with another child, and it turns out her mother had just died. Of course, I didn't know that, but suddenly I started speaking about reincarnation, which I really knew nothing about. But the information just started flowing through me. I talked about souls and how they come back and that there's never loss since we are eternal."

The child started crying really hard, she was in such grief. Tiffany had no idea why until she later learned that the child's mother had died. "I watched her fall apart. What should have been helpful was harmful." This was yet another kind of confusion for her. Years later, studying with Norm Shealy, M.D. and Carolyn Myss at Holos University, she learned the importance of "never committing psychic malpractice." This means, "just because you have information doesn't mean you should share it."

She came home and told her mom about what happened at school with her classmate whose mother had died. Her mom said, "Tiffany, you have ESP." Tiffany said, "I don't want it anymore," but her mom said, "It's a gift. You've always had it." Her mom then gave her a book about ESP. She said, "Keep it to yourself. It's more adult language but read what you can." Tiffany asked why, and her mom said, "If you speak about these things, you will lose the gifts." "In my head, I went ding, ding, ding! That's what I'll do." So she took the book to school and showed it to everyone she could, saying, "This is me." She wanted to lose the gifts.

That was the beginning of her willing it all away. "I remember so clearly thinking, 'I just really don't want this anymore.'" That was the beginning, and over a period of a few weeks, the clairvoyance and clairaudience closed down; all that remained was her strong intuition, the claircognizance, something she has remained very grateful for all these years.

The significant downside for her, though, was the suppression of these abilities eventually turned into high anxiety and OCD as she became a young teenager. It was a suppression of who she really was as a person, something she would come to realize years later.

Her first out-of-body experience occurred when she was seven years old, the day her father was arrested and taken to jail. He was taken out of the house in handcuffs. Tiffany was standing on the porch with her mother and younger brother, the three of them holding hands. "The next awareness I have is up and out. I'm looking back at the three of us standing on the porch, and then I floated out over my dad, just a little bit behind him. I was over his head and watching how they were handling him as they put him in the car taking him to Terminal Island, a federal prison located in San Pedro, California. I think I was trying to go there. Fathers and daughters, whether it's good or bad, have a lot of connection." Tiffany choked up as she said, "I loved him so deeply and didn't want any harm to happen to him." She said she was soon back in her body and told her mom she couldn't go to school that day. Her mom said okay.

Through her young years, she was often told by people in and around her family, "You should be a healer." But her observations of people who said they were "healers" caused her to think to herself, "Nope, I don't want that. These people are spacey, way too ungrounded." It wasn't until twenty-two years later, as a successful manager in the music industry, that she would receive the calling to become a healer, a calling she could no longer refuse.

She was living in lower Manhattan, across the street from Ground Zero, at "the second-worst time to be there," when the clean-up was happening. Tiffany was at the top of her professional career, working at a high-powered law firm creating state-of-the-art digital copyright solutions to fairly pay musicians for their work. She thought she was on her purposeful path of doing good in the world. She got very sick, however, from all of the debris being stirred up at the cleanup site. She dealt with the illness for two years, with no real medical answers coming her way other than being put on a mesothelioma medical watch list, which was the opposite of hopeful.

Sick with no medical solutions in sight, she decided to see a Medical Intuitive, hoping for some answers. "I told her I was sick and needed to find solutions." The Medical Intuitive closed her eyes and did her reading. When she came out of her trance-like state, she said, "You have three days to make up your mind to get on your real path." Tiffany emphatically argued back saying, "No, God doesn't work that way." The Intuitive replied, "Au contraire, it's like a job offer. It's on the table now and for the next three days." She added, "You're supposed to do what I do. In fact, the instructions I was told for you are, 'Do-One, See-One, Teach-One.'"

"My blood was pumping so loudly in my head that I had to get out of that office, so I walked out and headed down West 34th Street. I fell apart, damn it. Those streets of New York hadn't before seen so many tears from a single person." She said she "kinda knew what the Medical Intuitive was talking about but didn't want to hear it." She wanted to refuse the call.

Two days later, on a sleepless night before Thanksgiving, Tiffany peered out the window at an incredible full moon that had a gorgeous 360-degree rainbow harvest ring around it. The moonlight glistened off the pristine snow on the ground. "At that moment, I found the grace I needed. I relaxed and surrendered and said, 'Okay. I'll go into the unknown. I will trust in the divine more than anything I can see or touch.' Within two weeks, all of my symptoms had subsided."

Tiffany said it took a lot of work to get her gifts, all the claires, back online. "I can't imagine where I might be today if I hadn't surrendered that night to the unknown, to answer my soul's calling. Since I surrendered at that moment, it's just been so much easier to be in my own skin. I am on my path. I'm doing what I'm supposed to be doing and honoring my soul's journey. In that, I have a deep inner peace."

She went on to say that on the monomyth, the heroine's journey, "I'm on the ascent, you could say, of returning with what gifts I gathered so that

I can share and bring them back to help others. It's handing over your life to the divine. It's taking it from just the staunchly material way of existing to letting the divine lead. Yes, it means you're going even more into the unknown, but you're doing it in the light, not in the dark. It's the opposite of repressing energy—more accepting of the energy and the light. So I feel very fulfilled."

She added, "In fact, as a practicing Medical Intuitive, I've been able to help many people who were, too, at a crossroads, as I was years ago. I've since counseled thousands of people at this precipice of change, helping them hear and answer their own call." Today she is a published scientist, a Medical Intuitive, and a spiritual counselor—a "Guide on the Side," as she likes to say.

Mary Jo Bulbrook, in contrast to Tiffany, did have support for her claires, which included her family's Catholic devotion, of which she too was deeply devotional as a child. While praying, she said she would routinely experience Mother Mary and other saints, seeing and hearing them. She later wrote a book about those experiences. Hearing voices and seeing things fit her religious background, she just accepted them and thought the gifts might be for "being holy or special."

At that young age, those experiences seemed normal to her. She assumed others had them too as part of their Catholic faith. As Mary Jo got older, she became aware that others, in fact, didn't routinely have such experiences.

I asked her if, when she was younger, she ever resisted the voices and the guidance they would provide, as some people do. She said, "No, I never tried to ignore it. What I needed to do is figure out when and how to share it with others, that was my biggest thing." She said she had to learn to actively filter what she saw and heard, not just blurt out the information she would hear about a particular person. This was especially true years later as she started working as a therapist. She had to learn how to work with the information she would hear about a particular person's problems and if or how to share that information with them. This issue of ethics is always front and center for her.

To help her with this, she explained that early on she asked her internal guidance to no longer give her random information about someone, "but please only give me information that I am supposed to work with for that person and teach me how to do it ethically." To this day she follows this principle as it empowers a person to be in charge of their own life rather than be at the whim of others in a person's life journey.

At the start of our interview, Mary Jo wanted to clarify upfront that, although she trained in the sciences and has conducted research, she never viewed herself as a scientist because "that definition separated things, the body, emotions, mind, and spirit, until they were not a whole."

Mary Jo actually wanted to be a physicist and was enrolled in a college to start her education, but that was not to be. While sleeping one night, she was awakened by a voice that said to her, "You are to become a nurse." She thought to herself, "I've never wanted to be a nurse in my life. I never even thought about being a nurse." The voice even told her the college to attend for this, College of Mount St. Joseph's in Delhi, Ohio.

In the morning, she told her mother what happened, and her mom said, "What? We already have you enrolled in another school. Go back to bed!" She went ahead anyway and visited the college she was told of by the voice. There, she told the Dean of Nursing, a nun, that she wanted to attend their school and become a nurse. The dean said, "Sorry, but we've already filled this year's class, they're going to start next month." Mary Jo said, "Well, I want to be a nurse." The dean then said, "Tell me about yourself."

Mary Jo told her some things, including that she was in school plays and was in the National Honor Society.

The dean immediately said, "Okay, I will admit you," to which Mary Jo replied that she didn't have money to attend school. The dean ended up admitting her and placing her on a scholarship paid by the National Institute of Mental Health, which started her on a career trajectory of being a psychiatric nurse.

She eventually received her Ed.D. in College Teaching from North Texas State University with a minor in psychiatric mental health nursing. She went on to the University of Utah where she ran the psychiatric mental health nursing program where she had many training grants and also conducted occupational safety and health research as part of a large interdisciplinary grant.

Mary Jo also served on the university's Institutional Review Board research committee. She shared a story when the committee was once reviewing a proposal to permit lobotomies, which after a brief discussion, was approved. Mary Jo said, "I can't believe you guys question me when our team wants to do energy medicine work where we hold our hands non-invasively over the body and you think that's so ridiculous, but you can justify putting a needle into someone's brain to sever nerves for this other kind of research."

From Utah she was guided to St. John's Newfoundland Canada to head up a graduate program in nursing. Again, this was from clairaudient messages she received. About 6 years later when her mother died, she realized she had to move back to the U.S. to help take care of her father.

She then shared an unusual story. "After hearing the news that my mother died, I visually saw an angel take a bell out of the church tower at the bottom of the hill where I lived in Newfoundland, fly some distance, and then place the bell in a different bell tower. I said 'what does that mean?' and heard an inner voice say, 'You will interview at three places and the third place is where you are to live.'" She said she responded, "Can you give me something clearer than that?"

She went on to interview at three institutions and was, in fact, offered a job at the third one—a psychiatric nursing position at the University of North Carolina at Chapel Hill. Remarkably, the psychiatric nursing position at Chapel Hill came with an office with a view of the campus' bell tower, the Morehead-Patterson Bell Tower.

Another fascinating piece of this story is that Mary Jo was not initially offered the job at Chapel Hill, someone else was, but her internal guidance told her that the person would not take the job and in fact she didn't. Mary Jo was then offered and accepted the job.

There were other similar stories Mary Jo shared during our interview in which she would receive clairaudience guidance about her next steps in life, and things would always work out exactly as the voice said. Mary Jo said, "It's not an intellectual, ego-driven life I've led but a spiritually guided one; we must come from the heart."

THE STRESSORS OF LIFE AND DEATH

For some of my interviewees, it was their response to significant stressors in the home that started them on their journey. Julia Mossbridge and I met at the IONS. Her early experiences of the spiritual were related to family trauma. Her father, a theoretical physicist, had a case of extreme obsessive-compulsive disorder (OCD) and, as a result, couldn't hold down a job. As a child, she made a keen observation. She noticed that when he went into a scientific mode, with a deep commitment to studying something, his symptoms of obsessiveness and rigidity would go away. He would often come up with ideas for Julia and himself to go on a scientific adventure somewhere on their farm property, which she loved.

She shared that one day, her father said, "Let's go to the well and see if we can get the well to show us what kind of minerals it has in the water." She said that in those excited and creative states, his rigidity would just go away, and his creativity would emerge. He would be like a child enthralled with the universe. "I loved participating with him when he was in those states."

Her observation that her father's OCD personality traits would go away when he was in those discovery states led her to the sudden realization that "to be in a state of constant discovery and retaining the inquisitiveness of a child is the pathway to freedom." Julia shared that another great gift from her father was the ability to use science as a way to shift her consciousness to explore the universe.

Something I found fascinating about Julia's story is that she experienced a strong relationship with her future self, which helped her survive experiences of trauma in her childhood. When things were rough with her father in the home environment, her future self would tell her she would be okay and would get through it all. She learned to journal to her future self, and her future self wrote back to her. It was, in fact, her future self that guided her to become a scientist to explore the metaphysical.

What was unusual, though, was that, as a child, Julia imagined herself being a scientist. "In my white lab coat walking around in a lab, I was in a man's body because I was a scientist!" It wasn't until her twenties that she actually noticed during her fantasies that she was in a man's body. Upon realizing this, she thought to herself, "My God, that is how much the cultural gestalt had affected me. To be a great scientist, I had to be a man."

She went on to say there can be an upside to being a woman in science because "you're in the position of the immigrant. You're naturally a pioneer of sorts. The expectations are lower, so if you're doing well at all, people are like, 'Oh, you did well.'" She clarified she wasn't downplaying the difficulty of being a woman in science. At the same time, she noted that "it cuts both ways — the expectations are low, but also, you are often not really noticed if you succeed. Sort of like how you might respond if your cat talked to you one day. You'd be more amazed by the fact of the accomplishment than interested in what the cat said.'"

She remembered being a post-doctoral fellow when a fellow postdoc said to her, "Maybe there's a really good reason that so many women don't go into physics because their brains just don't think about things in the right way." She turned to him and said, "Or maybe physicists would have more questions answered if more women were in physics because their brain thinks about things differently."

Julia's experiences as a child with her future self guided her to get her Ph.D. and provided her deep insight into the phenomenon known as precognition, or future vision, the psychic ability to see events in the future. Today much of her scientific work is about precognition. "What we see right now is the tiniest tip of the iceberg of what exists," she relates. "The value of what we don't see with our eyes is profound, healing, and transformative. If you don't acknowledge what you can't see, and you don't do science on what you can't see, then why are you doing science?"

During the course of our interview, there were so many really beautiful gems and insights. I found that Julia has deep insights into the nature of the world and what science has the potential to offer us in terms of a kind of inner freedom, which is not how it's typically practiced today.

One of the things she spoke about has to do with the kind of relationship a scientist can have with themselves in order to further their ability to see into the world and guide meaningful scientific discovery. "The more I am connected to myself in the moment," Julia said, "the more loving and unconditionally accepting of myself I am. The more I feel connected to my past and future self, the more I'm able to easily have insight into creating the next experiment." She explained that "Science is a mystical path. An experiment is a mystical request to the universe. Egotism limits what the scientist can know, what they can receive from the soul." Egotism in science is a form of control, the belief that we can understand all we actually can't.

Rael Cahn had experiences of the spiritual through the trauma of his psychiatric challenges as a young man. He said he often left what is commonly called "consensual reality," where he experienced a dissolution of his ego and identified more as an unbounded consciousness with a oneness with all things. Rael said those experiences helped him see through the illusion of separation and provided a "washing away of the heaviness of the human drama. When I entered 'I am' I could see through the eyes of all beings. I'd feel less of an affiliation with the part known as Rael, more an affiliation with myself as the All."

During our interview, Rael readily acknowledged the reality of the significance of the "psycho-spiritual crises" he had as a young person. One of those challenges included how "my immature mind at that age couldn't put meaning onto those incredibly mysterious and intense experiences." The experiences also got him into the psychiatric system, "so he could return to the Rael my family knew me to be."

As he got older, he started reading teachings from meditation traditions as they understood the nature of the human being and consciousness. This was like a healing balm to him in that he began to understand the reality of those early experiences as gateways into his real spiritual nature, though acknowledging it was a traumatic and highly disruptive way to have such insights. Nonetheless, they helped him "understand that then, I was having a kind of awakening to something like my essential nature, this unbounded consciousness. At that young age, I was seeing through the illusion of separation."

As so many consciousness explorers come to understand, he realized that those experiences made it clear to him "that that view of reality is more real than our normal experience of ourselves as a mind and body. I was given insight into that world which aligned with what I felt intuitively to be really true."

Those experiences eventually led Rael to attend medical school at UC San Diego, which is where I met him. I gave a lecture to the medical students, and he and I chatted afterward. There he earned his M.D., as well as a Ph.D. in neuroscience. Based on his childhood experiences, he chose psychiatry as a path to help people know "they are not only the ego." Today, he is a practicing psychiatrist and scientist.

Rael shared with me how he brings the lessons he learned from his own psychiatric challenges into his practice. "When a person is going through a psychotic break, the personal meaning of their experience is extremely important for them to understand and come to terms with it. To recover and grow from the experience, it's important they just don't wall it off." Rael sees that, innately, most people do try to suppress the experience with the belief that it will help them avoid it again "rearing its head."

I told Rael that as long as I've known him, he's been moving the center point of what psychiatry can do for people. He said he's been on a path now for twenty-five years toward specifically looking through the lens of modern medicine and science with insights from metaphysical traditions of knowing and caring for each other through the type of conundrums he faced as a young man. I said it will be a welcome change because, oftentimes, in mainstream psychiatry, the tendency is to discount the experience of the psychosis and chalk it up to delusions, grandiosity, and paranoia as opposed to giving any validation to at least the possibility that the person may be experiencing insight into the greater reality of ourselves and the nature of existence.

Rael explained there's a good reason for this, as there is the need to get the patient grounded as soon as possible. It's important not to give extra validation to their internal world—their anomalistic experience which could possibly slow down the process of grounding them. Rael described that in his psychiatry work in the emergency room, this is his attitude toward his patients: to see their other nature, too.

I replied, "That's beautiful. You're seeing a human being in front of you really as a spiritual being. Even if you're not sharing this overtly with them, there's an energy there, an innate compassion. I hope that is perceived and transferred. I think it's remarkably healing and one of the foundations that we want to see more and more of in medicine." I also commented that UC San Diego has recently launched a new compassion institute to do just that—to find ways to bring compassion back into medical care.

Daniel Vicario was an academic colleague of mine at UC San Diego. His grandfather died when he was a young boy, and Dan lay there in bed with him after he had passed away, wondering, "What's the difference between being alive and being dead?" He said he felt his grandfather was still there, even though his body had died, but he didn't understand. "I had a sense about his spirit and body being different but no knowledge about that." Based on that experience, and a similar one with an uncle who passed away shortly thereafter, he wanted to understand life, death, and what happens in the afterlife. That desire led Dan to become a medical doctor.

During his medical training in Argentina, he found every topic of deep interest, but in particular, he always made a point to ask his "professors to share their insights into how patients actually heal." "They would tell me about the importance of being present with a patient, really listening and honoring their beliefs. All of that resonated with me."

Dan's first foray into meditation was during medical school. He took a Silva Method course on visualization and mind development so he could better focus on listening to patients. "It was about visualizing to help someone, scanning their body with my mind. It was very exciting." He also met with traditional healers and seers, "people who can see into the different dimensions."

Despite being in the Western mindset of his medical training, he saw no conflict with this at all. "It made sense to me; it all deeply resonated with me." Together, these different approaches helped him open up all of his senses to be of the best service he could to his patients. Essentially, Dan combined those approaches with his deep appreciation for the human spirit and admiration for the sciences so he could be skilled in both the art and science of being a healer.

After graduation, he landed at Stanford University for his residency training. While deep in his medical work, he continued his interests in healing, meeting, and learning from healers in the area. He started reading the work of Dr. Wayne Dyer. By then, he was thinking about pursuing oncology as a specialty. Reading Dyer's book planted a seed in Dan's mind of creating a new type of cancer center, one that would bridge traditional approaches to healing with biomedicine so as to better support people with cancer. I'll share more on this in a later chapter.

THE CALL DURING ADOLESCENCE

Dusana Dorjee started our interview by saying, "We have eighty percent of the population believing in the spiritual, yet we have science doing everything else but." This is a remarkable statement and, in many ways, one of the reasons for writing this book.

Dusana wanted to be a scientist as long for as she could remember: "Since I was four years old." She said that her "only interest in dolls was to take them apart. I just had this fascination with mechanistic things and trying to figure out how things worked." She recalled, "Even as a child, I had this sense of excitement of discovery with the pleasure of learning new things." Her thoughts of becoming a scientist caused her to go out of her way to read books and see movies about scientists.

Her grandparents, who were forest engineers, had a cabin deep in the woods. As an adolescent, she would spend her summers there, falling asleep watching the clouds while resting alongside the streams. Unbeknownst to her, she would go into a profound state of mindfulness under those clouds that made her feel more alive and boosted her energy. "There was a sense of freedom in all that." Later, as a young student, she found she had proficiency doing math and would go into a flow state. All these experiences were in some ways a harbinger of her later movement to Buddhism to cultivate what is called the "calm abiding" state.

A significant turning point for Dusana was a powerful dream she had when she was fifteen or sixteen years old. It was a "very, very vivid" dream about Nirvana. In the dream, she was outdoors in a meadow, along with many others. She didn't know them, but she was able to communicate with them with her mind; there was no need to talk. With that, she "felt an extreme sense of understanding and connection that I never felt with a person before." There was a silence, a complete sense of understanding that she had for them—and they for her. "There was just the deepest sense of contentment, calm, peace, and love that I never even thought existed."

Upon awakening, she realized it was a dream and started to cry, realizing she was back to her normal reality. "It's difficult to describe it. It was really a deep experience—an early experience of something that was close to the nondual kind of consciousness." She further explained, "In the experience, there was no heaviness nor luggage of all the assumptions, structures, motives, drama of what is me, what is you; it was not there. There was just complete peace and understanding, no need for explaining things, no games, no drama."

Years later, as an undergraduate studying psychology and neuroscience, she came across Buddhist writings with which she felt deeply connected. Sometime later, an acupuncturist friend recommended she try meditation, which she did, leading her to a deep dive into the Tibetan Buddhist tradition. Those meditation practices were very natural to her, almost familiar. As she continued meditating, her experiences accelerated—experiences of deep calm, deep peace, and transcendence of duality.

She explained that she had a certain unusual mental ability, such as she could affect the outcomes of rolling the dice. Demonstrating it would frighten and amaze her friends. Her meditation teacher told her, "These are normal consequences of developing the mind, and they are not much use unless you use your mind to help others." This was great advice for her to hear at that stage of her training and development. Often students can get caught up in certain abilities, or powers, that some meditation practices can foster. She clarified that these abilities are signposts at best and should not be allowed to divert the student from the goal or lead astray with their allure.

One day, Dusana was in her office on the university campus, "reading some deep teachings about unconditional compassion, trying to generate compassion for all beings. I remember having a deep experience while reading the book, then I walked outside to head home, still thinking about what I was reading. As I was walking along, my consciousness shifted. There was no longer a difference between me and the people I encountered.

I was looking at myself, and there was no distinction between me and the people walking past. I wasn't scared; I knew what was happening." She added, "I could also see what was behind some of the buildings, which I experienced as highly relative, not seeing them concretely as I normally did. I was thinking, 'Well, this is very interesting, but am I going to be able to function like a normal person in this state?'"

Indeed, she did learn to function, and much of her scientific work is researching and writing about stabilizing practices and the progression of the spiritual path. Of course, her own calm helps guide her work. Her work is devoted to these topics. She noted that we tend to lump all these experiences together, using terms like transpersonal and awakening. Her work seeks to identify the unique aspects of each stage of consciousness development and, significantly, how to integrate these experiences into everyday life. That is her work now as a practitioner and scientist.

I was moved by Dusana's story and the trajectory on which it has taken her scientific work. Having been in this arena for many years—that is, aware of people following a spiritual path of self-development—I've seen quite a few people get lost along the way, not able to successfully integrate transpersonal and other transcendent experiences into the psyche part of their life.

Historically, various meditation traditions had ashrams and monasteries where proper teachings and guidance by the experienced would be provided to the novice. In the West, there's been much fewer of those resources reliably available to the practitioner. In the 1970s, I saw MIU students have mental breakdowns—even psychotic breaks—and have to leave campus. It's not that there wasn't some support there, but the degree of oversight was not adequate.

I asked Dusana why compassion is such a powerful experience of the human soul, and why it is so powerful as a spiritual path to overcome illusions. Her answer made immediate sense to me. "It breaks down the duality. The moment you start caring for somebody else beyond your immediate circle, you are breaking down the typical duality of me versus other." She went on to explain that this naturally develops a deep compassion and connectedness with others, that deep state of being within us.

Dusana explained, "One of the reasons I'm doing this work is because, from the start of being involved in mindfulness, I've noticed that they borrow words. It's perhaps understandable, given that it's such a new discipline, but they do borrow words from Buddhist writings which can lead to confusion, and this can lead to people's stagnation on the meditative journey.

I asked her what she meant about stagnation. She explained that "along the spiritual path, there are many stages or states of coming into awareness that are vast, peaceful, and calm, where we're more of the witness. We're just seeing everything that's going on. We don't have any more sense of self, any more sense of personal involvement. It's hard to find a reason to engage, to move forward." Without the personhood and the desires and concerns that prompted that person to action, what can carry a person into activity in the world?

As she said this, I thought of what is commonly called "Shiva Consciousness." Shiva is one of the triune gods of Hinduism, the other gods being Brahma and Vishnu. Shiva is often described as the supreme Siddha, the one who has achieved spiritual perfection and the highest degree of enlightenment. There is a phrase that comes to one's awareness when residing in Shiva Consciousness: "Nowhere to go, nothing to do, no one to be, I am Shiva." That is, residing in such a perfected and peaceful state of consciousness, there is no motivation to pursue action, to engage in life's relative activities. There is no desire to take the cloak of individuality and enter relative life.

For Dusana, to help engage people, the element of compassion and Bodhicitta can and does drive us forward. They are inherent in the advanced states. She says, "If we are not here for that, what are we here for but to break down the illusion of duality?"

I am in full agreement with her, and her response reminded me of something I had read in the book *I Am That, Talks with Sri Nisargadatta Maharaj*.[44] The Maharaj (1897–1981) was a *Jnani* in the Indian Tradition, which is someone who possesses the knowledge of gaining liberation and fully lives the relationship between the Self and Brahman. In the book, he is asked the question, "The only difference between us seems to be that, while I keep on saying that I do not know myself, you maintain that you know it well. Is there any difference between us?"

The Maharaj responded, "There is no difference between us, nor can I say that I know myself; I know that I am not describable or definable. There is a vastness beyond the farthest reaches of the mind. That vastness is my home; that vastness is myself. And that vastness is also love." He often says in the book that it is his experience of himself as love that was the basis of his bringing forth his teachings. He was otherwise content to remain silently in Being.

I've always considered a Jnani a special breed among awakened individuals. Though they operate in the world as a seeming "normal" individual, their awareness of themselves as the totality of being is so established that their perception of their body and of others is entirely from the perspective of the one. In this same book, the Maharaj describes what a Jnani is: "The Jnani is the supreme and also the witness. He is both being and awareness. In relation to consciousness, he is awareness. In relation to the universe, he is pure being."

In the introduction to this book, I shared some of Maharishi Mahesh Yogi's teaching about different states of developing human consciousness, including the so-called four "higher states," which are characterized by unity and oneness. Buddhist teachings enumerate many more stages or possible experiences of human consciousness. The Maharishi was skilled at simplifying the Vedic teaching so they could be more readily accepted and understood by his students in the West.

What the Maharishi didn't much explain, however, are the challenges that a person can experience when they move into these higher states. Self-realization is awakening to a non-ego-self state where the identity has shifted to the unbounded Self. There are challenges here, including that the loss of the ego-self can be difficult to manage and adapt to. The closest Maharishi got to sharing this aspect of spiritual transformation with us was a story he told in one of his lectures.

The story goes something like this: Imagine there was a man living in a small hut in the forest. He lived there his whole life and was comfortable and familiar with every aspect of the hut and its surroundings. One day, someone came along and told him he was actually part of the local royalty, and it was now time for him to move out of his hut and into the giant nearby castle. So he moved into the castle, and it was beautiful—shining jewels, seemingly endless, vast and spacious, where he had everything he could want.

Now here's the kicker. The Maharishi then said something like, "Sometimes the man would remember the little hut he lived in, and he would miss it." That was his way of saying, "In the memory of the Self-realized person who has transcended identity with his former small self, or personhood (the hut), one can miss the lost self for the orientation and sense of familiarity it had provided the person in their day-to-day living.

The truth is that the missing of the lost individual self occurs only at the onset of realization and fades over time. The personality self becomes more and more a faint memory, harder and harder to recall. Over time, I would say impossible really.

If people are unprepared for these states, even just temporary glimpses of them, the outcomes can be significantly adverse. Most people in our modern cultures are not educated on these realities and thus have no understanding of what is going on if they happen to fall into such a state.

I asked Dusana if she is pleased with the progress of her research. She said she is pleased with how far their group's gotten with their mindfulness research so that people can manage themselves better. "Opening up the door to them, to be able to notice more about their mind and just opening to the possibility of understanding and self-managing a bit better." One of the products of her efforts is a beautiful book titled *Neuroscience and Psychology of Meditation in Everyday Life: Searching for the Essence of Mind.*

Dusana has a theory that I very much liked hearing about. "We all have a 'Who Am I?' instinct," she said, "which drives us to find ourselves since we are little. If we don't satisfy this instinct in healthy ways through self-exploration and altruistic behavior, we can slip into seeking harmful facsimiles of such states, such as addictions." These nondual states are fundamental to our existence and well-being and need to be understood and embraced. It's important to talk about these stages of development and further explore them and distinguish them. She emphasized this has nothing to do with the mystical, it is part of our nature.

David Muehsam was with the Consciousness and Healing Initiative when I interviewed him. It is a nonprofit collaborative of scientists, practitioners, educators, innovators, and artists seeking to advance the knowledge and practice of consciousness and healing.

David said that he had never felt a conflict between science and spirituality. "I never had that kind of conflict because science is a way of expressing something about nature and spirituality." For him, "Science has always been a way of studying nature, and spirituality feels that way too."

David said, "To describe the spiritual world, you really need a poem; it's almost too much to ask that of science." He added, "With the sciences, we do a very good job identifying things, naming them, but seeing the connections between them is harder. Science, to me, has always felt a natural art, one way to interpret nature."

I asked him why we elevated science as a dominant way of knowing, as it can't really fulfill those expectations. He said he's always had a sense there are limitations to the knowledge we can obtain, citing the uncertainty principle in physics and Eugene Wigner's essay on the unreasonable beauty of mathematics. "It's okay to lower our expectations of what science can do for us," he explained. "I'm fairly satisfied not knowing and not expecting science to tell me the way everything works."

He shared with me that when he was in a junior high biology class, they looked at the slices of onion root tips. "I could see the organelles and the different stages of the cell's division. I could see it all through the microscope and the sense of nature doing something that was just so awesome and remarkable. Its beauty brought tears to my eyes—the idea that I could look at nature in new ways and discover new things. I think a lot of what motivated me to embark on science, especially in the beginning of those early years, was just my love of nature, having insight into the magnificence of it all, the sacredness. There was a sense of wonder. It wasn't for personal gain or to save the world or cure cancer."

When he was fourteen, he learned TM at the local TM center in New Brunswick, New Jersey, and started having experiences of expanded consciousness. In college, he read the *Autobiography of a Yogi* by Swami Paramahansa Yogananda. "It was difficult for me to discount everything he was saying out of hand. By then, I had learned quantum mechanics and was studying physics. The world the *Autobiography of a Yogi* described was very different from the materialistic, dualistic perspectives of the Western sciences. I struggled with that stuff when I was younger, but at the same time, it fit for me with what Yogananda was saying about everything being connected and human transpersonal experiences, including my own—all the sorts of things that are more commonplace in other cultures."

He decided to learn more about it. "In the early 1990s, I was in a group of scientists discussing consciousness and parapsychology. I felt they were trying to fit science into something that it wasn't. I just got so sick of those endless debates, trying to define terms, etc., that I basically got on a plane and went to India to study yoga." He added that those discourses by his scientific colleagues weren't futile. "It's just not what I needed. I thought to myself that to be a good scientist, my spiritual life needed to be satisfied."

Once in India, he started studying yoga more deeply, which was "a homecoming for me, as is music." Since then, he's gone deeply into yoga and meditation practices, even teaching them. "I had a Christian mom and a Jewish dad. I wasn't going to sign up for either. I didn't want to join their religion. I had a spiritual sense of myself, and I had science."

David closed our interview by sharing that he "had a spiritual longing." I asked him what that meant, and he explained, "You know, a feeling that you want to be in touch with that. It's natural, like a moth to a flame."

PSYCHEDELICS, PLANT MEDICINE, AND SHAMANIC JOURNEYS

Psychoactive compounds played a part in the stories of several interviewees for this book. Cassandra (Cassi) Vieten's father was a University of California professor and an atheist who resisted spiritual explanations of life, but as Cassi put it, was a "very romantic naturalist in the sense of feeling awe with the wonder of creation. He had telescopes and microscopes; he would share with me his love of the natural world."

It was her mother, after her parents got divorced when she was young, that brought her exposure to various spiritual philosophies and ideas around raising consciousness. It was while still a young teenager that she started using drugs in the orange groves of Southern California. It was there that she had mystical experiences, including "dissolution of the boundaries of myself" and "experiences of oneness."

One night, while gazing into the stars, she entered into a Samadhi-like experience. Samadhi is a deep absorption where our true essential nature is experienced and can be known without distortion of the mind. She then knew clearly that "there's so much more about reality than I knew," and decided she needed to start learning about it.

She continued with drugs and alcohol, and at age seventeen, she hit bottom and needed help, ultimately finding a Twelve-Step program that helped turn her life around. She said she was extremely scared at the bottom, "and the only thing I could think of doing next, my only way forward, was to help other people." So at age eighteen, she became a counselor in drug treatment centers—helping others and teaching them mindfulness meditation techniques to help overcome their cravings and addictions.

Along the way, she participated in a three-year psychedelic shamanism program that was life-changing. She maintained a meditation practice the whole time. She joked that if she's having a really good day, it's an awareness practice, a nonduality meditation day; if she's having a tough day, it's a mindfulness meditation practice; if she's having a terrible day, it's a Christian practice day.

In a college class on Buddhism, the teacher described the four noble truths and the eight-fold path, and just like that, she had the realization that this was truth. "I felt that was the first true thing I had ever heard from a religious perspective." This information launched her on a path of meditating and deep learning on these topics. That became her impetus to pursue her Ph.D., which she did in Clinical Psychology from the California Institute of Integral Studies. There she learned Sanskrit and took classes on deep spirituality alongside studying Freud.

After getting her Ph.D., she focused on building her research skills and scientifically investigating the implications of spiritual experiences because she said, "Echoing my father, if there's no scientific evidence; it's just a fairy tale." She landed a job in a lab at the University of California San Francisco studying the biological and genetic aspects of alcoholism. From there, she changed to a position at IONS, which is where I first met her. She started there part-time, while still working at the University of California San Francisco, eventually going full-time at IONS.

While starting to work at IONS, Cassi also had an appointment at the California Pacific Medical Center (CPMC) where she obtained grants to research meditation. This was part of her interest to find non-drug approaches to help people with alcoholism. She also studied the effects of meditation on pregnancy and postpartum. She kept her appointment at CPMC active so she could have "one foot in the mainstream world and then the more far-out research at IONS." Eventually, she fully left CPMC to go full-time at IONS. When I first met her at IONS, she was Director of Research there and soon became President, which she was during the time of our interview.

During our interview, Cassi related a story about her father. Although, initially, he wasn't sure about her interest in a scientific career studying spirituality, when she first started working at IONS, which was founded by astronaut Edgar Mitchell, it was "kind of cool, since my dad is a big space fan, very into space and the space program." Eventually, she traveled with her father and Edgar to NASA to see one of the Space Shuttle launches, where her dad was in "seventh heaven."

Much of Cassi's early research at IONS was on major transformations of consciousness. Her early experiences with psychedelics gave her insights into transformative experiences for the better, but also transformative experiences for the worse with drugs and alcohol, which is when she hit rock bottom. She said studying transformative experiences was beautiful work. As part of that work, Cassi got to interview sixty-five masters and teachers across numerous transformative traditions—Ram Dass, Stan Ross, Mike Murphy, and "all kinds of swamis and spiritual teachers and priests and ministers and rabbis." She said, "It was like heaven; it was absolutely what I would be doing if I never had to make a dollar in my life."

For that project, which yielded the book *Living Deeply*, she tried to make it as scientific as possible. "I can't help it; that's just how my mind works." Today, Cassi is Director of Research and Associate Scientist at the Arthur C. Clarke Center for Human Imagination at UC San Diego. When we interviewed, she was the leader of the American Psychological Association's Division 36 Task Force, which was developing spiritual competencies for mental health professionals, a topic she has written a book on.

Joseph Tafur wanted respite from the growing depression he was experiencing while in medical school at UC San Diego. Depression in Western medical schools is more common than you might think, with recent estimates as high as 44%.

Joe is a sensitive and deeply spiritual person and needed something to help him navigate out of the depression and "spiraling negative thoughts" he was trapped in. A fellow medical student suggested they attend a weekend retreat at the Peyote Way Church of God in Arizona. Despite his initial skepticism, "I discovered a powerful ally in that plant medicine," which helped me "quiet my mind and connect to my heart." That ceremony helped him "reset." It turned him around. After months and months of mental suffering, he felt peaceful and joyful. That first "Spirit Walk" helped him find the strength to complete medical school. Some years later, Joe would come into my lab as a post-doctoral fellow. We conducted research on the effects of low-intensity light therapy on treating medical conditions.

Guided by his Spirit Walk experience and the realizations contained within it, Joe eventually spent many years in Peru training with indigenous Shipibo healers and learning to blend his medical training with the use of ayahuasca to support people through crises and self-development. He would live in the forest for many months at a time, undergoing purification diets while being guided by the Shipibo shaman, learning the plants and the other aspects of that tradition, and developing his own capacity to do shamanic work. Since that training, he conducts healing ceremonies around the world.

Joe wrote a book based on that journey, *The Fellowship of the River: A Medical Doctor's Exploration into Traditional Amazonian Plant Medicine*, which received wide acclaim for its insights into the nature of the relationship between spiritual and bodymind healing.[45]

Through that work, he came to a significant understanding related to his medical training and what would become his life's work: that consciousness expansion is integral to health and that spirituality and health are, in fact, the same thing. He noted that this is the perspective of the ancient traditional systems of medicine. He told me, "Health is about being in a kind of right relationship with yourself, with your loved ones, with your community, with your society, with your ecosystem, and with your spirituality." Listening to Joe, I found his insights particularly moving, which, too, provided me with more encouragement to continue this book project.

Through that work, Joe continues to experience his own healing and awakening. A foundational part of that journey is that, "in a metaphysical sense, access to spirit is through ourselves. I've come to realize that if I feel lost in life, it's usually because I'm not being friendly or linked to some part of myself. Once I slow down and see what's going on, see what I'm abandoning or rejecting or denying, and reconnect in a kind and loving way, then I find my center again."

Regarding whether we are loving to ourselves or not, I asked Joe how the Shipibo tradition understands self-judgment. He explained, "I think the shaman might put it simply that we have to approach with love and humility to receive the gifts of the spirit." He said the plants don't like people having a bad or negative attitude such as anger or resentment. It's a turnoff to the energy's ability to move. "There is a need to be humble, to accept your own sacredness. It's better not to have a contrarian attitude. Humbleness lets the spirits come through."

I asked him how all that indigenous training has affected his regular medical work at his clinic back in the U.S. He said it's helped him better understand the emotional roots of many illnesses. "I have more sensitivity to that possibility and can explore with my patients their patterns and behavior, ask them about their life. It's what I call psychospiritual counseling." He said he helps them identify those psychological aspects of their life, to see those patterns. "So basically, it's changed the way I see some health problems and then how I approach those problems."

That's the treatment approach he shares in his book *The Fellowship of the River: A Medical Doctor's Exploration into Traditional Amazonian Plant Medicine.* As part of that, Joe covers emotional biology, linking emotions to aspects of certain health complaints: "I keep coming back to things like forgiveness, self-love, compassion, and gratitude."

I asked Joe if, over the past twelve years or so since he started doing this work, he has noticed that people's capacity to enter into spiritual healing states has changed. Has there been a progression of sorts? He said, "My first response is that I see more and more people getting back to their origins. Whether they're using mindfulness meditation work to develop their personal experiences, whether its plant-based and psychedelic exploration, breath work, or some other kind of honoring or exploration of ancestral indigenous practices like the sweat lodge or some other kind of return to nature, there really has been a continued upward trajectory of that work across the society, and that is what I'm witnessing." He added that "the global pause, as a result of COVID, helped people do more practices, such that there's an increase in the tangibility of what some people might call the magic of nature."

Richard Hammerschlag's use of psychedelics was not in response to the stress and pressure of his academic studies but in pursuit of his academic research. Richard attended the High School of Music and Art in New York City. "If you were interested in science at that school, it wasn't too hard to be a top science student." The rude awakening came when he was later accepted at MIT, "which was really a mistake for me. I would have been much happier at a liberal arts college with a strong science department." "MIT was like drinking from a fire hose."

However, there was a big silver lining: "The best thing that happened to me during those years was taking a course on Eastern religions with Professor Huston Smith, who at that time had written one of the most comprehensive introductions to the world's religions." The course introduced Richard to ways of thinking to which he had never been exposed, either in his History of Science course or the Reform Judaism in which he was raised. Examples of the new ideas he was exposed to included "the Hindu concept of original virtue, instead of original sin, that we are born with Atman, a spiritual essence and that our goal is to merge with Brahman, the all-pervading spirit in the universe."

In those days at MIT all undergraduate students had to complete a thesis. Richard's major allowed him to combine any science or engineering course with either literature, philosophy, or economics. "I majored in literature and chemistry", he said, "which provided a balance that helped me survive MIT. When it came time for my thesis work, I was inspired by Aldous Huxley's experience with mescaline, which he described in *The Doors of Perception,* an assigned book from Professor Smith's course. For his thesis, he decided to compare the peyote-induced hallucinatory visions of Native Americans to those of Huxley and other non-native people, and he would "review current theories of the biochemical basis of hallucinations."

While in school Richard learned about a *Life Magazine* story that described how a New York City banking executive, R. Gordon Wasson, and his pediatrician wife Valentina, had encountered a mushroom cult in the mountains near Oaxaca, Mexico. As a culmination of the couple's long interest in ethnomycology, they were able to bring samples back to the United States. They also shipped some of the samples to the chemist Dr. Albert Hofmann at Sandoz Laboratories in Switzerland. Hofmann, over a decade earlier, was the first person to isolate, synthesize, ingest, and describe the psychedelic effects of lysergic acid diethylamide (LSD). He would also soon become the first person to isolate and synthesize psilocybin, the principal psychedelic alkaloid from the "magic mushrooms".

After finding that copy of the *Life Magazine* article, Richard expanded his thesis idea. He decided he would also compare the rituals surrounding the ceremonial use of the peyote cactus and the mountain mushroom by peoples for whom there was no recorded contact.

Richard wrote to Wasson, saying he was a student developing ideas about the mushrooms for his thesis, which he would like to speak about with him. Wasson wrote back, inviting Richard to join a planned mushroom discussion evening several months hence at his New York City apartment. Richard drove from Boston to New York City and greeting him at the door on the appointed date, Wasson informed Richard that he'd come on a very auspicious evening because samples had recently arrived from Switzerland that were believed to be the main psychoactive ingredient of the mushrooms. Wasson, his wife, and a number of other invitees were going to see how the effects of the presumed active ingredient compared with what they had experienced when taking the actual mushroom itself.

It was "decision time" for Richard since Wasson asked if he wanted simply to observe or take part in the evening's experiment. "On the one hand", Richard recalled, "I had no previous experience with any psychedelic substance. On the other hand, how could I not want a direct experience of what I was going to be studying?" Wasson offered Richard a measure of reassurance saying that his daughter, a nurse at Mount Sinai Medical School, and her fiancé a medical resident, were both going to be present and would serve as a medical support team for the evening's participants. Richard told me he pondered Wasson's question for a few long seconds before responding, "I'll take part!"

"My decision was strengthened", Richard continued, "when Wasson introduced me to the other guests, among whom was Robert Graves, the scholar of ancient Greece, historical novelist and poet. Graves was the one who had put the Wassons on the trail of the mushroom cult, in part from his reading of 16th century diaries of Spanish priests who, in the name of religion, had burned a path through middle America." The diaries described a mushroom ceremony, which previous scholars had speculated was actually a peyote ritual, with the dried mushroom caps incorrectly observed by the priests as slices of the cactus (now often called 'peyote buttons').

After relating all this, Richard then changed gears and was about to share a follow-up event to his mushroom journey when I jumped in and said, "Wait, wait, what happened, what was your experience that night?"

"Oh", he said, "I experienced kaleidoscopic hallucinations and felt I had received an initiation into the psychedelic state, including a liberating loss of self with an emerging sense of oneness." He explained that "once you've had that experience, your life is very different because you're able to glimpse that ineffable interconnection wherever you are." This was Richard's senior year in college.

After graduation he and his best friend were heading off to separate graduate schools and wanted to take a trip together before starting their respective Ph.D. programs. They didn't have enough money for Europe, so they headed south to Mexico for 6 weeks. While they were in the Oaxaca region, after visiting the Yucatan, Richard remembered that the 'mushroom village' was somewhere in the area, although Wasson had purposely not named the actual village in the *Life Magazine* article. After discussing the idea with his friend, finding the village became their primary goal.

After a week of searching, they found it and how they were able to find it was fortuitous. In one of the small villages they were looking in, they came across an American man dressed in a three-piece suit. Turns out he was from *Life Magazine* and was heading to the mushroom cult village for a story. He had a letter of introduction from Wasson for the Mazatec school teacher, the main contact to the village Curandera. Richard and his companion asked if they could tag along, and the man said yes.

Once in the village, they met with the teacher who understood enough of their minimal Spanish to agree to their request to take part in an already scheduled mushroom ceremony. A time was set to meet so she could walk them to the ceremonial site where she would act as translator from Mazatec to Spanish.

For the night of the ceremony, they entered a mud floor hut and met the Curandera and her daughter, who asked them their names to incorporate into the chanted invocation. After being shown to individual mats, they received a portion of dried mushrooms and were instructed "to chew slowly." They were also given some chocolate to ease the bitter taste of the mushrooms.

"That experience blew me away, it was far more intense than my New York City psilocybin encounter. I experienced a profound merging into the oneness, even as I delighted in a background village soundscape of babies crying, dogs barking and people calling to one another. I believed I had opened into the source of creativity for painters and poets, if not also for scientists."

"At this point there was still a 'me' to appreciate the experience. But then my relationship to the external world changed. I lost consciousness of everything around me. I could neither hear nor see anything. I couldn't communicate with anyone, yet I was very much awake inside. I remembered that psychedelics were also called psychotomimetics and concluded that I must be undergoing a psychotic episode. I eventually returned to 'consensual reality,' stumbled out of the hut to sit on the grass and share experiences with my traveling companions."

I asked Richard which aspects of that decades ago experience lasted into his future. "The mushrooms give me insight that there's something behind what we call reality, something more than we're seeing. That's what

I live with. There's a deeper way of experiencing that's always there. Even though I'm not experiencing it 24/7, there's the knowing that it's there."

From there, Richard returned to the U.S. and started his graduate studies in biochemistry at Brandeis University, where "I was thrown right into the deep end of materialism. We were taught that understanding the complexity of metabolic activity in living cells began by grinding up tissue samples and isolating as many individual molecules as possible." He said while on the one hand he deeply understood the limitations of the materialist viewpoint, he also had to focus on what molecule he would isolate to get his Ph.D. He completed his graduate studies in neuroendocrinology, an emerging area of research that was demonstrating communication between the nervous and endocrine systems, a step toward seeing the body as an integrated whole rather than isolated parts. This work and this way of thinking led him to accept a neurosciences-oriented post-doctoral appointment in London and then his first independent research position in a newly formed Division of Neuroscience at the City of Hope Medical Center in Southern California.

I asked him how he managed moving further and further into the reductionist mindset of Western science. "Did you need to do things to kind of bolster your well-being?" He said he certainly did, and the first steps of that journey came from a movie he saw. He had seen a documentary on China and "was enchanted by the image of people doing Tai Chi in the mists in the woods." He then sought out and found a studio where he learned Tai Chi. He said he would practice Tai Chi at the beach where he lived in Santa Monica. "Tai Chi helped get me in flow states, and in those states, I would get ideas on what my next experiments should be."

Unfortunately, after an injury, he developed sciatica. His Tai Chi teacher took him to the home of an elderly Chinese acupuncturist whom he said could help cure his sciatica. It was his first exposure to acupuncture, something he would eventually train in. He rose from the treatment table, surprised to be almost pain free. The elderly Chinese acupuncturist also gave him herbs to take. "The herbs smelled very bad when I prepared them, so I would take them to the lab and prepare them in the fume hood, so I wouldn't smell them." He laughed and said he got lots of flak from his lab colleagues, who would also ask him, "You don't really believe in acupuncture do you?"

Richard lived near the UCLA biomedical library where he began reading about acupuncture and its basis in a view of the body that differed markedly from the biomedical model. The Chinese medicine concept that health involves a balanced flow of energy appealed to his sense of physiological regulation. Research findings, which suggested correlations between acupuncture analgesia and endorphin release also drew his interest as a neuroscientist.

In the early 1990s he was introduced to two Chinese-American brothers who were starting an acupuncture college near his Santa Monica home. They had heard about Richard and his dual interests in Western sciences and acupuncture. When he described his first acupuncture experience, the brothers looked at each other and laughed because it turned out it was their father who had treated him years earlier. Little did Richard think at the time that this meeting would lead to a career change. "I only thought, I'm on some path here, and I accepted a part-time teaching position at the new college." At that point Richard's neuroscience research had been fully funded for nearly 25 years, but he felt drawn to this East Asian healthcare system based in energy, flow, and balance, and he wanted a new challenge - acupuncture research.

He eventually left that acupuncture school and took a position at the Oregon College of Oriental Medicine in Portland, Oregon, where he was asked to create a research department. His subsequent NIH-funded acupuncture research, in addition to his efforts to help form the Society for Acupuncture Research, his participation in the FDA workshop that reclassified the acupuncture needle as a safe and effective medical device, and his presentation at the landmark NIH Consensus Development Conference on Acupuncture, ultimately helped get acupuncture recognized and accepted in the West and incorporated into many academic Centers of Integrative Medicine.

After retiring, he helped launch the Consciousness and Healing Initiative (CHI), where I met him and where he continues to work today. As Richard explains, "CHI explores the concept of the 'biofield' to explain aspects of 24/7 physiological regulation that are other than biochemical. CHI also examines how Reiki and similar biofield interventions produce clinical benefit without physical touch, and, most importantly, explores whether all living systems exist within a subtle field of interconnection."

THE MYSTICAL AND UNEXPLAINED

While some of the stories so far have had to do with some kind of external event causing an insight or dramatic experience, what about experiences that occur in the absence of any such thing or that are spontaneous?

Eckhart Tolle, in his book *The Power of Now*, shares a deeply personal story about his own awakening. It happened while he was simply looking in the mirror. In another instance, in the esoteric tome *Thinking and Destiny* by Harold W. Percival, which I came across while in college and became fascinated with, he describes the massive spiritual experience he suddenly had while crossing the street in New York City.

William (Bill) Bushell had just such an experience while he was in his 50s. Bill had "bottomed out" on alcohol and drug abuse in his early twenties and started attending a support group where he first encountered spiritual ideas, which the group used as a foundation for their recovery. On the one hand, he recognized that the spiritual aspects of the program were important for members of the group; it was powerful for them. On the other hand, he thought it was odd "because God obviously doesn't exist, and these people just need to pretend that God exists to feel better." He thought to himself that "the spiritual effect was most likely some kind of placebo."

Ironically, it was this experience, this frustration and curiosity around why and how people experience spirituality, that made him decide to pursue science so he could study spiritual beliefs and their associated experiences in a multidisciplinary scientific and scholarly framework. He decided to study objectively accessible physical manifestations of the spiritual, things like people putting skewers through their body without feeling pain or bleeding, yogis suspending their breathing for extended periods of time, and apparent retardation of aging in some meditative traditions. Bill wanted to understand how spiritual beliefs enable people to do these things. He shared that he channeled his same obsessive-compulsive nature into his scientific research that earlier had driven him to seek the help of the support group. "My work became a consuming passion."

To enable that passion, he pursued his Ph.D. in Medical Anthropology at Columbia University. Later, he was awarded a Fulbright Scholarship in order to travel the world to conduct this research in the field. After obtaining his Ph.D., he went to Harvard University for a post-doctoral fellowship at the Center for the Study of World Religion. From there, he became affiliated with the Anthropology Department at MIT and continued his global research, along with his scientific and scholarly studies. The rub for Bill was that he himself had never experienced anything he would consider spiritual. He explained in our interview that he didn't even believe in it and wasn't interested in personally pursuing it.

With Bill's extended anthropology work around the world, the local people he studied saw him as sincere in his inquiries and desire to study their traditions, human potential, and spiritual lives. Things, however, remained curious for Bill in that, while he was increasingly successful studying these phenomena across cultures, he remained "in a certain key sense, only a spectator." Nothing in the form of a personally transformative experience had ever emerged for him. He said, "My life remained devoid of any semblance of a personal spiritual experience."

This changed suddenly and dramatically when he was fifty-eight years old. One afternoon, he lay down to rest. As he got comfortable, a holographic-type vision appeared. There was a beautiful woman floating above him whom he said he immediately experienced as a goddess or "Dakini." She was wearing a sari and was in a beautiful garden radiating its own light. She had liquid green eyes and smiled like the sun. She was radiant, with vibrations of endless light. "The word blissful doesn't do it justice—it was beyond that." In conventional time, it lasted for several minutes.

Bill subsequently came to understand that he had most likely experienced a "spontaneous Kundalini awakening," where the energy centers (chakras) of the subtle bodies spontaneously open up, including the heart center, leading to an intensive alteration in perception. He also came to realize that what he experienced had been "a profound Tantric experience," connecting the erotic to the other spiritual centers, and intensely integrating and amplifying the experience to an extreme degree. And yet he had never been a Tantric practitioner.

According to the discipline of Kundalini yoga, mankind contains all of the higher levels of consciousness as potential, a potential known in general terms as kundalini energy, which lies dormant in the unconscious. In the lowest state of Kundalini—the state wherein it initially slumbers, waiting to rise upward toward higher levels—it is represented as a serpent said to lie coiled at the base of the spine, the site of the lowest chakra. Man's potential for higher consciousness starts out at that lowest chakra and from there can evolve or awaken to successively higher centers of awareness, into truly superconscious states, ultimately vanishing into the unity of Atman.[46]

Bill also came to understand that, along with it, he had an experience of what is called "immanent divinity," an intensely intimate experience that reveals to us our own divine nature. Immanent divinity as a phenomenon is more commonly understood and accepted in India and in the esoteric Eastern Christian traditions. Through that experience, "For the first time in my life, I felt like there must truly be divinity, and that we seem to be human and divine at the same time."

Following the experience, he was stunned and disoriented for hours and even days. He realized that it was literally the most important thing that had ever happened to him. One of the gifts of the experience was that, afterward, his decades as a medical anthropologist and his field work into spiritual phenomena had begun to transform into a profound and more transcendentally personal journey.

Bill also confided that, while he had always had the highest integrity in his science and scholarship, he hadn't experienced it fully in other aspects of his personal life. After that experience with the Dakini, he was changed and began to experience integrity in all other aspects of his life.

He shared, too, that the transformative experience caused him to feel anger at Western science for so long ignoring the study of human potential through spiritual phenomena. He said, "Our arrogance aimed at metaphysical traditions has cost all of us the potential for a much greater quality of life and participation in it." Indeed.

Dean Radin, like Harold W. Percival as described in his book *Thinking and Destiny*, had a transformative experience while walking—not in New York City, but across his college campus. I met Dean at IONS. Dean is a rock star in the world of parapsychology research. He's authored many scientific papers on the subject, plus a number of award-winning books.

Dean started our interview by saying that one of his big complaints about our educational process is that "it pretty quickly narrows down what we're able to study. We're shuttled into majors." He said that, as a young boy, he asked "why" all the time. He had a curiosity to "learn literally everything." One of the ways he sought to overcome that effect of our educational system was to "buy a multitude of different kinds of those science kits for kids and follow the Amateur Scientist column in *Scientific American*."

Despite his intensive interest in the sciences and mathematics, up until around the age of eighteen, he planned to be a concert violinist. He played the violin up to four hours a day for twenty years. He described it as a "moving meditation." He told me that he had attained such a degree of expertise with the violin that it enabled him to actually do other things while he played. In a kind of automatic way, his body and mind would be fully engaged in the process of playing so that he only had to pay a little bit of attention to it. He found that he could even read a book while playing.

That got him wondering what else he could do with his mind. To help answer this question, he started learning about hypnosis and psychic anomalies associated with yogis. He began to read the stories of the masters of the far East. He then learned that there was a discipline called parapsychology that used scientific methods to actually study psychic anomalies. This really caught his attention because he realized he could now take the tools of science to learn about the frontiers of the mind's abilities.

For Dean, understanding psychic phenomena, commonly referred to as "psi," touches upon the very best that the human intellect and spirit have to offer. It's a domain "suggesting that our minds are capable of much more than we've been told, or at least that we've been told in school." In his psychology courses, he learned that academic belief in psi was uniformly negative, but at the same time, there was a very high public belief in it. He knew that people believe in things because of their own experience, that they had experienced "something weird," so why the negative attitudes toward it? He said he was always disturbed by the lack of tolerance of science in the official public story, e.g., "Oh, that phenomenon is a delusion." For those reasons, Dean became interested in the sociology of science as a result of seeing these contradictions. His rebelliousness, in part, drove him to study psi.

He thinks part of the reason for the divide with science, in a historical sense, had to do with the enlightenment period, which was driven by the desire to pull away from superstition that was sustained by religion. "Superstitious ideas were running rampant, essentially, but Francis Bacon, Galileo, and others began to develop ways to ask questions of nature that you could actually get answers to. For that to happen, there had to be a split between the religious authorities of the day, who would allow it to happen only if someone didn't touch certain areas such as things having to do with the mind and spirit. 'You can take care of the physical world, and that's okay. We won't bother you too much.'"

On his twentieth birthday, Dean had what he calls his first "step-up awakening" experience. It was 8 a.m. and frigid cold as he walked across his college campus in Massachusetts. He was walking through a cold fog. He walked past the campus' Gothic cathedral which is where they would hold his orchestra practice. He clearly remembers thinking, "I'm twenty today, and this is my fifth birthday [he had been born on February 29th].

I need to remember this." And with that thought, he said, "I simply woke up. There is no other way to describe what it was. I was able to look back at everything that happened prior in my life as though I had been completely dreaming the whole thing."

He saw that everything in his life up to that point as "Dean" had been some kind of continuous dream. It's as if his life up to this point had been as foggy as the fog he was walking in. He remembers feeling both delighted and shocked at the same time. The moment left him with a new level of clarity of consciousness that never went away. He calls these experiences "step functions in our experience of consciousness." Since then, he's had other similar step function experiences that have been lower in intensity but have resulted in continued incremental increases in the clarity of his awareness.

His description reminded me of a "step function" experience I shared in the introduction. The language I used was different, but the overall description was very similar. For Dean, it was a recognition he had been living his life in a fog prior to this realization, whereas I experienced it in terms of a kind of darkness. After that step function, to use Dean's term, I was no longer living in a dimly lit consciousness but one that was bright and self-illuminated.

As I reflect upon this further and consider what the other scientists shared with me, it's clear that the circumstances under which one has an awakening experience are as varied as there are people. I think the underlying mechanism is that our innate primordial Self-awareness is always looking for a way to come forward into our outer self-experience, to break through the typical ego-mind consciousness. When an opportunity arises, perhaps through a distraction or an anomalous thought, it can break through and reveal itself to us.

When I asked Dean if he had ever meditated before, he said he had learned TM just a few months before the experience, and "perhaps it was related to the experience occurring." He said that, while practicing TM seemed pretty natural and easy for him to do, within a month of starting, he was experiencing "all kinds of very strange perceptual distortions and weirdnesses that I didn't particularly like." He went to the TM teacher and told him what was happening, who told him, "Yes, this sometimes happens. Don't worry about it." But Dean was worried about it, so he quit practicing TM. After the step function awakening on the campus, he began to study transformative experiences so he could understand what he had experienced and what factors are known to cause them.

This reminded me of some of Ken Wilber's writings on the self's stages of development along the spiritual awakening journey.[47] At each stage, or level of development, the self is faced with certain decisions and tasks, and how the self negotiates them determines whether it winds up relatively healthy or relatively disturbed. At each stage, the self starts out identified with that particular stage and must accomplish certain tasks appropriate to that stage. In order for development to continue, however, the self has to let go of that stage, or dis-identify with it, in order to make room for the next new and higher stage.

Dean then posed the question, "How can we use quantum mechanics in that realm of reality to think about consciousness and psychic phenomena?" He went on to answer the question, saying, "The closest thing you can get is perhaps the descriptions of an entangled holistic reality where everything is interconnected at very fundamental levels throughout the universe, throughout time and space. In which case, we are made of that stuff, too, and are interconnected. We most often experience, 'I feel most of the time like my mind is related to my head and my brain.' But that's a local illusion, part of the processing of the brain to sustain the illusion of independence.

If you take quantum mechanics seriously, then that really is an illusion. Occasionally, you can predict that not only do I act like a localized particle, but sometimes I'm wave-like, in which case my consciousness gets very expanded—that's the wavelength aspect of nature. So I can think about such phenomena in quantum mechanical terms, admitting it's more analogy. In general, this is a way of staying within materialism and still accommodating these transformational phenomena."

PERSPECTIVE

Once these scientists—and the other scientists I cover later in the book—heeded their call, what was ahead for them? For some, as you've read, their desire to use science to better understand their experiences propelled them to earn their doctorates. Many would later make major discoveries and contributions to traditional and nontraditional scientific fields. Others chose paths less traveled, seeking out less traditional academic institutions for their educational and scientific careers, ones that provided them the freedom to more openly pursue academia and their spiritual path. Their paths were incredibly diverse, but the common element among them was that these first spiritually insightful moments have remained as a guiding light through their lives and scientific careers. Another common and important feature among them is that they never doubted the veracity of their transpersonal and metaphysical experiences.

EMBARKING ON THE JOURNEY

At the end of our interview, I asked Tiffany Barsotti what kind of advice she would give to somebody who is reading this material and learning about people who've been on this journey of awakening. Perhaps they're feeling some rumblings within themselves, but they don't know what it means or how to really respond. I asked her how she would advise them to listen internally to potential promptings and guidance.

She said to "go back to your very first passion, the dream that was inside, the thing that you knew when you were a young child that gave you such profound interest, where so much time would go by that you wouldn't even know what was going on around you. You would immerse yourself and lose all sense of anything else. What was it that made you feel so alive, so connected? Recall, rekindle, and follow that."

She added that answering the question of *how* to rekindle that isn't really our job. Our job is to identify *what* we want and *why* we want it. We need to learn what that impassioned thing is. "Yes, the ego will fight back because the ego is operating out of safety and will want to stay in the familiar. The ego will ask the question of *how*, and it's a very seductive question." The mind might say, "I need a map. I need a compass. I need to know how to read this terrain." But rather, we need to focus on the *what*. When we do that, the forces of life will organize around you to manifest the *how*.

She went on to say that there are a lot of teachers out there who teach us to annihilate and destroy the ego as though it's the adversary. "I can understand how they got to that conclusion, but it's a bad idea to try and annihilate any part of ourselves. We want to include it and integrate it because the smarter we get, our ego gets smarter—like book smart and carnal knowledge and all of that." "We need to show the ego that it's safe to come along for the journey to the unknown. 'Follow your bliss,' as Joseph Campbell would say. Just surrender because the universe has the most incredible strategies to bring us what we need. Just free yourself of any notion of the *how*."

In the book *No Man is an Island*, the American Trappist monk and mystic Thomas Merton addresses the human journey and how it unfolds. He wrote, "In the last analysis, the individual person is responsible for living his own life and for finding himself. You cannot tell me who I am, and I cannot tell you who you are. That is something you yourself can only discover from within."

Spotlight

CONSCIOUSNESS EVOLVES TOWARD WHOLENESS AND UNITY

Robert Atkinson, Ph.D.

The spiritual lives of the scientists in this book express vividly and convincingly how innate potential is fulfilled. Without a deeper experience of life, we miss out on something that can become so central to those who do have this. In fulfilling our potential, we achieve wholeness while discovering that, to get there, we have followed a timeless pattern underlying all versions of a universal transformative process.

In focusing on scientists who have found a spiritual perspective in their lives, Professor Paul Mills has wisely chosen to structure this book upon a framework that follows the pattern of what Joseph Campbell called the *monomyth* because this identifies the one pattern, or formula, that the mythological adventure of the hero follows, making the pattern universal—everywhere the same. I say wisely because this is the pattern found in all the world's myths that not only centrally locates transformation within its core but also highlights the awareness of the responsibility that comes with having experienced it—to serve others in whatever way one is best suited for, i.e., "giving back" in some way. You can see this entire pattern in his book's chapter titles. Yet this pattern goes well beyond mythology.

In my book *A New Story of Wholeness: An Experimental Guide for Connecting the Human Family*, I explore how this pattern can be seen as an avenue of transformation in various expressions from myth to mysticism to initiatory rites, as well as psychological development. The pattern describes a life lived deeply and, at the same time, confirms that all stories following this pattern are reflecting a sacred pattern found everywhere in the cycles of Nature.

We're familiar with patterns like the cycle of seasons. The rise and fall of civilizations as well as the cycle of spiritual epochs also share repeating cycles of growth, maturity, decline, and renewal. All governed by one natural law, or as Pierre Teilhard de Chardin says, by "a single energy at play in the world," these cycles, and the common pattern they follow, illustrate how evolution in all realms is tied together, as all of creation is an indivisible wholeness.

This ubiquitous pattern illustrates ultimately how everything on the micro level reflects everything on the macro level. The nature of transformation is the same on all levels of existence.

As Plato noted, "Perhaps there is a pattern set up in the heavens for one who desires to see it, and having seen it, to find one in himself." We discover this pattern in ourselves through all varieties of guidance and practice meant to assist us in the journey to our fullest potential.

In mythology, Joseph Campbell pulled together the archetypes of the world's myths to form the pattern consisting of *departure, initiation, return*. This has become well known as a pattern of transformation—breaking away from the familiar, entering the unknown, encountering difficulties, undergoing a symbolic *death*, followed by a *rebirth* and *renewal* resulting in the restoration and rebalancing of the whole.

In ritual, Arnold van Gennep identified the pattern that all rites of passage follow as *separation, transition, incorporation*. This closely parallels the process of myth while also being designed to guide the young person from a dependent state through independence and on to interdependence within one's community.

In mysticism, Evelyn Underhill, who did for mysticism what Joseph Campbell did for mythology, but fifty years earlier, describes *the mystic way* as a universal, androgynous journey of spiritual transformation following a pattern that leads from *awakening* to *purification* to *illumination* to a *dark night of the soul* to *union* or living in harmony with the whole.

In psychology, C.G. Jung called this the "individuation process." It consists of the conscious experience of the archetypes we are born with, embedded in our unconscious, bubbling up from within, released by life experiences, making us aware of their innate existence, and enabling the merging of opposites into a new whole. Though the individuation process involves great struggle, the experience of these archetypes helps direct our thoughts and actions in new ways that reinforce our understanding of wholeness. In psychological terms, this is a process involving the *birth of the ego, death of the ego, birth of the whole self.*

All these variations of the same pattern emerge from the familiar storyline of beginning, middle, end. Yet what they all represent on a deeper level is *beginning, muddle, resolution*, another variation of the same pattern which takes us through all manner of possible *muddles* in life for the sole purpose of bringing about a transformation that prepares us to understand and accept the *resolution* that will eventually emerge.

What makes the process of personal transformation even more interesting is that it is intricately connected to the process of collective transformation. Both are aspects of the same process, one leading to the other. Individual transformation carries within it a collective function, that of acting in the world in a way that leads to the betterment of the world—what Kabbalists call Tikkun Olam, the work of repairing the world or restoring the world to wholeness. This comes naturally to those who have consciously gone through this pattern.

The good news is we are all capable of experiencing this depth of becoming as well as this commitment to wholeness. As Evelyn Underhill said, "The germ of the transcendent life, the spring of the amazing energy which enables the great mystic to arise to freedom, is latent in all of us, an integral part of our humanity."

The accounts here of the spiritual lives of these scientists attest to this. The transformation of consciousness, the dynamic and dramatic unfolding of a systematic awareness of ourselves in relation to others and the world—and how we make sense of it all—is seen by the Parliament of the World's Religions as what is most needed now in the world.

We can identify the rough outline of a blueprint that this ageless pattern takes us through in our lives, as follows. The first archetype, a *call to wholeness*, begins with waking up to something beyond what we've known before. This propels us on a journey to a higher consciousness, signaling destiny unfolding. Entering unknown realms, we discover we are being guided and assisted, most often by a mentor who comes along at just the right time. New challenges appear, but as we proceed, we become more comfortable and assured of our path. If the world becomes a distraction, we may turn inward, slow down, withdraw, or retreat from our everyday experience, as a desire to fully develop and evolve in the world grows in us.

The second archetype, the *path of purification*, opens us up to greater challenges. As our consciousness continues to expand, we meet these challenges and get glimpses of a new reality, as we are drawn further into fulfilling our innate potential. We encounter more guidance and assistance and become more intent on cleaning up, integrating, unifying, and healing all parts of ourselves to reclaim our innate wholeness. Though we continue to live in the realm of dualities, our consciousness of wholeness grows, our values solidify, and our fear of letting go of old ways of being dissolves. This completes the dying to the limited self, and we are renewed and reborn with a full consciousness of wholeness.

The third archetype, a *return to wholeness*, focuses us on sustaining the unitive consciousness we are convinced of as the highest reality. We embrace showing up for all this demands of us, giving back to others what we have been given, lifting others up along their journey, and linking up with others to serve the evolutionary impulse—all this contributing to the betterment of the world. We remember throughout that we are always in the process of becoming, interconnected with all others, striving to live as the whole being we are.

This book contributes a great deal to understanding how this pattern is expressed in the spiritual lives of these scientists and beyond. So many of the motifs and archetypes mentioned above are found in the life experiences of the individuals here who have contributed so much to the world through their science as well as other ways.

This underlying pattern guiding all things in a greater unfolding of a purposeful evolutionary flow further implies that there is a hidden wholeness to all things. That we see this central motif in our own lives is a testament to us being an integral part of this wholeness as well as to the ultimate union of all components of the whole.

This book also illustrates, just as importantly, that science and spirituality are not only representations of this wholeness, but they are also equal partners in this wholeness. The lives of these scientists show how both partners contribute to a fuller, more complete understanding of reality, giving a deeper sense of order and meaning to the universe.

Reality is wholeness, and the awareness of this dawns when the subject-object split is healed. "Wholeness is our birthright and our destiny," he says. This book conveys, in a way very few others have, how the personal experience of wholeness heals the divide between science and spirituality, reason and faith, separation and union. Here we have a rich collection of unitive narratives that tell how we can experience wholeness when we are open to all aspects of reality.

CHAPTER 3

MENTORS AND OTHER ALLIES

I am not a teacher, but an awakener.

– Robert Frost

A ll professions need mentors, and science is no exception. Science relies heavily on the mentor-mentee relationship as it expands the young scientist's exposure to ideas, methodologies, and potential future collaborators.

Following graduate school, newly minted scientists typically seek a post-doctoral fellowship with an established scientist in the field they wish to enter. Mentors guide and, in many ways, help their students choose what research they will ultimately engage in. A successful fellowship is crucial to helping land research grants and find a suitable academic or industry position. For me, I was fortunate to find Joel E. Dimsdale for my training in behavioral medicine. He was and remains an outstanding mentor.

Many professional scientific societies establish special trainee activities as their annual meetings. The American Psychosomatic Medicine Society, for example, in which I was involved for many years, including serving as president, actively sought established scientists to match with young members of the Society to create mentor-mentee pairs. Special events for the pairs would be held at each annual meeting, plus the mentor would make themselves available during the course of the year for any support that may be needed.

How many young mentees are wanting to explore spiritual topics during the course of their scientific training? There aren't studies on this topic but there are studies on religion in scientists, which reinforce the common thought that most scientists are secularists. However, these same scientists don't see an innate conflict between science and religion.[48]

While it's difficult to determine rates of interest in spirituality in young scientists, it's certainly easy to determine rates of actual spiritual and metaphysical research in traditional versus non-traditional academic institutions. The more traditional the institution, the lower the chances a young scientist could find a mentor who would encourage any kind of spiritual inquiry. For the nontraditional academic settings, the likelihood is much greater. For those students with a more spiritual orientation, a suitable and supportive mentor is vital to their development into a successful scientist so they can remain true to their developing inner life. This chapter explores how mentors shaped the interviewees' careers, in positive and, at times, not-so-positive ways.

Meeting a mentor is a highly significant stage in the monomyth. By definition, the "call to adventure" brought the individual out of the world they knew and were comfortable with and asks them to enter into a new way of being. The mentor shows up at that crucial stage in the journey to help the person temporally bridge the two worlds—the known with the unknown—then successfully leave the old one behind and embrace the new one. The mentor provides the needed knowledge for them to be successful with the transition and keeps them on the path when they might feel discouraged or ready to turn back. Sometimes the mentor emerges for only a brief encounter, but the effects are long-lasting.

THE SEARCH FOR SPIRITUAL TEACHERS AND TEACHINGS

Across cultures, there are innumerable stories of children hearing the call at a very early age and leaving home in search of God. Maharishi Mahesh Yogi told the story that his guru, Swami Brahmananda Saraswati, left home at the age of nine, wandering from town to town in search of a teacher who could help him find God. The rub with searching for a mentor, though, is that it's not always like the movies where, for example, Obi-Wan Kenobi showed up for Luke Skywalker very early during his quest and was, in fact, his perfect mentor. Luke was spared from having to wander for years from one potential teacher to another until he found his destined teacher.

The young Brahmananda Saraswati, born in the village of Gana—near Ayodhya in Uttar Pradesh, India—wandered for five years before he found his master, Dandi Swami Krishnanand Saraswati, in Uttar Kashi, India. Maharishi Mahesh Yogi said that during those five years, the young ascetic roamed from village to village looking for his teacher, asking potential teachers probing insightful questions about their knowledge and wisdom, and rejecting one after the other until eventually meeting Dandi Swami Krishnanand Saraswati and, according to custom, surrendered to him as his master.

Early in his training, he entered a cave, resolving not to emerge until he had attained enlightenment. This was to be where he ended up spending the majority of his life, living a solitary life in caves in the forests and jungles. In 1941, after repeated coaxing from local citizens over a twenty-year period, he left the caves at the age of seventy and accepted the seat of the Shankaracharya (spiritual head) of the Jyotir Math monastery in northern India, which under his leadership became a famous center for

Advaita philosophy. He held the position of Shankaracharya until his death at the age of eighty-two.

In seeking his own spiritual mentor, Maharishi Mahesh Yogi was fortunate. Maharishi told the story that at the age of twenty-three, he accompanied a friend to a lecture he was told was to be given by a saint, Swami Brahmananda Saraswati. Maharishi said he was immediately overwhelmed by the Swami's presence and asked him if he could stay and become his disciple, to which the Swami responded, "Go complete your college studies, then return to me." By the time the Maharishi completed his studies and went to find the Swami, he had become the Shankaracharya, so Maharishi joined the Jyotir Math monastery and eventually became the Swami's assistant.

In India, the search for a spiritual mentor and a spiritual path is a normal part of society. In the West, embarking on such a journey is not always understood or supported. For some, family members are obstacles; for others, family are supportive. Despite any doubts I had about pursuing an education at MIU, it seemed the best direction for me to dive deeper into meditation and spiritual development while simultaneously getting an academic degree. Fortunately, my family was supportive of my decision.

Thomas Liu's search for a mentor mirrored that of Swami Brahmananda Saraswati's, in the sense that it took him many years of going from one potential mentor to another before he finally found what he was looking for, a spiritual teacher who was actually resonant with their own teachings. I first met Tom at the UC San Diego Center for FMRI where he is Director. I immediately felt comfortable with his presence. UC San Diego is a high-pressure, competitive institution, and it's unusual to encounter a faculty member who exudes a relaxed and peaceful state. Tom is a gem and a beacon of light on the UC San Diego campus.

Tom didn't grow up in a particularly spiritual environment, although he shared that his mom started attending a Baptist General Conference church while he was in high school, "which meant I attended too." In a college essay describing the experience he wrote, "My mother had heard the church was a good place where one could find God." There he participated in a church youth group and attended religious study groups. While participating in the church community, he experienced the first of what would be a handful of repetitive and disappointing experiences. In a spiritual environment, his expectation was that everyone would get along and exemplify the loving ideals of the religion. He learned that is often not the case: "the church ended up blowing apart due to personality and internal political issues." Even at that young age he began to appreciate the difficulties inherent in spiritual communities.

After college, while in his early 20s, he began studying Aikido and immersing himself in the practice of the martial art form. The philosophy was appealing to him, including the sense of energy and flow and being in harmony with the life around him. He also learned to relax and to "unify his energy." He considered Aikido, "the start of my deeper spiritual practices." After some time there he unfortunately again experienced what he had experienced in the church community. "While the Aikido dojo was a very renowned and respected place, there was the human element, and I began to see surfacing the human issues of conflict and politics. It was a disconnect, as here was a place to learn the harmony of energy, yet people couldn't get along." Part of his insight with this was that "Aikido is essentially a body practice with a limitation that it is missing a way of dealing with one's thoughts. There's no effective mental training that accompanies it." Although he continued to practice Aikido, he had the sense that he needed additional forms of practice.

About a decade later he and his wife were in a bookstore in San Diego. She found a book titled *Everyday Zen* by Charlotte Joko Beck and suggested Tom buy it. She was aware of his prior experience at the Aikido dojo and thought an "everyday" approach to a spiritual discipline would presumably be more realistic in dealing with day-to-day human emotions and interactions among people. At that time Tom had two young kids and was a relatively new assistant professor at UC San Diego. He realized he needed to learn something to help him to "better manage his emotions as the stress of academic life was getting more intense."

When he read the back cover of the book, he saw that Joko Beck's Zen center was located right in San Diego, in fact in the community of Pacific Beach not far from where he was living. Shortly afterwards, we went to the Zen center and joined the community, learning the Zen meditation techniques taught there. Joko was in her late 80s by then. "She would give a dharma talk every Saturday and occasionally I would get a private audience with her." Tom said the main initial teaching at the school was "to label your thoughts." Through daily practice, the weekly meetings, and occasional week-long intensive retreats (sesshins), Tom found that his sense of self was gradually shifting and he noticed that he was less emotionally reactive.

At the same time, he started to sense things "weren't quite right at the Zen center; I'd hear comments here and there." At the time, Joko had two dharma heirs that she had been training and grooming for many years to take her place, but she suddenly disowned them and revoked their dharma transmission. In Zen Buddhism, there is a long tradition called "dharma transmission" in which the teacher passes their spiritual wisdom to their disciples in order to keep intact the specific lineage of that teaching. Tom said that revoking this transmission "was a very big deal" at the center.

Towards the end of her life, Joko appointed another student to be her primary successor. It became clear to Tom, however, that he couldn't stay and study with the successor who had received Joko's dharma transmission, so he decided to move on again. Tom said he had "internal warning bells" about the new teacher, which ended up being validated by his own experience as well as that of other students. He recalled, "The new teacher would say one thing and my intuition was telling me something else."

Before Joko passed away, she recommended to Tom that he read the book *The Impact of Awakening* by the nondual teacher Adyashanti. He did purchase the book and spent the next few years meditating more on his own and gradually listening to Adyashanti before beginning more intensive study with him through in-person retreats and on-line courses. These studies helped Tom to clearly see the fundamental non-reality of thought. Today, he continues to follow the teachings of Adyashanti while at the same time studying with other spiritual and martial arts teachers to deepen his understanding of embodied spiritual teachings.

Tom said these practices also help his scientific work. His "greatest satisfaction is working at the intersection of pure engineering and pure science, applying mathematical clarity and simplicity to things which are otherwise just very qualitative." For him "the best part of science is always the discovery. It doesn't have to be very big, but just something you can understand that no one has understood before." He noted that spirituality is about, "trying to see things in new ways with new insights and perspectives."

At the same time, he has found that basic scientific knowledge of the workings of the mind and body has helped him to better understand the "why" and "how" behind various spiritual and mind-body practices. His experience is that fundamental insights into the nature of mind and body come more readily when he approaches his practice from an inquisitive and scientific perspective as compared to blindly following a set of prescribed routines.

Jeffery Martin, like Tom Liu, experienced a mismatch between what religion preached and what he observed it to be, and this led to a crisis of faith. It was also the start of a many-year journey of self-exploration and consciousness development which ultimately came to fruition. Jeffery said he came from a family with "a Theology professor and another one who was a dean, various other pastors and Christian educators, the whole bit. Plus my mom was a Christian TV show host for a missionary interview program." He was raised as a very conservative Christian, yet during his teens, that "belief system rug got ripped out from underneath me." He recalled, "It's just one of those things that happens in life." He said that it's amazing he survived; it was "a deep crisis of meaning."

How did it happen? Basically, as he moved into his later teen years, he began seeing "the disconnect between people who professed to be Christians and what they were really like." He was working in a Christian television station with people at the highest levels of Christianity, some of them famous pastors with some of the largest ministries in the world. These were "pinnacle Christians," as he called them, running these megachurches, yet many of them were in fact "kind of horrible people," not Christian in their real ideals or behavior. He said these individuals were not congruent with what the Christian system said it would produce. Jeffery then had to admit to himself that all the Christian knowledge he was raised with "just might not be true," which precipitated his crisis. He also thought that the broader religious system that he and others were taught to model their life on might not be true either.

He attributed his survival through that crisis to someone he met at the Christian television station where he worked. Jeffery acknowledged that all of us, "at some point, stand on another's shoulders in our lives." There was a broadcast engineer at the station, a man who had been through the Vietnam war and in the intelligence community. This mentor, this ally, had taken Jeffery "under his wing," teaching him the ropes about engineering, how to maintain the equipment, and how to run a TV station. He noticed that Jeffery was in the midst of some kind of crisis and asked him what was going on and if he could help. Jeffery then asked him point-blank, "Do you believe in Christianity?"

The engineer paused and then responded, "What if it is true? How does it hurt me to come here and go through the motions with this stuff?" He continued, "If Christianity is true, then fine, I'll go to heaven forever. If it winds up being true, how smart it was for me to get up every Sunday, go to church, and go through the motions. If it isn't true, though, did it really kill me to attend all those religious services?"

Jeffery reflected on this. It wasn't a Zen Koan, but the rationale didn't sit right with him. It was either true or not true, and he didn't want to "sit on the fence the rest of his life." Their conversation led to a discussion of the upsides and downsides of religious indoctrination and the choices one needs to make along the way.

During the conversation, Jeffery had the sudden realization about the religious indoctrination he had experienced and of the fear that had been driving his lifelong religious beliefs and behavior. He realized that fear had been the underlying motivator of his life. He also realized that there are plenty of people out there like his engineering mentor who weigh the odds and risk factors that it might be true—or it might not actually be true—but with eternity in heaven at stake, why be stupid and not follow through on the practices?

Jeffrey then made a firm commitment to himself that he wasn't going to live a life based on fear and threats. This realization propelled him to seek a path to develop his own consciousness and well-being, one that's independent of any religious belief or indoctrination—not something based on fear but on a sincere desire to experience the sort of well-being of which a human being is fully capable and deserving.

He then decided to begin an intensive search for a religious system where there is indeed congruency between what is believed and what is practiced. He wondered, "Might I actually find one?" He went on a multi-year, multinational search studying different religious systems to find congruency, to see if the adherents were living a life consistent with what was being professed. He systematically examined every other faith philosophy he could, even practicing many of them.

After about ten years of searching for a religion where the practitioners were actually living in alignment with the principles of the religion, he gave up and decided to take a step back. With all that he had learned, he started thinking anew about his life. He realized that "the only thing we really have is our moment-by-moment well-being. Is this moment as good or better than the last one or not? If not, why not?"

He started taking courses on personal development, trying to focus on developing his sense of well-being because he eventually wanted to teach about it. This was in his early thirties, but he hit a ceiling of sorts. It was clear to him that he met people who seemed a lot happier than he was. He wasn't necessarily unhappy or "miserable or suffering in some existential crisis or some black hole or something," yet he realized there were people that were definitively happier than he was. "Being a Type A person," he decided on a new goal, "to become the happiest person out there!"

He had already done exhaustive research on self-help and even funded someone else's research project on it. Ultimately, those efforts led him to pursue his own research to better understand and maximize his own well-being. He wanted to do it independent of any religion or particular philosophy. He wanted to learn whether he actually could develop what he would later call "fundamental well-being." He even wondered if his genetics or his nervous system could support such an experience.

To accomplish this, he went back to school to obtain his Ph.D. so he could conduct professional quantitative and qualitative research on well-being. "In truth, it all started as a selfish pursuit of happiness domination for myself," he explained. But in the end, his work became a much larger endeavor, one that would ultimately have positive effects on the lives of tens of thousands of other seekers of well-being.

Over the next two decades, Jeffery conducted the largest international study on what is often called "persistent non-symbolic experience," which includes the types of consciousness commonly known as enlightenment, nonduality, the peace that passes understanding, unitive experience, and oneness. He literally traveled the world finding and interviewing hundreds of people who professed to be in persistent non-symbolic experience states. He learned from them what such states were like and how they achieved them. One of the fruits of his work was to develop a pan-tradition classification system for these diverse types of experiences.

Along the way, he recognized that these were psychological traits that have been known and adopted for thousands of years by many cultures and belief systems. "They were not inherently spiritual or religious, nor limited to any given culture or population, and could be molded in many ways to shape the experience." He eventually founded the Center for the Study of Non-Symbolic Consciousness, an independent academic research center dedicated to the study of ongoing forms of persistent non-symbolic experiences. The Center's research is secular and focuses on the psychology, cognitive science, and neuroscience underpinnings of these states of extraordinary well-being and highly beneficial experiences.

TEACHERS IN INDIA AND TIBET AND THE SEARCH FOR SELF

At the start of our interview, Tawni Tidwell said, "My entire journey has been about mentorship. Each and every step of the way, I've been supported by key mentors at key times." Tawni has a fascinating story and trajectory, and her education took her around the world, with much of those travels being in India and Tibet. She studied Tibetan medicine at Men-Tsee-Khang in North India and Sorig Loling Tibetan Medical College of Tso-Ngon (Qinghai) University in Eastern Tibet, with a year internship in gastroenterology at the Tso-Ngon (Qinghai) Provincial Tibetan Medical Hospital. In July 2015, she graduated with a Kachupa-level degree in Tibetan medicine.

Like Tom, I first met Tawni at UC San Diego. She was interviewing for a post-doctoral fellowship, and we had a chance to sit down and chat. I found her background and training remarkable and had many questions about Tibet and its medical system. As fate would have it, the following year, I next saw her in Tibet. My wife and I had been invited to Lhasa, Tibet, to speak at a conference celebrating the 100-year anniversary of the Men-Tsee-Khang, or Tibetan Medical and Astrology Institute. The Institute was established in 1916 by the 13th Dalai Lama. Given her fluency in the Tibetan language, Tawni served as a translator for several of the Westerners presenting at the conference. We met again a year later in Boston, MA, where a group of us had organized a conference on Tibetan Medicine at Harvard University. Attending were many Western scientists like me, along with Tibetan physicians from around the world.

Tawni's life of travel began at the young age of two when her family moved to Korea. Her father was a U.S. army orthopedic surgeon and was sent there for duty. She stayed there until five years old when they moved to the front range in Colorado. Once back in the U.S., she had a series of remarkable and eccentric teachers in a small suburban Colorado town who gave her free range to study what she wanted from grade school through high school.

Ultimately, she heard a keynote lecture at her high school by Candy Vallado, who had been a systems engineer for spacecrafts on the NASA Apollo, Viking, and Magellan projects while working at Johnson Space Center and Jet Propulsion Laboratories and coordinating with Goddard, Kennedy and Ames Space Centers. At the time, Candy was at Martin Marietta working on shuttle robotic systems and transitioning to Bell Labs.

She was living in a canyon in Colorado, in a small cabin with a compost toilet and a wood-burning stove. Though it was a lecture presumably focused on engineering, Candy focused her talk on featuring all the latest findings from space missions. Tawni was hooked. "She was a hippie and a rebel," Tawni said, "A passionately curious mind." Candy got her an opportunity to shadow some engineers and scientists at Lockheed Martin and Bell Labs and told Tawni, "The world is your oyster. Just ask good questions and go!"

In high school, Tawni was a dancer, was student body president, played lacrosse, and got involved in environmental organizations. She did very well and eventually made her way to Stanford University to study physics so she "could understand how things work." As an undergrad, she approached a professor and told him she wanted to study the molecular basis of thought, "how thoughts translate into biochemical cascades in the body." He was dismissive and told her to come back when she was a graduate student. Eventually, she switched from her intended physics major to an earth sciences and geophysics major and spent a summer in the Pacific Ocean studying the Mariana trench, the deepest trench in the world at 36,201 ft.

While a sophomore, she had an opportunity to travel to Dharamshala, India, which proved to be a life-changing event. She was there for six months and studied with senior teachers at His Holiness the Dalai Lama's Institute of Buddhist Dialectics, where she also received teachings from him. The year before, while still a freshman at Stanford, she knew the India trip was a possibility, so she started studying the Tibetan language. It turned out her Tibetan language teacher's father was a Tibetan physician, an influence that ultimately affected her choice to become a Tibetan medicine physician.

While at Dharamshala, she met one of the Buddhist nuns who took her under her wing. The nun lived up in the hills in a cabin, away from the busy town. Tawni was able to visit her there on occasion. "As I got to know her, see her in retreat, her devotion just blew my mind." It was a form of spirituality that Tawni hadn't encountered before. She said, "It all felt really familiar to me, kind of like coming home. It deeply resonated with me."

It left such an impression on her that she went to speak with the Dalai Lama's senior English translator in Dharamshala who lived adjacent to His Holiness' residence about becoming a nun. He explained to her that she would have to study Buddhist philosophy for 16 years before she could even begin learning meditation and that she would have to give up music and dance, both of which were important in her life to that point. In her mind, becoming a nun also meant distancing herself from her family and friends, which, though ultimately untrue, became the primary reason she decided not to take vows. Her hopes had been quickly dashed. "I was devastated," she said. She realized that she "wasn't ready to renounce the world, at least not in the way they said. I cried all the time for weeks."

While on the trip, she had an opportunity to live with a Tibetan family high up in the mountains of Ladakh, a Buddhist region ethnically related to Tibet in the state of Jammu & Kashmir. It was there she had perhaps her most significant spiritual experience. As the many weeks went by living there, she realized she "was in a place so far removed from anything that reminded me who I was." The family called her by the name "Tsering" because it was easier to pronounce than her English name and Tsering was the mother's name so it showed a degree of adopting Tawni into the family. She wore all the local clothes like her new family. They all slept in the same room and ate all the same food. She helped with the farming, herding the goats and sheep, and milking the *dzomo*, a female yak-cow hybrid.

At one point, a strange thought occurred to her, "There is no way I can ever go home." She explained, "It wasn't like I couldn't intellectually think about the feasibility about going home, that I couldn't imagine the steps to get on a plane and go back to the States. For some irrational reason, something in my psyche firmly felt I would not be able to return home because I had entered a new life, a new identity there." She pondered to herself, "Who am I? Who is this new person?" Her life there "felt so familiar and easy." She wondered, "Where is my go-getter and dancer being?" She reflected, "All these things were stripped away from me, and I was really trying to figure out my identity as if it was an energetic signature. With all the previous relationships, roles, and conditions stripped away from her, what was left? Who was left? She wondered if anyone could even see her as her in this new context. Would they be able to identify the person she had once been, would they be able to pick up her energetic signature, so to speak, or was she literally a new person. She asked herself, "Did I literally become a new person?"

"It was like a Zen Koan. I don't really know how to describe it, but that experience of trying to figure out this new identity was really frightening." She had spontaneous crying episodes, yet, "I'm not a big crier. I just felt such incredible grief. I would stand there and churn butter and sing Amazing Grace because I was trying to remind myself that I knew English. She had only been speaking Ladakhi with the family. She'd say to herself, "'I know

English, that is my language.' It was the weirdest thing. I don't know why, but I was worried about forgetting English. I can't explain it much."

As someone who has had what are commonly called 'past life experiences,' I couldn't help but think that a prior life of hers as a Tibetan was actively resonating in her current consciousness as Tawni. Few in the West believe in the phenomenon of reincarnation, but it's commonly accepted in much of the world. Of the several profound experiences I have had of being another person in another time and place, I don't know how else to interpret them other than as a prior life of "mine," my soul, of being another person. There are many fascinating stories of young children knowing languages they've never been exposed to. Some people interpret these experiences not as evidence of reincarnation but of the existence of no time—that all past, present, and future as we understand them are all occurring simultaneously.

In any event, I asked Tawni how she came back from that. How did she transition out of that self she had become and return to her Tawni self? She said she struggled with depression for a few years. She said she didn't know where she was going. She was "kind of devastated with academia, felt that the kinds of questions we were asking were not right, and what was being done with that knowledge was irresponsible. It took five to six years of wandering before I came to know myself again."

As part of her searching, she found herself on 40,000 acres studying at Tom Brown's Tracking and Wilderness Survival School in the Pine Barrens of southern New Jersey. After a year living in the woods and working for the school, she went back to India to teach in the Emory University Tibetan Studies Program to reconnect to Buddhist teachings there because during her time in the woods she found the teachings of Buddhism had "come alive" for her.

After returning from India, she taught children survival skills for a related wilderness organization, looked into naturopathic school (even took a firefighter exam to consider a path that could fund naturopathic medical school), and then moved to San Francisco to join a group of her college friends as a renewable energy engineer for a large engineering firm in the financial district. All the while, "I just felt really lost. I was trying to figure out my path." On one of the engineering projects designing a landfill-based biofuels project in Lima, Peru, she ended up leaving the firm to work for one of their subcontractors to implement the project. "I felt inspired by the people there – their way of living and their view on life and the land."

While there, she taught cultural ecology to college groups as a contractor for a study abroad organization working closely with communities in the Amazon and Andes of both Peru and Bolivia. She had the opportunity to meet *curanderos* of the Nacion Q'eros and the Machiguenga tribe and she realized that they had a vast knowledge of the land and medicine, but that it was being rapidly lost. Don Alberto, one of the master *curanderos* of the Machiguenga tribe, had told her that his grandmother knew 4,000 plants, but that he knew only 2,000, and his niece, who would carry on his work, was probably going to know only a few hundred. "I could feel a sense of desperation, and I recalled all of the written knowledge in Tibetan medicine and the methods Tibetan medicine has to propagate precious knowledge like this natural medicine tradition." She suddenly realized, "I want to be a Tibetan doctor."

Over the next many years, she traveled back and forth to India and Tibet studying Tibetan Medicine and working in communities in both regions. She applied to several Ph.D. programs at Harvard, Columbia, and Emory Universities, eager to do her doctoral work in an interdisciplinary science program. As she was hearing back from programs, she met comparative human biologist Carol Worthman who asked her why she didn't pursue a Ph.D. in anthropology since biocultural anthropology is an interdisciplinary field of science. She suggested Emory since it still retains a more integrative approach. Tawni shifted her application to the anthropology department and was accepted by the end of summer.

It was serendipitous as well since as for many years Emory University has had a program called the Emory-Tibet Science Initiative, which works at the intersection of science and spirituality. After graduating with her Ph.D., she landed a postdoctoral fellowship at the Austrian Academy of Sciences where she could study Tibetan pharmacology, and then at University of Wisconsin at Madison working with Professor Richie Davidson bringing a Tibetan medical perspective to their neuroscience studies on meditation research, while simultaneously establishing her private clinic practicing Tibetan Medicine locally.

Tawni's passion for Tibetan medicine is to bring Tibetan medical paradigms to research and health contexts to help shed new insights on how we understand the body and mind and spirit opportunities to cultivate health and wellness from those understandings. She wants people to learn about this system of healing, "to learn about why this healing system is about connecting spirit and mind and consciousness as well as supporting our sustainability as human beings."

FINDING TEACHERS WITHIN FAMILY

Sometimes, we are fortunate to have close family members be our much-needed mentors on the spiritual journey. Wayne Jonas' exposure to religion and spiritual concepts came from his father, who was a military chaplain for thirty years, having gone through three wars. Wayne said, "My father was a chaplain, so I'm a preacher's kid growing up in his benign, benevolent shadow." Years later, his daughter trained in hospital chaplaincy at Yale, then took a job at Johns Hopkins before moving on to become Head Chaplain at Goucher College in Baltimore. Wayne joked, "In my world, I've been sandwiched between chaplains."

As a child, Wayne was immersed in church work as a gopher, handing out bulletins and running whatever errands his father needed. During the church services, he'd sit in a pew and "doodle on the back of the bulletins, wondering what his father was talking about." What he did absorb during those years was insight into this deep sense of faith his father had, the "place out of which real abundance and peace arises, regardless of where you go in life or what you do in life."

Wayne would go along with his father to the hospital for his chaplaincy work and ask his dad what he did, to which he replied, "I help people heal." Wayne said, "But that's the doctor's job. What do you do?" His dad replied, "I take care of the spiritual side of the people." Wayne's interest in healing and medicine came from "absorbing by osmosis" his father's dedication to helping others, which ultimately led him to medical school.

Very early in medical school, he realized that healing involves a lot more than he was learning—that he was "learning a sliver of something through a very narrow lens." This was because his father's work in the military then involved living in multiple foreign countries, where Wayne had observed other approaches to healing with thousands of years of tradition behind them. These approaches incorporated their version of spirituality into the normal part of health and healing, including medicine.

"Ayurvedic medicine is based on consciousness and Chinese medicine is based on energy. These traditions incorporate the non-material as a core part of what a human being is and what the world is all about." A beautiful byproduct of those observations was that it confirmed for Wayne that "my dad was a spiritual healer!" His medical training "only looked at one layer of what a human being is, and that's all they're teaching me. So if I don't get outside that box and really begin to learn more about this area, I'm missing the majority of what it takes to be a healer."

During his training, he encountered another mentor. Although it was a brief encounter, it literally gifted Wayne with a crucial insight that has guided much of his career. This occurred during a six-month hospital chaplaincy program that he took. As a twenty-three-year-old student chaplain, Wayne related to me the experience he had with his first patient. The sharing of this profound moment for Wayne was one of the many honors I had conducting the interviews for this book.

There was a patient who was dying of bone metastases and in severe pain. He asked the hospital for a chaplain visit and was told there was only a student chaplain available, to which the patient said it was okay. Wayne said he went to the man's room, but when he got there, he was asleep, so he sat by his bed. He said he was extremely nervous as he wasn't really sure what to do, yet he knew he was there to somehow minister to the man. Sitting by the bed, Wayne decided to close his eyes and wait for the man to wake up. After a few minutes, Wayne felt the man's hand on his hand. The man had opened his eyes, saw Wayne sitting there, and reached over and touched him. With tears in his eyes and deep emotion in his voice, Wayne told me that when he felt the man's touch and opened his own eyes, the man looked at him and said, "Son, you're going to be okay."

Wayne said, "Oh my God, that was remarkable. I suddenly saw what healing was all about—it's about deep connection! It was not me to him. It was bidirectional, and he had tapped into it, and he was healing me!"

To help round off his exposure to the more spiritual side of healing, Wayne went to the Esalen Institute in Big Sur, California, and took a month-long residential course on world healing traditions. This further opened his mind. During the course, he was exposed to Tibetan, Chinese and Ayurvedic medicine, among others, which all helped confirm and fill in the details of his observations while he was a young man living in those other parts of the world.

These broader perspectives on healing and medicine would get him into trouble, though, while still a medical student. "I had my hands slapped several times and even had to repeat a rotation because I had suggested something that the chief of pediatrics didn't like. Anything outside of molecules was seen as quackery. I recall being looked at cross-eyed by surgeons after suggesting that acupuncture could help with post-operative care."

He reflected, "I knew not to even mention homeopathy, although I had been studying that too." Years later, while doing a rotation in Germany, he would learn that homeopathy was routinely used in their medical system for pain management. Wayne would later take these tools into his family medicine practice after completing his medical training. As a result of his broader vision of what medicine can be, in 1995, he went on to become Director of NIH's new Office of Alternative Medicine during a time when the office was still trying to define itself. Prior to that, he served as the Director of the Medical Research Fellowship at the Walter Reed Army Institute of Research.

During those early years at the Office of Alternative Medicine, he found a very important mentor, the then NIH Deputy Director Ruth L. Kirschstein. "She was fantastic. She got it. She understood the importance of it—the whole person, body, mind, and spirit." Wayne said she really supported and helped guide him.

There were, of course, difficulties as most people "weren't quite ready for that kind of holistic thinking. They were stuck in the more reductionistic approach. Western reductionistic science is great for generating knowledge but it is not very good for applying that knowledge." Today, the Office of Alternative Medicine has evolved into the National Center for Complementary and Integrative Health with over a $138M budget for 2022. The Center has matured enough so that its current Director, Helene Langevin, M.D., has advocated Whole Person Health as part of its official Strategic Plan, helping to bring fruition to Wayne's vision of the then-fledgling Office. He recalls, "I tried to define it in these holistic terms." After NIH, Wayne became the Founding Director of the Samueli Institute, which would help bring that holistic vision for health into the DOD and VA hospital systems.

Melinda Connor's first mentor, like Wayne Jonas, was a family member—her grandmother. Melinda is a biofield scientist and Buddhist priest whose mother was a psychiatrist and father a Harvard-trained scientist who developed the field of highway safety. As a young child, Melinda displayed certain psychic gifts that were unfortunately not encouraged in her home. When she was a young girl, she mentioned to her mom that she could read her thoughts. Her mother said, "Telepathy is not real. It's not possible." Melinda went on to "read" her mother's thoughts and proved to her that she could, in fact, read her thoughts.

The next day, her mother took Melinda to the hospital where she worked as a psychiatrist and had her sit in the hallway outside the room where the electroconvulsive therapy (ECT) was done for patients with psychosis and depression. Melinda watched the patients go into the room and sometime later come out "remarkably changed for the worse." They were worse not only in terms of how she perceived their general state of being and mood, but she could clairvoyantly see the energy of their biofield which was now highly distorted. Melinda said, "It was extremely frightening to me."

After several hours of watching patients going in and coming out transformed for the worse, her mother came out and sat down next to her. After a few moments passed, her mother said, "Melinda, this is what happens to people who hear voices. Better that you don't hear people's thoughts." This was, of course, a demonstration to discourage Melinda from further speaking about anything psychic to her mother or anyone else.

Melinda was fortunate, for despite what her mother wanted, her grandmother actively but quietly supported her in furthering her education about and development of her psychic gifts. Her grandparents had a farm where Melinda spent most of the summers. Her grandmother taught her energy work and how to further develop her clairvoyance and other gifts. Her grandmother was from a line of royalty, whose mother, Wilhelmina Von Wrengle Rudi Hawk, was a German princess. The family had to flee Germany during the Franco-Prussian war. On the farm, Melinda learned how to speak with plants. "They would tell me when to pick them and what they were for," she recalls.

There has been a long lineage of psychic and clairvoyance gifts in Melinda's family. She said, "My great-grandmother had it, my grandmother had it, I have it, and my daughter Catlin has it." Curiously, her mother didn't get it—hence the innate challenges posed for Melinda as a child. The truth is, many children have such abilities, but through the process of socialization, their abilities are actively discouraged from being used. Since they are not reinforced by the people around them, the gifts are not attended to and slowly go away. With my own work in these areas, it's clear to me that faculties such as clairvoyance and clairaudience need to be practiced and attended to, or they do go inactive, unseen, and unheard in the background of our perception.

Despite the challenges in her home about how Western science and medicine viewed things like psychic gifts, Melinda went on to pursue a scientific career. Despite the disconnect she knew existed in Western science, she felt immensely comfortable around it, having spent a childhood "with endless series of interns, residents, and post-doctoral fellows coming and going from our house." She went on to obtain her Ph.D., knowing that because of her perceptions and experiences of the non-physical world, her views would be very different from her average classmates. She relates, "I knew the boundary where science stopped its explorations. I knew what was beyond that boundary, and I wanted to pursue a scientific degree and conduct research on that spiritual side of the boundary."

With her Ph.D. in hand, she went on to study at the University of Arizona's Center for Integrative Medicine under the tutelage of Andrew Weil, M.D., and Iris Bell, M.D., Ph.D. Melinda was the first non-M.D. admitted into that integrative medicine program where she started conducting energy medicine research, which she continues to this day. She described how the gifts her grandmother helped her cultivate in the garden of the farm enabled her to successfully pursue biophoton research on plants. The plants tell her how and when to do the biophoton measurements.

As Melinda was describing this process to me, I immediately thought of Johann Wolfgang von Goethe, the great German scientist, poet, and novelist who wrote *Faust*, who described the methods he used to communicate with plants. He described scientific approaches to plants where they "would allow themselves to be seen." Goethe was ahead of his time, a pioneer in the sense that his approach to science was to work with the conscious wholeness of nature, and, through that approach, he made his scientific discoveries. There are scientists today who actively use Goethe's approach to science to discover and understand the primacy of the wholeness of nature.[49]

Melinda's research efforts have also been used to standardize the qualifications of energy healers. She created a testing set of measures that scientists can use to determine the competence of energy practitioners, with the aim of identifying qualified practitioners who can be employed in improving research accuracy.[50]

THE TEACHER WITHIN

Many meditation traditions speak about the "guru within," stating that our journey is to find our own Higher Self so that it can be both our guide and destination. Rajnish Khanna consciously begins each day by setting himself in a place where he is spiritually and scientifically aligned. He went to St. Xavier's School, a British Catholic school in Delhi, where there was a church on campus. He visited the church once in a while and was exposed to the Catholic catechism. At home, he was exposed to the Bhagavata Gita. "Growing up, I didn't believe in religions specifically. What I did believe in was what all these religions were trying to tell me." He thought doing science could give him insight into this.

Rajnish moved to the United States in 1991 to pursue his graduate studies and a career in plant sciences. In 2019, he went to Brazil to conduct plant research with the N,N-dimethyltryptamine (DMT) molecule to see if it would alter the light response of plants. He had come across the book *DMT, the Spirit Molecule: A Doctor's Revolutionary Research into the Biology of Near-Death and Mystical Experiences* by Rick Strassman, and it gave him new ideas about his own plant research. He knew that tryptophan is important for what is called dark elongation in plants—the growth of seedlings when there's no light. He soon was able to meet Rick and share with him his ideas, which further motivated him to pursue the work. Rajnish thought that Rick was an experience of "good karma" for him along his life's journey. His new ideas also led him to conduct research on the effects of anesthesia on plants. His research group found that "if you give anesthesia to plants, they can't orient themselves to light."

He had the thought that psychedelics share the tryptophan pathway, so he asked, "What are the mechanisms that consciousness has in terms of signal transduction?" He wondered if plants would have a parallel response to humans. He had gotten his Ph.D. in plant biology years earlier, followed by a post-doctoral fellowship at UC Berkeley.

Years earlier, when he first told his father he wanted to study plant biology, his father worried whether that was the best choice for him and if that path would lead him to become a farmer. In those days, the top career choices in India were Medicine, Engineering, Indian Administrative Services, Chartered Accounting, or similar highly competitive career tracks. The next thing Rajnish knew, his father had signed him up for a military service interview at an Officer's Training Camp where he was shipped off to another city for training and evaluations. Ultimately, he wasn't accepted because he told the military leadership at the final oral interview for acceptance into the program that he didn't actually want to be in the military but wanted to be a scientist.

Years later, he decided to pursue ayahuasca ceremonies in Brazil. Ayahuasca changed him. He said that as soon as the ayahuasca took effect, "There was a voice, and the voice sounded like my own voice. But what was crazy was I didn't feel like I knew what it was going to say." The voice said," Thank you for letting me in." I said, "You're welcome." The voice then asked, "Would you like to chat, or would you like some visions?" Rajnish was scared about visions, so he said, "Let's chat." Rajnish then asked, "Who are you?" and the voice replied, "We'll come back to that in a later ceremony."

Pondering what to chat about with this voice, Rajnish asked, "What is life?" Throughout the remainder of the ceremony, he received answers to this and other related questions. In a later ceremony, when he again asked the voice, "Who are you?" he was then "shown the universe, floating out in the vastness." At that moment, he realized who he was speaking with. It was that "vastness beyond the universe, the oneness of it all." He realized then that we are all the oneness.

"When we speak to someone, we are speaking with ourselves as that totality in human form. I realized I was speaking with just this one thing that I am, what everything else is." He concluded, "When we're speaking to someone else, we're actually speaking to ourselves." He said, "Everything is just one, and it is so big, bigger than the universe."

As he said these words, it reminded me of the famous reply by the Indian Saint Ramana Maharshi. Someone asked him, "How should we treat others?" to which he replied, "There are no others." In that state, everyone and everything is seen as Self, but not in a personalized way—in a deeply experiential way that is known foundationally.

This experience changed the way Rajnish saw his science, the science that he had been conducting for the past many years. He said, "That experience changed me dramatically, and when I came back and started looking at my science, which was about signaling and consciousness, I now understood it from a completely different perspective." Going forward, he decided to have his science work be from that perspective, a living conscious universe that he himself was. He told me about his work and his theories on physics, consciousness, and the universe. I encouraged him to write a paper on his ideas and submit it to a scientific journal. He's been developing that paper in which the principles go back to Vedic principles of Akash, the primordial ether.

One of the things I so appreciate about Rajnish is that, like Melinda Connor, he is an admirer of plants. He said that "plants are so beautiful, creating and recreating themselves endlessly, adapting to pretty much every climate on this planet." He said it's sad to him that many people have "plant blindness," which means, "so many people don't see plants. They ignore them." He appreciates, too, that plants have the same source of consciousness as all of us, but their level is very different, and they have a different experience of time.

He said we have flexibility that plants don't have. "We can stop being in one profession and take up another profession." A fruit tree that makes apples can't change what it makes. "We humans have some additional abilities but that doesn't mean that we are fundamentally of a different source. It's just that the information is different, and how we interact with it is a little bit more sophisticated."

He reflects, "All life forms have signaling mechanisms. We just need the right antenna to pick up pieces from the vastness of information that we need to move forward in life, both physically and mentally." What he would like to show with his work is that "consciousness, our awareness, is an interaction with that information," and he thinks this is doable because biological systems are mechanical. He works with plants for his research because he admires them, and "it's easier to do these things with plants."

CHANCE ENCOUNTERS WITH A TEACHER

Bill Bushell told me a mentor story that greatly influenced his commitment to his work and the direction he needed to take. He was traveling the Ethiopian desert as part of his medical anthropology work. Hours into the desert drive with his translator, they encountered a hermit monk alongside the road. They stopped and spoke to him. While Bill had learned some of the language, his skills weren't good enough to fully understand what was being said. The translator related to him that the monk said he had walked a long distance to get to the road, "to come out of the desert after a year-long meditation retreat because he had a dream about him [Bill] coming to that spot at that time." He said he knew Bill had questions, and he came to answer them.

"The hairs on my neck stood up; he was like St. Anthony in the desert," Bill said. "He was supposedly very old, nearly a century, but didn't look even forty." Bill's questions for the monk had to do with what he was studying there—exceptional people who seem to resist the aging process by using esoteric prayer and meditation techniques. Turns out there were remnants in that particular region of an ancient Christian sect of monks who have lived in the desert for centuries. "When I saw and spoke with this monk, which was just before the war that chased me out of Ethiopia, I was thrilled, and the experience greatly inspired me about the potential truth of my scientific and spiritual model.

Bill added that for many years—before and after that particular experience with the monk—he had devoted his career to studying advanced meditation practitioners across many cultures and learning about the effects those advanced techniques had on their lives and biology, especially regarding aging, longevity, and potentially super-longevity. But Bill had not been much concerned about his own physical health, especially later in midlife—until the "visionary" experience with the Dakini I shared earlier. It was only after that experience that Bill became personally open to the desire to actually obtain longevity as that allegedly achieved by the Ethiopian hermit, though not through such an ascetic path, but rather through a Tantric path inspired by that vision.

Rudy Tanzi began to professionally write and speak about his interests in the metaphysical and consciousness, including speaking at the Science and Nonduality conference. "My work started to expand into integrative medicine, holistic medicine, that whole side of things I had never pursued, including consciousness development."

There was a significant experience that Rudy had that prepared him to pursue these longstanding areas of interest. The experience took place on

September 11, 2001. He was scheduled to be on United flight 175 to attend a medical conference. The night before the flight, he attended the Boston Film Festival where he met Steve Martin, speaking with him into the early morning hours about music and banjo playing. "Bottom line, I drank way too much, more than I had ever before or since. Frankly, it was the most I ever drank in my life." He missed the flight due to a hangover.

When he woke up the next morning, he learned what happened to that plane he was scheduled to be on. He went to the beach for a long walk with his wife. He told her, "This is bonus time. What are all the things I'm not doing in my life that I've been wanting to do?"

He had long wanted to get back to his music, which he did. They had wanted a child and pursued that, and they now have a beautiful daughter. He started working with the New England Patriots football team, helping them take better care of the player's brains. Before that missed flight, he would have said to himself, "I don't have time to do all these things." But now, going forward, he made the time. After 9/11 it was, "Do it. Do it. Do it." He then started taking bigger chances, including with his research.

I asked him, "Are there still remaining frontiers that you want to pursue but haven't?" He said, "The metaphysical becomes physical and scientific as soon as you can measure it and predict it." A remaining goal of mine is to learn how to do that, to measure the metaphysical, and then work more in that domain.

One of Rudy's approaches to his own self-development is practicing lucid dreaming. Many years ago, he read Carlos Castaneda's book *The Art of Dreaming* and became proficient in it. Rudy said that no one has had as great an influence on him as a teacher of the mystical as Castaneda.

When Rudy was a young boy, he started keeping a journal, writing down his thoughts on the metaphysical, and juxtaposing those thoughts with the physical world. "When I had the opportunity to write the book

Super Brain, I picked up every one of those journals and used all of the notes going back to when I was ten years old. He shared, too, that much of the content in those journals came from his astral traveling during lucid dreaming and the information he would gain on those night journeys. "The book became a New York Times bestseller; it was a dream come true."

He shared with me a lucid dream experience from when he was on the island of Kauai, HI, some years ago. One night, he was lucid dreaming and found himself in a wooden building, like a temple—he was flying around. He then encountered a person in their spirit form, a monk in an orange robe and white beard who told Rudy he couldn't be there and to get out, which he promptly did.

The next day, he and his wife Dura went for a long drive on the island and ended up at a Hindu monastery. While walking around, he saw the very building he had flown up to the night before while lucid dreaming. Some moments later, a monk approached in an orange robe and white beard. Rudy instantly recognized him from his lucid dream. "Right there, I think I had the stomach flip, the same as I used to have playing that game as a child with my sister." He leaned over and quietly told his wife, "This is the guy who kicked me out last night."

Rudy related, "All realizations are personal; nobody can make anybody realize something. Over the past few years, I've become much more convinced that the physical material world does not exist. It's all consciousness. The number one rule is love and service."

"What I have trouble with is whether there is life after death," he said. He's had no experiences about that per se, so it remains an unknown. "Without a brain when I am dead, will I still be able to astral travel as I do during lucid dreaming?" He said he would love to get to the level of scientifically exploring these questions, to be able to turn those mystical questions into the known through science. For example, when he was twenty years old, he did research with the team that discovered several genes responsible for Huntington's Disease. Centuries earlier in Salem, MA, if you had Huntington's Disease, you were considered possessed or a witch and burned at the stake. So here, science made what was once considered metaphysical into the physical, the scientific.

Cassi Vieten's experience shows how a mentor can sometimes shows up in our life for just a brief moment, but that moment turns out to have been highly decisive. When Cassi was wrestling with the idea to attend the California Institute for Integral Studies—a nontraditional school, which her parents and college professors said wasn't a good idea—she randomly met a woman at a coffee shop and was prompted by an inner voice to ask her for advice. Cassi asked: "Should I get a more traditional education? Am I going to make a big mistake by going to a nontraditional school? Will I be making myself more alternative and not potentially able to be a professor someday?"

She reminisces, "I'll never forget. This woman looked me in the eyes and said that she had changed her name when she turned sixty. That she had waited until she was sixty to do everything that she had always really wanted to do her entire life and would strongly recommend that I don't wait." Something clicked inside Cassi, and she knew it was time to enroll at the California Institute for Integral Studies, a decision that set her life's direction and has enabled her to successfully pursue her dreams of integrating science, the transpersonal, and the metaphysical.

INNER GUIDANCE SHOWING THE WAY

Much of Mary Jo Bulbrook's nursing career has been in mental health, having served as a psychiatric nurse and nurse educator for many years in multiple countries. She told me the story of how Virginia Satir came to be her primary mentor, colleague, and friend.

Virginia Satir was a renowned psychotherapist and one of the foremost leaders who helped reshape how we approach mental health. She is considered the 'Mother of Family Therapy'. Virginia was visiting the University of North Texas where Mary Jo was working on her doctorate. Her clinical professor had suggested she study the work of Satir who came to the university to do some teaching. Mary Jo decided to attend the meeting where Virginia was demonstrating her patient interviewing methods. Mary Jo heard an inner voice say, "You are going to be very involved in Virginia's life. You are meant to work directly with her." As she watched Virginia demonstrate her interviewing methods, Mary Jo thought to herself that it was a totally different approach from how she had been trained as a psychiatric nurse and what she knew about psychiatry. "It was a whole new way of dealing with people."

At the time, Mary Jo was working on a book titled *Development of Therapeutic Skills*, which was to teach people how to become therapists. After the demonstration, Mary Jo approached Virginia and asked if she would write a chapter in her book, to which Virginia said yes. "A year passes, and everybody else had written their chapters for my book but Virginia had not, so I caught up with her." Mary Jo asked her about the chapter, and they started chatting about energy work. Mary Jo realized Virginia also was very interested in the hands-on energy therapy known as Therapeutic Touch. Their mutual interest in extra sensory experiences was the foundation of how they connected and the role they eventually played in each other's life. Their bond grew closer and closer each year. Mary Jo told me she dreamed of Virginia's death before it happened.

Mary Jo recently became President of Akamai University to advance energy medicine education and research. She said she was nudged by Spirit to do so. She holds space for students to follow their voices, spiritual nudges, and development to become fully guided by higher sense perception rather than ego based mental thinking. This is fully accepted and joined by other faculty and students to embrace this innovative way of being in the world.

There, she's created the Integrative, Multi-Dimensional Empowerment, (IME) Model of Care and Caring for Health Professionals to help guide, direct and launch solutions in academia and provide unique opportunities for individuals to realize the hidden truths of their inner knowing and inner direction.

CHAPTER 4

FROM TRIALS TO COMMITMENT AND TRANSFORMATION

If you know you are on the right track, if you have this inner knowledge,
then nobody can turn you off... no matter what they say.
— Barbara McClintock, Nobel Prize in Physiology or Medicine

After completing each of my interviews for the book, I said, "Before the book is published, I'll send you all the text I'll be including from our interview so you can review it to make sure you're okay with the content." Neil Theise immediately responded, "I doubt that will be an issue. I own my crazy." I immediately laughed out loud and told him I really admired his sense of freedom. The point is, he made the choice years ago to proceed with his own transformation and be open about his interests in the metaphysical and mystical, something he's fully okay with.

This gets to the issue of one of the most important stages of the monomyth journey: How does the seeker deal with the inevitable trials and decision points that come along the way? Some trials are minor while others are more significant. Does the seeker choose to proceed along the journey as planned, make a minor course correction, or perhaps abandon the journey and return to the known? For Neil, how people might think about him is not much of a consideration.

For many academics though, there is the career side of income and tenure to consider. Neil followed up by saying, "If I were dependent on grant support for my career, this could be dangerous stuff for me to talk about. But I have my clinical practice, and no one questions the diagnosis of a pathologist because they talk about weird things in their off time."

The earliest scientists faced more significant trials, potential life and death trials by the Catholic Church which stated they couldn't continue their science if it contradicted church teachings. Giordano Bruno was burned at the stake in Rome in 1600. The church brought a philosophical charge—as well as a scientific charge—against him, for which he was executed. In 1616, Galileo Galilei was issued an injunction not to teach his heliocentrism model. He was later tried for heresy and persecuted by the church. These were tough times for scientists who were faced with the decision to pursue their inner inspiration of discovery or turn away from it to ensure their safety. Many of these early scientists had deep spiritual convictions about the nature of the world, and I wonder what it was like for them to have to compartmentalize that, to put aside those convictions in terms of expressing them in the context of their research.

In addition to the type of work, scientists must decide what institutions will be a match for them. While there are many private and public institutions and universities to choose from, only a minority of them are open to conducting more metaphysical research. Institutions such as the IONS, the California Institute for Human Science, and the California Institute of Integral Studies are examples. But outside of such places, there are challenges—typically in the form of other scientists—to being able to pursue these nontraditional areas.

Challenges can come from family and friends too. Recall Thomas Brophy, who had a family member negate his psychic experiences. Most often though, trials come in the form of one's own internal struggle and decision-making. Following a transpersonal transformation, does the seeker proceed further into the unknown and the new life that has been glimpsed, or do they retreat back to the familiar, known, and comfortable?

TRIALS FROM MATERIALIST SCIENCE

Cassi Vieten shared a poignant story in this regard having to do with her interest in spiritual research and her interest in pursuing a nontraditional academic institution for her next steps. She had choices to make. While working at an alcohol research lab at the University of California San Francisco, the NIH put out a Research Funding Announcement to study spirituality and alcoholism. With excitement, she went to the lab head, who was then Chair of the Department of Neurology, and told him she wanted to apply for the grant. He responded, "Not only can you not apply for this grant, but I'm starting a petition with all of my colleagues to protest the use of federal dollars for this ridiculous bullshit." He also thought it was a violation of the separation of church and state.

Cassi stood up to him for the first time ever, telling him he was wrong. She said that even if it's "imaginary bullshit," it does help people quit drinking. "Even their perception that there is a God is making more change than any other treatment ever has, including your research to find a drug to treat alcoholism." She noted that NIH was spending $130 million to try to find drugs to treat alcoholism, and the spiritual approach has already saved many more people than this research has. He didn't agree, and she didn't agree with him, so she started looking for another job because "I realized I was in the wrong place. I couldn't work there anymore and pursue the type of work I wanted to."

That decision was what led her to IONS, where I first met her. She told me after her conversation with the Chair, she put the words "spirituality and science" into her internet browser, and up popped IONS—it was only ten minutes from her house. She went there to find out what they were doing and met Marilyn Schlitz, then President of the Institute. She started working there one day per week, then two days. When she went to her University of California San Francisco job the other days of the week, she said, "My stomach would hurt. I realized I was happy working at IONS." But then another challenge arose in that her University of California San Francisco mentors and other research colleagues told her, "You are crazy to go to IONS. It's a pseudo-scientific cover for fraud, complete pseudoscience. You will destroy your scientific career."

While she described this, I thought of the material I wrote about in the first chapter of this book, the deep sense of materialism in science, the scientism, the inability to have an open mind to the possibility of there being anything worthy in these pursuits. The COVID pandemic has made many of us more acutely aware of the phenomenon called "group think," or worse, what is called "mass formation," where entire groups of people adopt an unreasonable mental and emotional position on a topic that has no basis in rational thinking—or at times is even contrary to the science behind it. I felt a sadness with this thought, the thousands and thousands of scientists our system has created who have no idea what is beyond the known, where they could potentially find a more meaningful life and their very Self.

I asked Cassi, "Why the strong negativism among scientists?" It often actually offends them that other scientists pursue this type of research. I found her answer very interesting and provocative—something that perhaps resonates with C.G. Jung's and others' work on the collective unconscious.

She said, "I think there's kind of a collective post-traumatic stress disorder from the dark ages and the inquisition, when all of society went crazy. It's kind of like an addict hitting bottom. It's hard to believe how bad it was, how much they were relying on the perversions of religion to make these terrible choices about the crusades and the inquisition, with the terrible things that were done to people in the name of religion."

She continued, "Once we finally started to enter the enlightenment, there was a whole collective agreement that we're never going back there and never going to be that blind again. I have compassion for people who see it as a danger to civilization to pursue these areas of study. They think it's their job to make sure that never happens again."

She concluded by saying that there is fault, too, on the part of some people doing alternative research in areas like energy medicine, where there have been "snake oil salesmen and a lot of fraudulent claims that have harmed people." She raised the valid point that there are many examples of corroborated fraudulent claims on the part of pharmaceutical companies that, too, have harmed and even killed people, so why isn't there as much animosity toward them?

Cassi did overcome her doubts and took a full-time research position at IONS. She eventually became president of the organization.

INNER TRIALS FROM ADAPTING TO A NEW WAY OF BEING

Recall the experience Dusana Dorjee had upon leaving her university office to walk home. She had been deeply engaged with a reading and practice of unconditional compassion. The practice induced her into a nondual state of consciousness of oneness, where there was no longer any difference between her and the people she encountered. She found the state profound and, at the same time, wondered if she would be able to function like a normal person in this state. Should she proceed further with her meditation practices and transformation or turn back?

Soon after her nondual experience, Dusana decided to attend a Summer Research Institute conference. These are weeklong immersive programs providing opportunities for students and faculty to engage in deep dialogues across disciplines and support inquiry through first-person reflection and contemplative practices. They are hosted by the Mind & Life Institute, which brings science and contemplative wisdom together to better understand the mind and create positive change in the world.

At the Institute, Dusana met a Buddhist teacher. It was a brief but fortuitous encounter and turned out to be very helpful. She described her nondual experiences to the teacher and asked him questions about how to function in these states. The teacher told her, "This is Nyam, part of the normal path." She asked him, "What do I do now?" and started to cry. He said, "Go with it." She replied, "But how can I function like this? He said, "You have to make a choice. Do you want to further explore these things or accept the conditioned reality?"

That was a powerful moment and a choice point for Dusana. She thought to herself, "There is some security in knowing the conditioned reality, but then comes the challenge of shattering our identities." Can we have true transformation without the commitment it takes? In the end, she better understood her nondual experience and experiences of unconditional compassion and deep peace and abiding. She deepened her conviction to proceed on the journey. The phoenix bird must commit to entering the fire to be reborn from its own ashes.

She told me about "Bodhicitta," which is translated as "awakening mind," the mind that strives toward awakening, empathy, and compassion for the benefit of all sentient beings, a commitment to walk a path of awakening and dedicate oneself to the liberation all beings. Bodhicitta is born of compassion. It's one of the foundations of the Mahayana tradition. Dusana said this path has a deep resonance for her.

Dusana's story reminded me of a similar experience I had when attending the Summer Research Institute years earlier. I then wondered how many scientists have been helped over the years by these summer conferences that Mind and Life has been hosting for over nineteen years.

I attended back in 2008, having been invited by the leadership because I was mentoring students at UC San Diego who had previously attended the summer conference. At that time, I knew very little about Buddhism, other than there was the Buddha who had achieved a state of enlightenment and whose teachings of compassion subsequently spread throughout the world.

While asleep in my room the second night I was there, I found myself transported to a room where, in front of me, on a dais approximately one foot off the floor, there sat four identical-looking Buddhas. My vision immediately went to their heart chakras, which I found very unusual in that they were diamond-shaped, not circular in the way I'd previously seen. Each was an identical blue color, darker than the sky but not a dark blue. The blue-colored energy was swirling around in their diamond-shaped chakras, and there was an endless depth to them.

The other thing I noticed was that while each of the four Buddhas looked identical to me, the one on the left was seated slightly farther apart than the other three, who each had identical spacing between them— perhaps one foot apart. As I gazed upon them, I heard a telepathic message in my mind: "You can give us your self or not." I had a deep emotional response to this statement and immediately awoke. I began to cry until eventually falling back to sleep. In the morning, I pondered the experience, particularly why four Buddhas, as I was under the impression there was only one historical Buddha.

In the morning, after breakfast, which we all ate in silence, I walked past a Buddhist teacher and monk in one of the hallways. I stopped and asked if I could ask him a question. We found a quiet place to sit, and I recounted the experience to him. I then asked what he thought it was about and why there were four Buddhas.

He said, "I think you had insight into the four immeasurables of the Buddha." I asked what that meant. He said the four immeasurables are love, compassion, sympathetic joy, and equanimity. He further explained that these are recognized as the four primary features of the Buddha, what were his essence.

I then thought to myself that perhaps the "equanimity" aspect was the one on the left who sat a bit apart from the others. Equanimity in spiritual traditions is akin to Christ's teaching to "be in the world but not of it." One partakes of the experience of life but isn't overly identified with them, being more of a witness. I felt grateful to the monk and thanked him for his time.

He then told me he came to the conference from China and that he lived at a temple outside a small village where he goes to teach on occasion. He then asked me if he could tell me about a dream he had before he left China to attend the summer institute. I felt enthusiastic and said yes.

He shared that one evening at the temple, he was preparing a lecture on the topic of the Diamond Sutra. He was going to head into the local village the next day to give a talk on it. The Diamond Sutra is an ancient Buddhist text, major themes of which are non-self and liberation without attachment to the world.

The monk shared that the night before he was to give the lecture, he dreamed he was seated in a movie theater. He was the only one in the theater and was seated three rows back, right in front of the screen. He watched the movie scene of a man screaming and crying wildly about the loss of someone in his life—a young woman, if I recall correctly. After some time with the man going on and on, the monk got up, climbed over the seats, came right up to the screen, and said to the man in the movie, "Why are you so upset? You're not even real." The man on the movie screen then stopped his ranting, looked down at the monk, and asked, "If I'm not real, can I still experience liberation?"

"Amazing," I thought to myself. "What a wonderful dream!" I hadn't heard of the Diamond Sutra before, but learning about it later, the dream made much sense to me and is at the heart of many enlightenment traditions. The "small self" or egoic self is recognized as not being real, even though the vast majority of people experience themselves as a separate self and very real indeed. The text of the Diamond Sutra is to help bring that realization to the person by practicing non-attachment to everyone and everything, which eventually helps loosen the attachment to the false self. Then liberation can be realized.

The "unreal" man in the movie, akin to our own unreal nature as an ego-self, was awakened enough by the monk's statement to make the inquiry about enlightenment and liberation. After the conference, we stayed in touch for a while. At one point, he sent me a manuscript he was writing for publication, asking if I would check the English, which I did.

Neil Theise, like Dusana, had a kind of decision point come to in the form of surrendering to a new way of experience and understanding things or sticking with his old way of seeing things. He'd been meditating for around fifteen years at this point, mostly Zen, and realized that his brain "just started to be insistently creative in a way that did not feel real to me. It was very uncomfortable. I initially wanted to turn it off but couldn't."

I asked him why it was uncomfortable. "Well, it felt that this stuff—all these ideas about how bodies work—was downloading, and I just couldn't stop it." He said when you meditate that much, it changes the way your brain works. Eventually, he overcame his resistance to it and committed to proceeding. Over time, he got more and more comfortable with it.

Another beautiful insight he described is that Zen practice is very much about constantly coming back to no judgments. "What you are experiencing in the present moment without making names, without making categories, without assuming principles, just what is there. This is the so-called Zen mind, the beginner's mind."

The Zen priest, Shunryu Suzuki, who went by the name Suzuki Roshi and who founded the San Francisco Zen Center, would say, "In the beginner's mind, there are many possibilities, but in the expert's mind, there are few." Zen practice taught Neil how to approach things that were routine for him, such as looking at things under the microscope, as new. That practice did something to his brain. He said, "Zen practice paid off. Zen has allowed me to be more open to all my experiences. Because on an emotional, intellectual, and experiential basis, I've lost a reason to say no to things. That's what came out of the incense realization."

Like my colleagues in this chapter, a significant choice point came for me approximately a year before graduating from MIU. As I related in the introduction to this book, one afternoon, I sat to meditate and upon closing my eyes heard a strong and clear voice state, "You can no longer serve two masters." The voice was not my own. I had an immediate visceral response to hearing those words. A part of me knew what was about to happen— I was to release the Maharishi as my teacher.

During my years at MIU, I became a teacher of TM. The curriculum was set up in the "block system." Two consecutive months are an immersion in whatever academic courses we were enrolled in at the time, and the final month is a "rounding month." Rounding was a curriculum where we spent many hours per day meditating and practicing yoga asana and pranayama (breathing) exercises. A minimum of three of these rounding months constituted the first requirement to become a teacher. The second requirement was to attend another three months of rounding that included learning all of the material needed to teach TM. Along with around one hundred other people, I completed this latter requirement at a hotel in the mountains outside of the city of Nice in France. Maharishi taught the main portion of the coursework.

The challenge was that Maharishi was important to me. I looked at him as my spiritual teacher. I had a choice to make. Could I make it? Could I release him as my teacher? I did release him, and it was that decision that opened a whole new life for me. The decision was accompanied by a transformation of my experience of consciousness that remains to this day.

A rub at that time was that I didn't have outlets to speak about the experiences I was in the midst of, there were no Summer Research Institutes then. I did share with my then-wife, Kim, and she was good at listening and very supportive. As far as speaking with any of the leadership at the university, that was not an option. The TM movement, at least at the time, like many other organizations that seek to foster self-development, discouraged sharing of such experiences. If people did speak openly about such transpersonal experiences, they were told they were "unstressing" and to go practice more yoga asanas to "get grounded." While I appreciate the reasons for this, such policies can also work against the progress of what people are striving for.

As with other similar settings of intensive meditation practices, there were instances at MIU of people becoming imbalanced psychologically and needing professional help outside the organization. I've often wondered if there had been more supportive venues for members of the community to openly share and explore their transformative experiences, would there have been less need for eventual professional mental health intervention.

I'll add a curious piece of information about that transformational time back at MIU. Some days after the transformation, I had a clear knowing of what the next step would be for me in terms of a further change in my experience of consciousness. I desired that change to happen right then. It wasn't meant to be, however, because when that desire arose in my mind, I heard an inner voice promptly say, "That won't occur until twenty-five years from now." That was not what I wanted to hear but I accepted it. I realized it was not something I could force or in any way control.

Over the years, I forgot about that prediction. One day, that next step of experience I had desired did emerge. It's a topic I'll cover more in Chapter 6, but it has to do with what is often called the no-self event, a dropping away of the center of consciousness. As it occurred, I realized what was happening. I soon recalled the prediction of "twenty-five years from now." The predicted timing had been accurate.

CHAPTER 5

GIVING BACK AND THE NEXT GENERATION

You cannot transmit wisdom and insight to another person. The seed is already there. A good teacher touches the seed, allowing it to wake up, to sprout, and to grow.

– Thich Nhat Hanh

N early all scientists help train the next generation of scientists. The mentee becomes the mentor. There's a stage in the monomyth when the hero or heroine returns to their community to teach what they have learned. They've had some sort of illumination, a transformation on their journey, and now are compelled to share it with others. Some of my interviewees give back through teaching. Others give back through their medical practice. In each case, their new insights greatly influence how they approach their work with others.

Marilyn Schlitz grew up in the 60s and 70s when "Detroit was literally on fire." In her early teens, she felt a deep inner impulse to do something about it. "I wished I could change the world then, but I was an impotent white girl in a largely black city, and what was I going to do?" As a teenager, she had taken psilocybin and "merged into the godhead," which significantly influenced the type of teaching and scientific work she would eventually pursue.

She had been ready to head off to the University of Michigan where she had been accepted for her undergraduate work. Then her father died so she couldn't leave home; she was needed to help manage things. As a result, she attended Wayne State University in downtown Detroit. She said, although it was a major disruption to her college plans, "it turned out to be the best thing that could have happened because I enrolled into a program at Wayne State called Monteith College. The program was based on the work of Thomas Kuhn and *The Structure of Scientific Revolutions*, with a focus on getting science into the humanities."

Marilyn said, "Getting into that program blew my mind open to the idea that maybe revolution is actually possible." She was acutely aware that we live within a certain worldview, a social paradigm, one that Detroit was evidence of, but she soon began to understand through the program she enrolled in that paradigms aren't absolute and can be changed. "So suddenly, this disempowered girl from Detroit had pretty lofty ideas about being part of a paradigm shift. It caught fire in my imagination!"

While at Wayne State, she did an internship at the medical school with a neurophysiologist named Robin Baracco. They often spoke about the mind, the brain, and consciousness. At one point, he gave her the book *Psychic Exploration: A Challenge for Science, Understanding the Nature and Power of Consciousness* by Apollo 14 astronaut Edgar Mitchell, "and that completely changed my life. I decided I wanted to be a revolutionary in consciousness, to show that we have all these human potentials that can transcend the material world as we know it. I know, I was really grandiose at that point."

She sought out and landed a summer internship in a lab doing psychic research, including remote viewing, and there she learned that she, too, had certain psychic abilities that could be measured in the lab. It was an exciting revelation for her. Upon graduation, she landed a research job "studying the science behind the mysteries of consciousness, healing, and transformation." She eventually headed to Stanford University in Palo Alto, CA, running the parapsychology program there. Although it was a great position, and at Stanford no less, she soon found herself as Director of Research at IONS. She remained at the Institute for twenty-seven years, including being president. "It was a great place to grow up." Her next stop was at Sophia University in Palo Alto, where she is now Chair of the Transpersonal Psychology program.

Over the years, Marilyn published numerous scientific papers and books on the mind, consciousness, and psychic abilities. I asked if any of her research test subjects ever experienced an enhancement of their psi abilities by participating in the research. She said she wasn't aware of this, but she was aware that the research affirmed for many of the test subjects that their prior psychic experiences were indeed valid. "To be in a scientific lab being studied with established experiments was affirming for many."

As part of her giving back, Marilyn created scalable educational programs that bring these ideas on consciousness and human potential out into the world and, in particular, serve health professionals. She's also made films on these topics, including *Death Makes Life Possible*.

RECONCILING ONE'S SPIRITUAL TRADITION AND THE CALL TO SCIENCE

Shamini Jain is a clinical psychologist, scientist, and social entrepreneur. Her desire to reunite science and the spiritual led her to create the Consciousness and Healing Initiative, a nonprofit collaborative accelerator that connects scientists, health practitioners, innovators, and social entrepreneurs forwarding the science and practice of healing. Much of the inspiration for her work came years earlier when, as a child, she would spend afternoons in her father's library, amazed at the images she'd see in his religious books on Jainism. She would climb the circular stairs to his library, which was a "quiet place in our home with large picture windows."

Jainism is one of the world's oldest religions in practice today. Her father, Surendra Jain, a highly successful chemist, inventor, and businessman, had a library of Jainism and other metaphysics books. "I remember even at that young age reading that stuff and wondering, 'How do they know that?'" It was odd because she also felt that what she was reading was true. "Even then, I understood it to be important information about the human being, but I wasn't learning about it in my school." Years later, as she formally entered into science, the same questions arose as to why the things she read about as a child, such as the energy body and chakras, weren't being explored in Western science.

During her undergraduate and graduate education, it was striking to her when learning about psychoneuroimmunology that she would recall those early childhood books that spoke about yoga and pranayama (breathing) practices balancing the sympathetic and parasympathetic nervous system and supporting the immune system. "I said to myself, how did they know that back then?"

I met Shamini at UC San Diego. I was her adviser and then dissertation chair for her Ph.D. Although we had many conversations over those five years when I was overseeing her graduate education, it wasn't until I interviewed her for this book that I asked her to share with me more of her personal journey. One of the reasons I'm grateful for writing this book is that I finally had a chance to sit down and ask some of my former students about their spiritual life, including Shamini, Joseph Tafur, and Christine Peterson.

Shamini told me what brought her to science was a "deep desire to heal what I saw as a fractured worldview." She said there was always a sort of knowing that we needed to heal these so-called disparate understandings of science and spirituality, that they were not separate and needed to be brought back together to allow us to better understand our humanity and to apply that knowledge to heal suffering. "And quite frankly, why not? It's super exciting; it's a frontier. Why would we not want to explore consciousness from a scientific perspective? Science is all about uncovering the unknown with curiosity and reverence. So what better thing to study than consciousness from my point of view?"

Being the daughter of a chemist, her father always told her that "'everything is understandable through chemistry,' yet he believed in the soul and in reincarnation." She thought to herself, "There's this one side of life that is very linear and this other side which is mystical and metaphysical. There's no one really connecting these dots." That was what motivated her to enter into science. She knew it was important to explore these areas from a scientific perspective.

She went to Columbia University to study neuroscience and behavior, working in a lab and learning how to conduct EEG and ERP research. She wanted to apply those tools to study the effects of sound on the brain. Shamini had trained as a professional singer and knew from personal experience the effects of sound on the mindbody. From there, she went to Stanford for a couple of years to learn more about EEG research. The rub, though, was that she couldn't get any traction from the lab's principal investigator to support her interest in studying the healing effects of sound. She was feeling at a dead end when she heard about the University of Arizona's program where faculty were studying many aspects of consciousness and healing. The program there resonated deeply with her, so she packed her bags and went to the University of Arizona to get her master's degree in

Integrative Health Psychology, conducting research on sound and healing as well as mindfulness meditation.

Her interest in the human biofield had started some years earlier while she was in Santa Cruz, California. She had a session of a hands-on healing modality called Reiki. It was the first time she felt a "healing vibration" running through her body. "I understood that this moving of energy was removing a kind of stuckness in me that was related to certain thoughts and emotions I had about myself and my own power." She thought the energy experience was amazing, and the experience brought her thoughts back to what she had read in her father's library those many years earlier.

Based on that experience, she decided to learn Reiki and train in how to do it. When a person learns Reiki, at the end of the training, they go through a kind of initiation process, often called an attunement, which the teacher performs on the student. For the initiation, the teacher told her to bring to her mind and heart the highest source that she felt connected to when she thought of spirit as the representation of God. So Shamini thought of Mahavira, who is to Jainism what Buddha is to Buddhism.

As she thought of Mahavira, the teacher put her hands above Shamini's head, "and all of a sudden, I felt this beautiful golden light and waves of bliss. It lasted for over an hour." Afterward, she wondered to herself what had just happened. It was profound and beautiful, and she wanted to understand it more. That spiritual experience further whetted her appetite to be a scientist. She wanted to know what had happened and how it happened. "The Reiki teacher didn't even touch me."

To further her education in energy healing methods, Shamini began studying intensively with Rev. Rosalyn Bruyere, an internationally acclaimed healer, clairvoyant, and medicine woman. Rev. Bruyere was then collaborating with physicians and scientists on research on energy healing at the University of Arizona. Shamini has been studying with her since then, completing many of her workshops and trainings.

Early on, while training with Rev. Bruyere, Shamini had what she called a "kundalini experience." In an earlier chapter, I wrote about Bill Bushell's experience of kundalini with the Dakini. Kundalini is a Sanskrit term for the energy and consciousness that resides at the base of the spine. Shamini experienced it as a strong energy that began at the base of her spine and rose up her body to her head. It was characterized by an enormous amount of light and a shift in her consciousness. She said, "Everything was all of a sudden a hundred times brighter and my body was absolutely alive and tingling with every cell's vibration. I had been plugged into the universal light socket."

I asked her how, with all of these transformative insights she had experienced over the years, did she now understand the nature of consciousness and her own consciousness at large? She said she still struggles with her egoic mind, the so-called small self, and how she interfaces with her larger consciousness Self.

She said her experience these days is "more of connecting with spirit and knowing that I'm not separate from that spirit, but there's still a sense for me as sort of the personification of certain elements of consciousness and to learn more about the devotional process as a process of surrender to that big Self-consciousness." She said she recognizes, too, that we as human beings are filters, conditioned beings, and yet we are looking for that totally unconditioned big Self-consciousness.

I said I appreciated this very much and that it reminded me of something the Maharishi Mahesh Yogi used to speak about—the Sanskrit term is Laisha Avidya, which translates as "remains of ignorance." He said that even the most advanced, highly enlightened sages all have some remains of ignorance, i.e., some remains of their conditioned small self.

If they didn't have that, they would not be able to stay functioning in our three-dimensional consciousness of the material world. The experience of themselves as the totality of awareness is there, the experience of Brahman, but yet some remnants of their former egoic self experience remains, serving as an anchor to operate here in the world.

When it came time for Shamini to pursue her doctoral training, she interviewed at my institution, UC San Diego. During our interview for this book, she recalled meeting me during the doctoral interview process. UC San Diego's joint doctoral program in clinical psychology with San Diego State University is one of the top programs in the country. Hundreds of applicants are screened, and typically two-dozen are invited for interviews to fill six to eight slots, depending on the year.

Shamini reminded me that during our interview for the Ph.D. program, she pulled out this large stack of paper with outputs from statistical analyses she had done at the University of Arizona for her master's degree. "I showed you some kind of ridiculous factor analysis output for spirituality being one factor, psychology being another factor, and biology being a third factor, with all these arrows connecting all the dots. I was trying to emphasize how important the findings were. I was so excited." She laughed and said, "You sat there taking it all in and eventually said, 'Yeah, okay, interesting.'"

Shamini was accepted into several Ph.D. programs around the U.S. As part of her decision-making process, she decided to give me a call to ask a question as a kind of follow up to our earlier interview. She said, "I called you and said, 'I really want to do my dissertation work on energy healing. I already know that's what I want to do. I know that that's not a popular thing to do in academic settings, but I'm just wondering if you would be able to support me in doing something like that?'"

As she shared this story with me, I recalled that conversation and was pleased when she choose to attend UC San Diego. During the third and fourth years of her doctoral work, Shamini was good for her word when she designed and conducted an ambitious and landmark, gold standard, randomized trial of a hands-on biofield healing energy modality versus a mock hands-on healing energy modality versus a waitlist control of women who were suffering from chronic fatigue as a result of chemotherapy for breast cancer.

She managed to land funding for it from the National Institutes of Health and the Samueli Foundation. She completed the study and published the findings in the journal *Cancer*, the flagship journal of the American Cancer Society.

Shamini recently published her first book, *Healing Ourselves: Biofield Science and the Future of Health,* which presents a new vision of health and healing, including the biofield sciences.

As part of her Consciousness and Healing Initiative, Tiffany Barsotti and I traveled to India with Shamini and a group of scientists and clinicians to study consciousness and healing. During the trip, we conducted studies on the effects of spiritual places and meditation practices on the brain, the heart, and the human biofield.

I noted earlier that, when younger, Shamini had trained as a professional singer. After getting her Ph.D., in addition to launching her clinical and research career, she started a rock band called Nuns and Moses, which played regularly in the San Diego region. I have fond memories of Shamini and her band members dressed up as nuns belting out Guns and Roses tunes.

GIVING BACK THROUGH THE WISDOM OF ANCIENT TRADITIONS

I first met Tamara and Mike Goldsby at UC San Diego, when Tamara started her post-doctoral fellowship there. Tamara and Mike had started their doctoral work in clinical psychology and computer science and mathematics, respectively, at Oxford University, but needed to return to the U.S. after 9/11 to complete their degrees.

Tamara started her spiritual journey while at Oxford. She found a yoga studio there and, while practicing yoga, had a spiritual opening. "It all deeply resonated with me," she recalled. Upon returning to the U.S. and settling in North County San Diego, she resumed her yoga and meditation practices at a studio near Paramahansa Yogananda's Self Realization Fellowship compound in Encinitas, California. With her ongoing yoga practice, Mike noticed that "Tamara transformed into someone else, someone she is supposed to be." She said that during her yoga sessions, she gets deep insights and information of what to write about for her articles for groups such as *Psychology Today*. "At times, there's so much information being downloaded I can't keep up with it."

I asked Tamara why she got into psychology. She said, "People get into psychology usually because of their dysfunctional families. Mine was rather dysfunctional. I needed to learn about all that."

Mike, too, had a stressful childhood. In addition to being poor, his mother and father divorced when he was a baby. Mike shared that there was a lot of gang activity and violence in his school, so he had to "learn early on how to deal with socially challenging and sometimes dangerous situations. I had to learn survival techniques." The upside of it was it gave him a better understanding of social problems, including what drove kids to become gang members, such as being exposed to violence in their own homes. "All of these challenges in my life made me the person I am today: empathetic, compassionate, and open-minded. It oriented me to try to make a difference in the world, to help people."

For Mike, the entry point to becoming a scientist came when he was five, while watching watermelon seeds he had planted grow into seedlings. "It was so amazing to me. I wondered how this happens. I needed to know." His father, who was very disconnected from him, was a geneticist working for Monsanto Corporation. When Mike was six years old, his father sent him a full set of Encyclopedia Britannica's as a Christmas present. Mike studied all the photos, eventually reading the books, "further satisfying a thirst to know." He said learning helped him "feel happier." In subsequent years, his father sent him chemistry and rocket sets, all furthering his learning and interest in science (as well as almost burning down the garage). A final nudge to pursue science came from Albert Einstein himself. While still a kid, Mike read a quote by Einstein about "intuition and science that deeply resonated with me" and set him on a path to be a scientist.

Part of Mike's spiritual journey included starting to surf when he was age eight. He lived in Long Beach, CA, where he found an old broken board in a trashcan, fixed it up, and taught himself how to surf. "As a child, surfing was a way for me to escape the chaos at home. It was safe being out on the waves. But more so, it was highly calming for me."

Within a few minutes of meeting Mike, our conversation stumbled upon surfing. We each surfed, and we became surfing buddies—for many years, we surfed weekly at numerous reef and beach breaks up and down North County San Diego. We'd typically finish our surf sessions with a big breakfast and discuss life and new research ideas to pursue. Mike shared what he called "a spiritual moment," which he recalled vividly. Around the age of nine, he was very upset as he told his mom, "I'm getting old. I can't believe it's going to be 1970 next year.' It was very emotional for me." He felt he had a sudden shift in his perception of time and the relationship of his life to it. His mom comforted him, didn't downplay it, and simply said, "It's okay. It's okay."

Tamara's yoga eventually led her to learning about singing bowls, often called Tibetan or Himalayan singing bowls. The bowls are made of either metal or crystal and, depending on the size of the bowl, they vibrate and produce rich resonant tones when played. Historically, such bowls were used by monks as part of their meditation practices. More recently, wellness practitioners such as massage therapists and yoga teachers use the bowls therapeutically for benefits including relaxation and improved sleep. Tamara eventually started playing the bowls for other people, even for large groups of people as a kind of group healing meditation.

The inspiration to find a way for Tamara and Mike to work more together and give back socially came while visiting Peru a couple of years ago. Tamara started playing her bowls on the beach where there was a group of autistic kids on a school outing. One of the children had Down's syndrome, which reminded Tamara of her brother named Scott who had Down's syndrome. She actually never knew Scott because he had passed away before she was born. Tamara shared that she often feels his presence guiding her in life. At the beach, "When I started to play the bowls, the children were immediately drawn to them. I could see their transformation. I could see it in their faces. Their constant agitation started to melt away." One of the children even started crying. When she stopped playing the bowls, the children motioned for her to continue to play.

"We're just now starting to understand what these bowls can do for people. We really need more research on their effects for people of all ages." Tamara and Mike purchased EEG and EKG equipment to examine the effects of the bowls on the brain and cardiac systems at the same time. They are living in Costa Rica now and collaborating with a faculty member at the

Universidad de Costa Rica who is an electrical engineer with expertise in sound research. Mike used to work for Qualcomm Corporation designing mobile medical devices, where he learned to manage research data. Together, they make a good research team with all the needed expertise. Mike shared that a recent study by scientists at the Salk Institute in La Jolla, CA, identified proteins in the brain that respond to sound, a field of research called "sonogenetics." He figures that might be the explanation for the bowl's positive therapeutic effects on mood and the sense of well-being.

Several of my interviewees have given back by helping to transform how Western biomedicine is taught and practiced. Some of them, like Dan Vicario, started medical clinics that were spiritually based. After Dan completed his internship at Stanford University, he headed to UC San Diego for further training in oncology. While there, getting deeper into the world of oncology, he began to reflect on his training in the healing and visualization arts from when he was in Argentina. Those memories and the lessons learned led him to open the first Integrative Oncology Center in the San Diego region. In addition to providing current state-of-the-art medical oncology treatments, his Center also provided massage, acupuncture, meditation training, Reiki, and art therapy for cancer patients.

Dan added that all of the complementary therapies were offered for free by trained volunteers. "This was a must for me, I could not see myself being an oncologist and not offering these modalities to the patients." The Center attracted cancer patients throughout San Diego and Los Angeles counties, helping to transform their lives for the better while they navigated their oncology therapies. Dan said, "Some of the patients had healings just through the presence of the Center's nurses and staff who believed in them and listened to them and supported and spent time with them in a deeply loving and caring way. It all happened so naturally."

I visited the Center several times and, indeed, found the entire staff—from the receptionists to the medical providers to the lab technicians—all oriented to genuinely recognizing and supporting the patients' spiritual and medical healing. Listening to them, hugging them when appropriate, helping them to know they were safe. The Center was indeed a healing environment.

Dan added that many of his oncologist colleagues at UC San Diego would roll their eyes upon hearing what his Center was doing. He found this a remarkable response. He would tell them that these therapies were being offered to support the patient's sense of well-being, to support them spiritually, which also helped them get through chemotherapy with fewer side-effects—and, often, they do better medically. Still, that didn't register with his colleagues. It was like a foreign language they literally couldn't understand or couldn't connect with. Such can be "group think" in biomedicine. I have seen it many times.

As Dan shared this, I recalled attending a meeting with Dan at the UC San Diego Moores Cancer Center. Dan was there to give a lecture on what his Center was doing and share some of the medical outcome findings. Many of the oncologists purposely didn't attend, and those in attendance were on their phones during much of the presentation. Dan shrugged it off, saying, "Once they come to understand that the nurses are happier, the patients are happier, they'll see the importance of Integrative Oncology. The patients are the wise ones in helping rewire the brains of oncologists."

I asked Dan what he learned about the spiritual nature of people through his years at his Center that he hadn't already known. He said he learned more about the miracle of healing and what circumstances support that kind of healing. He saw so many examples of the effects of positive thinking and belief—how a healing environment can radically turn people's lives around and help them gain healing and live longer, happier lives than was to be expected from their diagnosis. He said that many of the patients became depressed after they were finished with their chemotherapy or other medical therapies because they would miss everyone at the Center. So, Dan and his team encouraged them to come visit. "We'd say, 'Please come visit, get a hug from the nurses.' It was a ritual coming in and being nurtured by the nurses and staff."

Dan continued, "Every person is an infinite being with an infinite capacity to heal. When I say heal, a lot of people are cured but never get healed, and many people who live with life-threatening illness do feel healed; that's what helps them live longer with a better quality of life."

I asked Dan about patients who didn't get better physically, despite the healing environment at the Center. I said, "Perhaps on a spiritual level, you witnessed them become strengthened, and that helped them better meet their destiny." He responded, "Yes, absolutely that's true." For many of them, he did witness them become strengthened on a spiritual level. He witnessed many come to a place of real peace about their life and its outcome. Dan was and remains a pioneer in medicine.

A few years ago, the UC San Diego medical system purchased Dan's Integrative Oncology Center and, unfortunately, stopped providing the extensive complementary therapies to patients.

Jeffery Martin, as I shared earlier, went on a ten-year quest searching for a religion where the practitioners were actually living in alignment with the principles of the religion. He eventually gave up and decided to take a different path to find the well-being he was searching for. His work in consciousness development and non-symbolic experiences now includes teaching around the world.

I asked him, based on all the knowledge he'd gleaned from his own experiences and research programs, what the best way is to give back, to disseminate what he's learned. He described that people spend their lives building up who they are, a sense of identity based on their thoughts, emotions, and history, but with the transition to non-symbolic consciousness states, that prior sense of self largely goes away. It is replaced by something much larger. He said, "The field of Positive Psychology has helped a lot because, through it, people have learned to take more control of their life. It's helped people think of themselves a bit more reflexively, to understand there's a change that occurs in their nervous system away from that moment, that there is a contentment in the moment-by-moment sense of things as they are really."

Jeffery then mentioned the "Roger Bannister moment," when Roger broke the four-minute mile. Before that, people said it was not possible to break the four-minute mile because human physiology can never run faster than that. Similarly, "there's this belief in our psychology that there's this ceiling on well-being, but that doesn't turn out to be true. There's a whole other level to be achieved. We just need to get that message out there."

Jeffery said he's looked at this for years now, and he's at the point where he realizes the mistake in the past has been to try to convince people that they should want it. But now, success will be just letting people know that there *is* a third choice. "We have to just get it considered as a choice, even if it's in the back of their mind, just so it penetrates more into mainstream society as an option, that people have at least heard of it."

Pertinent to this point, Jeffery shared a story from when one of his research team members traveled to Myanmar. She had heard about a famous Buddhist teacher and wanted to speak with him in the context of their persistent non-symbolic consciousness research. While there, she met several very old Buddhist nuns at a local convent. She started to chat with them, telling them about the research, and they then shared an interesting observation. The nuns said that historically people would come to the monasteries to attempt enlightenment. This was of course a Buddhist culture; people would come to the convent and intensively meditate for some weeks and have deep insights into or transition into a higher state of consciousness.

In their earliest years as nuns, if people didn't soon achieve a higher consciousness, the nuns would have them stay a little longer, giving them more attention, until they did. The process was so reliable that if they hadn't transitioned by the end of the second week, the nuns became suspect. If someone hadn't transitioned by the fourth week the nuns called the army because they assumed the person must be there to hide out!

As the years went by, the nuns noticed that it started taking longer and longer for people to transition to a higher consciousness. By the time the researcher met with then, it had become quite rare. The nuns weren't completely sure why this was but had noticed it get worse each time a new form of media, including radio, movies, television, and most recently the internet, became common in the country.

The nuns believed that there was a change occurring in people's cognitive and attentional systems that was simply making their teaching methods less effective. People seemed to have increasingly less capacity to wake up to an expanded consciousness. This was a highly significant change they observed over their 70+ years as nuns.

GIVING BACK THROUGH EDUCATION

Neil Theise described how his approach to teaching students has changed over the years as a result of his transformational experiences of consciousness. He related a story from when he was recently invited to the University of North Carolina at Chapel Hill to present to graduate students on the topic of "how complexity theory applies to stem cell biology." In the past, when he gave similar lectures, he would include in it his thoughts of how the topic related to his understanding of the spiritual. More often than not, that approach didn't lead to further discussion on the spiritual aspects and even turned off some of the students.

This time, he decided to approach the lecture very differently. "I decided I would speak about complexity theory and the idea of the universe as a self-organizing system, but not give them the spiritual implications." Instead, what he did was include several guided meditations during his presentations for the students. "I thought it was likely that for the majority of the students, their initial impulse to pursue science didn't come from them wanting to cure cancer or win a Nobel prize. Rather their impetus came from that same felt sense connection I had when younger that drove me to science." He wanted them to feel again that "intimacy in the world that turned into curiosity, which led them to a scientific impulse.

To accomplish this, some minutes into the lecture, he told the students, "We're all going to meditate together. They looked at me like I was crazy." He asked them to sit up in their seats and start by following their breath. He said that many of the students sat "with their arms folded and their legs crossed and highly skeptical, rolling their eyes." But he pressed on. "Bit by bit they eased into it." By the end, the students were engaged in the meditation, with one of them even crying at the end of the meditation. He said that "several students came up to me after the lecture or sent me emails after, thanking me because they had forgotten." He then mused, "I wonder how many scientists out there have forgotten. They squelched that inner impulse, which is actually a spiritual impulse to begin with." I agreed with him; the scientific culture teaches us to squelch it. For those who keep it alive, their work is more meaningful to them and, I dare say, creative as well.

Neil said he's noticed that students today are more open than their forbears were on metaphysical topics, but that so are some of his older colleagues. Due to his consciousness videos on YouTube he has a reputation now in these areas. People who know him personally come to speak with him about metaphysical topics.

This occurs at more and more academic institutions where he gives lectures, including grand rounds. When visiting academic institutions to give a lecture, it's typical for the organizers to arrange meetings with faculty who share common research interests. Their meetings are to share ideas and perhaps create a new research collaboration. Neil said, "It used to be I'd meet them in the faculty member's office, and they'd say, 'Tell me about your research,' but now the door closes and they say, 'I hope you don't mind me asking but tell me about meditation,' or that they knew someone who died, and they think they experienced them after their death. This sort of stuff happens all the time now."

I said I appreciated this insight into the nature of so many people in the sciences. We all have the same nature, but it's been tucked away. That is a significant thing. Neil said that "Einstein, Heisenberg, Bohr, many other science greats spoke about this stuff all the time. It's just all fallen off the radar of what is acceptable."

Julia Mossbridge's efforts to give back fall into several domains. Firstly, there are her thoughts about how science is now taught and the type of person that is attracted to science. The way science is taught now attracts people "who want to be in control of the universe, who want to think that they can understand something and that understanding will give them some control; it's about control and domination."

She went on to say that when that happens, it's "a dying of the soul. I don't think the universe connects well with a scientist like that. I think the information they're receiving and the scientific work they're doing suffers."

She went on to say that if such a scientist is in a position of teaching, then what they're showing to students is that the way to make progress is to "kill your soul and let your ego be in charge." "That's incorrect information for those students, and it destroys their capacity to be a true scientist. If I didn't have my father's example, I would not have known that science is something completely different than how it is now presented."

Julia is therefore a strong advocate of changing how science is taught. She said you can be smart and discerning in how you teach science but have a discernment without being judgmental. Being judgmental is removing love from one's discernment. Her dream is that people love themselves enough so that their ego doesn't take over their work.

"I think the way forward is teaching science from an unconditionally loving standpoint where possible. I don't mean having teachers who are always unconditionally loving. Good luck finding that. I mean the professor can say, "In this physics course we're going to learn about optics and we're going to learn about it with an unconditionally loving stance, which means our goal is to try to unconditionally love ourselves and each other as we learn it." She says, "I think a lot of really beautiful learning could take place like that."

She added, what if the professor said, "By the way, science is about you. It's about the nature of your mind and its relationship to the rest of the universe, and it's about physical objects and how they work in relationship to your mind and the rest of the universe. It's about your intuition, and it's about checking your intuition and developing a relationship with God." She said, "If it was taught like that, who's it going to attract? What kind of minds?"

Part of her efforts in this regard led to creating her own Institute, The Institute for Love and Time, a nonprofit that conducts research and creates technology to support our understanding of both time and unconditional love. The Institute works in areas of experimental psychology and psychophysiology, artificial intelligence, the physics of time, and cognitive neuroscience.

Dean Radin told me he receives emails all the time from students wanting to get into parapsychology research and seeking his advice. They say, "I want to do what you're doing; how do I do that?"

He tells them, "The truth is there is no career track in parapsychology and probably won't be for another generation of scientists." What he does tell them to do is to get into consciousness studies, psychedelics, and meditation research. These are very acceptable disciplines these days. "Go learn neuroscience, learn something about physics, learn everything you can about statistics and so on. It's beginning to be possible."

Dean explained one of the more far-reaching projects that he and others are working on could more quickly change academia's attitude toward parapsychology and related fields. They want to "figure a way of breaking out of the taboo of studying esoteric topics in the academic world, without fear of reprisal." Their solution is to create a multi-university consortium.

"There will be an administrative hub for perhaps six to twelve universities around the world, each of which will have a senior faculty member who is part of the consortium. These faculty members will be charged with essentially looking at advances in consciousness studies to address some of the more flagrant holes. Holes, for example, as how do we understand the nature of genius? How do we understand the nature of spontaneous savants and acquired savants? How do we even begin to understand how people can do this? How do we understand the extreme psychophysiological effects that can occur in things like multiple personality disorder? How do we understand terminal lucidity?"

He explained that there's a whole collection of such things which are telling us there's something about our assumptions in terms of human capacities that we've got wrong. If we did understand these things, what would it tell us about who and what we are and what we're capable of?"

Dean added that such a consortium would provide a certain degree of cover for graduate students and faculty who would like to be affiliated with it and use it to pursue work in these areas with more acceptance. Ultimately, the consortium would provide recommendations for what we think needs to happen at this point, such as new research projects and how best to disseminate the material.

GIVING BACK THROUGH TRANSFORMING HOW MEDICINE IS PRACTICED

Both Wayne Jonas and Rael Cahn created opportunities to give back while they were still in medical school, sharing their knowledge and experience. Wayne began what would be a lifelong effort of teaching and giving back. He started retreats for medical students to be exposed to healing traditions, creating the program "HEART," which is still active today and is supported by the American Medical Student Association. It's open to fourth-year medical students to get immersed in yoga and other mindbody practices and learn the scientific evidence base for these practices. They also learn leadership skills. The program take place annually in the California Redwoods, typically before students begin their residencies. Wayne continues to teach in the program each year. Recall the deep insight Wayne had with the man dying of metastatic cancer, his deep recognition of what healing really is, that "there is no healer and healee—there is only healing."

As part of Wayne's commitment to teaching medical students, he describes his understanding of healing and teaches students how to get into that space where healing can occur—to enter that relationship out of which a mutual experience of healing emerges. He said when a person takes the time to pay attention to it, by pausing and stopping, then healing can emerge. He teaches his residents and students how to open that up, to be ready for a transcendent experience so that they can have a readiness for healing, to increase the probability that those kinds of experiences will happen.

He said, "Yes, you look at the facts, you look at the lab tests, etc., but all that gives you is information usually based on probabilities. Healing comes from another source. It's through the pausing and connecting with people that insight occurs. If we take our scientific tools and all its discovery modes and focus them on our consciousness, an entire world that we can only imagine now reveals itself. This I endeavor to impart to my students and help guide these efforts."

He also teaches what he calls the "HOPE note," which stands for healing, oriented practices, and environment. It's a set of tools to cover all aspects of the human being—physical, mental, environmental, and spiritual. It's to identify the personal healing plan for the person. "Everyone in the healthcare systems should have a minimum of proficiencies in these areas." Wayne has written a book on this work, titled *How Healing Works: Get Well and Stay Well Using Your Hidden Power to Heal.*

I asked Wayne broadly about consciousness and medicine. He described a dear lifelong friend of his, who recently passed away, who saw the spiritual aspects of everyone he met and their innate value in the world. Wayne said, "He knew the connectedness of us all, that we're really one foundationally through consciousness, through the heart." I wholeheartedly agree.

I told Wayne that he has accomplished so much in his career and asked him if there is anything he has yet to accomplish. He laughed and replied, "I'm sure about 75% of what I'm to do, I've yet to discover." He also shared that he became a Quaker a few years ago because he can "express a deep silence, experience the merger of his scientific mind and the spiritual world."

Rael Cahn, while still a medical student, started HI-MED-- an integrative medicine special interest group for UC San Diego medical students, which continues to this day. My Center of Excellence at UC San Diego routinely sponsors lunch and learns for the HI-MED students, bringing in integrative medicine healthcare providers as guest speakers to help round out their learning. Rael is now a Clinical Associate Professor of Psychiatry at the University of Southern California. He continues to give back by providing mindfulness meditation classes there.

Rael's neuroscience research focuses on how expanded consciousness interfaces with the human brain and how meditative practices change brain activity to support the experience. He said that, in many ways, his scientific and clinical work is an attempt to unravel these issues. He also explained that, as a result of his experiences as a young man and subsequent experiences, his sense of self is no longer as it once was—there is no longer an egotistical perspective. He posed the question, "What is me, and what is not me?" His research has been focused on how we can change our relationship to the mental concepts of our minds.

Some of his research work has been in psychedelics, which are known to support transpersonal, spiritual experiences. When he first started that work, it wasn't legal in the U.S. so he went to Switzerland to conduct the studies. "Both meditation and psychedelics share a commonality in that they alter our perception of the outer world, giving us access to know things more as they actually are, apart from our constructed meaning, and perhaps support a new way of seeing each other with compassion and meaning."

I told him I loved that vision and that these technologies can get humanity to a new place, as we really do need such a new place with more compassion for each other and an understanding of our true essential nature as awareness itself.

While Rael was speaking, I was reminded of one of Joseph Campbell's books where he speaks about those things that make us more transparent to the transcendent. He speaks about the role of the Shaman that helps people make the transition to the transcendent. Campbell also writes about how, historically, individuals in such cultures ended up becoming a Shaman— it was typically those who had more psychic sensitivities. I wondered to myself if Rael had been born into a more ancient culture, would he have been called to fill that Shaman role because his sensitivities make him more transparent to the transcendent? Being a Shaman is not necessarily easy; they have to be different in order to bridge the spirit and regular worlds.

When I asked Rael this question, he said, "That is an interesting question, one I, too, have thought about over the years. It seems to me that had I been born into such a traditional culture with the types of experiences I had as a young man, I would have been viewed very differently. Nonetheless, I would have gone through some kind of trial period to see whether the Shaman path was for me."

Given his history and who he is, Rael's research over the years has aimed to understand how the human psyche—particularly for those who are more fragile in terms of their ability to interface with our materialistic cultures—can correctly interface with the larger Self. I asked him how his clinical work is going in this regard. He said his focus is much more on research and the need to get grants to support that research and academic position. This keeps him from having the time to merge his clinical practice more fully so it can have a greater focus on his spiritual insights and have a bearing on his work with patients.

Gita Vaid, a psychiatrist like Rael has, over the past few years, been using psychedelics in the context of her psychotherapy work. She finds that, for the right person, it can be incredibly beneficial for their healing journey. She notes that "these medicines can help people go to some states of ego dissolution, a kind of spiritual field in the sense of unity."

As human beings, we have these dual natures. One nature identifies our temporal relationship with our mindbody and the lives we lead. The other is our identity as a vast universal consciousness. "We're always trying to come into a resolution of them, a harmonization of them. That is the journey of healing, to have them co-exist; the journey to wholeness is the journey to healing."

I asked her when she is with a patient who has gone into the state of ego dissolution, beyond constructs and egoic projections, what her understanding is of the nature of the human being before her.

She said her work has opened her up to "more of the mysteries of consciousness, of that dual nature of the human being." While she has always been a meditator, her work with psychedelics has taken it to "a deeper kind of space where one can really untangle from—in the fullest way—the narratives we live in, the things that pattern one's mind."

As far as the true nature of the human being, as the patient moves beyond the mental narratives to realize "that was not I," and goes further beyond the emotions, realizing "that too is not I," to further realize in a very clear way those aspects of experience we all know and attribute to our identity, and finding "still this is not I," that "at each step along the way, that awareness remains, the sense of I in consciousness. It is beautiful to witness."

"I can't even tell you how many times I'll do sessions with people, and at the end of the session, they will say, 'Oh my god, you are genius Dr. Vaid.' And I will say, 'Well, actually, I'm really good at listening. Everything you think is genius came from you.' They'll stop and think about it, then say, 'You know what? You're right, I did it!'"

She said she learns so much from these sessions, "because the knowledge that comes through is absolutely stunning. It gets into this kind of universal knowledge that we all possess in our DNA." She further clarified that it's not from their minds, and that most of the time, we just don't even recognize that universality of consciousness in us because we're too busy trying to think and figure things out with our head. "It's actually gnosis," she says. Gnosis is innate knowledge, wisdom beyond the books. She shared, "I feel very honored and privileged to receive that knowledge, to be almost a conduit for people to actually learn from their own experience through their own inner intelligence. I can reflect back to them what their knowledge is telling them. I can serve as a translator."

With that in mind, Gita said, "My healing work is spiritual encounter after spiritual encounter, miracle after miracle, which actually is what life is about. We just need to surrender to it again and again."

As she said this, it reminded me of her experience she shared when she was a child. I asked her if she recalled where she went during that experience as a child. She said she basically "exited some of the constraints of her mind and arrived at a child's mind, knowing we can indeed do anything. I slipped into that space of knowing I could do things and then was able to do it."

Joe Tafur's giving back includes opening the Nihue Rao Center in Peru, which, in the Shipibo Tradition, offers healing ayahuasca retreats with master plants and traditional Shipibo diets. As part of the Center, Joe included a medical clinic to accommodate people with medical and/or mental disorders, such as PTSD, so they can be cared for properly while participating in the healing ceremonies.

This work is described in his book *The Fellowship of the River: A Medical Doctor's Exploration into Traditional Amazonian Plant Medicine.*[51] Joe also created a non-profit, Modern Spirit, which is dedicated to demonstrating the value of spiritual healing in modern healthcare by illuminating the intersection between biology, emotion, and spirituality.

In addition, he's been working to get the U.S. government to recognize and legalize the use of traditional medicines such as ayahuasca. Joe explained that the challenge is that the government wants to call these traditional medicines a sacrament. "They say we can't call it medicine, that's inappropriate." But Joe points out that this is what they've been called for thousands and thousands of years. Whether it's the ayahuasca tradition or the peyote tradition, or their respective prayers and songs that go along with those ceremonies, they are medicines for healing, spiritual healing, and supporting overall health and well-being. "That's the purpose," he said. "The reason we do these things is to take care of ourselves and to nourish our health and spiritual well-being. We're not going to call it what it is not because the Supreme Court or attorneys say so. Rather, we're going to be honest and honor the traditions of where we live here in North America."

As Joe was explaining all this, he said he wanted to read something to me from Pierre Teilhard de Chardin's book *The Phenomenon of Man*. It had been his father's book, who passed away some years ago. Joe's father was a psychiatrist whose presence and work I greatly admire. Joe read the following, "So far as I understand the struggle in which I found myself involved in, it seems to me that it's prolongation depends less on the difficulty that the human mind finds in reconciling certain apparent contradictions with nature, such as mechanism and liberty or death and mortality, as with the difficulty experienced by two schools of thought and finding a common ground." Joe added, "This is a poignant description of what I've been facing these past few years, explaining to governmental agencies about the nature of psychotropic plant medicines." I thought to myself, "Joe's father must be very pleased with his efforts."

Joe then made what I thought was a beautiful observation, one we can all benefit from. He pointed out that the sacred sciences of meditation and yoga that came from the East were never separated from their innate role in supporting health. In those ancient traditions, they are part of their medical systems and are considered foundational to health and well-being. I reflected on this and fully agreed. Mindfulness and meditation, for example, entered the U.S. very much in the realm of health and well-being. Today, there are over 2,000 studies on mindfulness in relationship to supporting health and well-being. It's taught in hundreds of medical centers around the U.S. and Europe, including major academic medical centers (https://imconsortium.org/).

In that same way, the traditional uses of plant medicine should be honored and accepted into our Western medical culture for supporting health and well-being. Joe said that "spirituality is practical and health-oriented; a spiritual practice is spiritual healing. The entire purpose of spiritual practice is spiritual healing, which means to nourish our spiritual well-being."

Joe added another important insight to his work: framing these arguments in the broader context of ecological sustainability. "The only way we're going to resolve this environmental crisis is through a universal spirituality. There needs to be a common ground of spirituality.

Sustainability is the same thing— it's the health of the society. We must acknowledge the more holistic, spiritual approach to the environment, to identify the multidimensionality of each individual being and their relationship to the whole. Otherwise, it's not sustainable."

Spotlight

ADVICE TO SCIENTISTS THINKING OF PURSUING THE JOURNEY

Eben Alexander, M.D.

My father was a globally renowned neurosurgeon with a deep respect for the scientific method in seeking truth, but he was also religious. His extraordinary influence inspired my own life-long pursuit of science.

Having trained in medicine and neurosurgery at Duke University, followed by over fifteen years on faculty teaching neurosurgery at Harvard Medical School (as well as appointments at the University of Massachusetts Memorial and the University of Virginia), I was devoted to the reductive materialist, or physicalist, point of view—one that hypothesizes that only the physical world is real. Given my thorough education, I thought I had some understanding of the mind-brain connection and the fundamental nature of consciousness.

That all changed dramatically in November 2008, when I spent a week in a coma due to a severe case of gram-negative bacterial meningoencephalitis. The surprisingly vivid nature of my near-death experience (NDE) and of my most unanticipated full recovery over two months after awakening, is reported in *Proof of Heaven* (2012), the book I wrote about the experience.

Six years later, my claims were validated by a medical case report written by three physicians uninvolved in my care who were fascinated by my full recovery, unprecedented in the medical literature. They reviewed over 600 pages of medical records and pointed out in their peer-reviewed report the well-documented damage to my entire neocortex (which should have disabled all but the most rudimentary of conscious experiences) and the miraculous nature of my recovery.[52]

The spiritual experience I witnessed during my seven-day coma was multi-layered and complex. The seeming ultra-reality, persistence, and stability of the memories of it align with several scientific papers on the remarkable qualities of NDE memories. An unusual ingredient of my experience was an amnesia for my life before the coma that included loss of language, knowledge of humanity, earth, or this universe during the time of my spiritual journey.

This *tabula rasa*, or empty slate, allowed for some profound lessons and understanding of the nature of consciousness that I continue to unravel through meditation and analysis almost fourteen years after my coma. The astonishing return of memories over two months post coma was so complete and inexplicable that it led to a fuller understanding of memories and how they don't seem to have any permanent storage location in the physical brain. This remarkable and revolutionary observation was addressed in my third book about the experience (*Living in a Mindful Universe*, 2017), co-authored with my life partner, Karen Newell.

After awakening from the coma, my initial assumption based on discussions with my physicians was that it must have been a vast hallucination. But as I reviewed my medical records, lab values, and scans, it became very clear that my brain during this illness was too incapacitated, with documented damage to my entire neocortex and brainstem, to have generated the exquisitely complex, detailed, and meaningful experience I had. Something more was required to explain it all, and this is the most important lesson of my experience for other scientific minds—opening up to what that *more* might be.

In the early months after my coma, faced with such experiential challenges to my physicalist worldview, I came to see that true open-minded skepticism is one of the greatest gifts in coming to deeper understanding of something as fundamentally important as consciousness. Some of the scientific minds who questioned my story and announced themselves as "skeptics" were more like pseudo-skeptics. They are the opposite of what true scientists represent, ignoring (or debunking and denying) the evidence and dismissing rational argument. I found I had to question everything I had ever believed in order to rebuild a more comprehensive worldview from the bottom up.

The greatest advice I can give others interested in pursuing science is to look at such challenging evidence without prejudice and with as open a mind as possible. When faced with experience and observation that defies one's understanding of the world, we must be open to unexpected explanations and possibilities outside the conventional narratives. These challenging jagged edges at the boundaries of our knowledge often represent essential clues that might suggest our theoretical models are incorrect and offer specific clues as to their rectification. The greatest leaps forward in the history of science have involved those who questioned the main authorities and conventional narratives, embarking on an open-minded exploration fueled by child-like curiosity.

It's crucial to point out the essential nature of anecdotal accounts when our ability and knowledge of experimental pathways of inquiry fall woefully short of the mark. Some of the most profound human experiences cannot be neatly assembled into a form that is openly subject to experimental modeling. At this stage in the game, it's crucial to accept these numerous accounts and study them *en masse*, as many researchers in the modern science of consciousness studies have done in addressing NDEs and related phenomena.

Given that direct experience is so powerful at informing worldviews, it is clear that a major benefit would result from an effective mechanism of *driving* experiences, like NDEs on demand, to facilitate their deeper investigation. It is certainly not morally or ethically practical to put people in a physiologically near-death state in order to model the experience, as depicted in the 1990 movie *Flatliners*.

Some researchers have observed similarities in the linguistic descriptions of NDEs compared with certain psychedelic drug experiences (notably ketamine, *Salvia divinorum*, and serotonergic substances such as dimethyltryptamine, or DMT, the active principle in ayahuasca), and suggested using such substances as drivers. Based on my personal experience, such entheogens do offer a glimpse into spiritual realms, but the glimpse is as if through a small keyhole, as opposed to the panoramic penthouse vista to those realms offered through a full-bore, deep, and transcendent natural NDE. Institutional Review Boards might still balk at a study intentionally exposing subjects to such agents, but such issues are not insurmountable given that these substances are relatively non-toxic and non-addictive.

Given my personal successes with a daily meditation regimen over the last decade, and what we have seen in our meditation workshops, I believe a prolonged regular program of meditation is superior to psychedelic substances in simulating the benefits and lessons of an NDE. Psychedelic substances create a large biochemical splash in the brain, offering too much noise around the spiritual signal that I believe is much better amplified and refined through meditation. However, we have a long way to go in designing very specific meditation protocols that would best simulate NDEs.

Meditation and exploring of one's own consciousness are absolutely essential in any attempts to more deeply understand the mental realm. But, in fact, a regular process of "going within" provides tremendous benefits for health in general, and certainly in accessing creativity and insight far beyond what we can normally accomplish.

The key to such successes in meditation involves acknowledging that the voice in our head, our egoic mind or linguistic brain, is not our only mode of accessing knowledge about the universe. Cultivating a mindset in that hypnagogic space between awake and asleep has offered creative energy to some of our greatest scientists, philosophers, artists, and musicians. Albert Einstein, Thomas Alva Edison, Robert Louis Stevenson, and Salvador Dali all had ways of getting into that dreamy hypnagogic space to enable uprushes of creative genius and insight that fueled some of their greatest contributions.

Meditation, or any mode of exploring the mental realm of possibilities beyond the standard pursuit of logic and rational chains of argument furnished through our linguistic brain and ego mind, will often provide novel insights that prove indispensable in achieving creative success in investigating the most vexing of mysteries.

NDEs provide the tip of the spear in the modern scientific study of the mind-brain connection and the nature of consciousness, not to mention that of reality itself. Consilience of information from widely disparate fields of inquiry (neuroscience, philosophy of mind, quantum physics, and parapsychology) all support the philosophical position of objective idealism or a top-down causal principle from a shared mental and/or spiritual realm that better explains our notions of free will and of human agency in manifesting our emerging reality. The brain thus serves as a filter of this primordial, unified source of consciousness.

Note that this concept of idealism is one that naturally defines spirituality to include two main ingredients: a sense of connection through our mental processes ("sharing the dream of the one mind") and a sense of shared meaning and purpose.

Coming to a deeper understanding of the mind-brain relationship and the ultimate nature of consciousness remains one of the deepest unknowns facing modern science. NDEs and this broader line of investigation into the leading edges around our understanding of the world-at-large are leading to a worldview that will be far more harmonious and peaceful than the current status quo that has been wholly based on the false sense of separation inherent in materialist thinking. The deepening science of consciousness leads the way into this refreshing worldview that demands more open minds from its investigators.

CHAPTER 6

PUTTING INTO PERSPECTIVE THE CONSCIOUSNESS DEVELOPMENT JOURNEY

Evolution of consciousness is the central motive of terrestrial existence.

– Sri Aurobindo

A t the close of my interview with Dusana Dorjee, she gave me a message, a charge really. "Paul, in your book, ask the hard questions. Go beyond the wow factor of simply saying scientists have these states. Be sure to emphasize there is a reason for these states and that we need to build self-regulation before getting too deep into the practices."

I very much took Dusana's message to heart, and this chapter endeavors to address these issues. What are the reasons for transpersonal, metaphysical, and mystical states? Why bother to pursue them? What do they tell us about our nature, and how do we best honor that nature? How can we best prepare for and understand them? What type of life might they be directing us toward?

SOME TEACHINGS ABOUT THE JOURNEY OF DEVELOPING CONSCIOUSNESS

In this book's Introduction, I started addressing aspects of these questions when I briefly reviewed Maharishi Mahesh Yogi's teaching on the evolution of human consciousness beyond the states that most people are familiar with—that of waking, dreaming, and deep sleep consciousness. I reviewed what Maharishi called the higher states of cosmic consciousness, divine consciousness, and unity consciousness, with its refinement called Brahman consciousness. These states describe a refinement of perception and experience of consciousness and awareness as the person moves away from being identified with the egoic personality self to the experience of Self as pure awareness.

Along this journey, there can be many types of metaphysical and/or mystical experiences, some of which have been described in this book. What I've covered here is but a small sampling of such experiences that have been described and cataloged across Eastern and Western cultures for millennia. Ken Wilber has written deeply on the intersection of the enlightenment teachings of the West with the enlightenment teachings of the East, each contributing key elements to a spirituality he calls Integral Spirituality.[53] Included is his "spectrum of consciousness" work, where he traces human development from infancy to adulthood and beyond into those states described by mystics and spiritual adepts. It is essentially a model of human growth and development, integrating the systems developed by psychology with those of the contemplative traditions.[54] In The Atman Project, Ken presents seventeen basic levels, or structures, which make up matter, body, mind, soul, and spirit, which comprise the Great Holarchy of Being.[55]

From my own experiences and those of countless others, it's reasonable to say that while these types of experiences have common foundational features, how they manifest and are understood across people are highly inter-individualistic. Ken Wilber's Special Commentary addresses the "why" of this fact. While descriptions of metaphysical and mystical states are as varied as there are people, descriptions of the state of oneness are uniform across people and across cultures. The reason is that the person describing that oneness state is no longer doing so from the perspective of an egoic personality, but from being the true Self as the dominant experiencer. Since that Self, referencing as Awareness, is common to all, the description of the experience is much more uniform. Put another way, while there are as many "small selves" as there are people on the planet, there is only one "true Self" common to all perception. Living from that state is for the most part uncolored by the filtering of the mind and personality. Awakening to a state where duality and separation don't exist is perception of our true nature and of the world itself.

In addition to the Maharishi, there are of course innumerable teachings on this perspective. Over the years I've been particularly fond of the teachings of Sri Nisargadatta Maharaj, as expounded in the book *I Am That*, and Ramana Maharshi, of which there are many books, including *Who Am I*.

More recently, through long evening conversations with my dear friend Dr. William Reed, M.D., I've learned about Tibetan Buddhism's Mahamudra teachings. These teachings are parallel to the Vedic nondual traditions, with Mahamudra meditations placing a special focus on the mind and its relationship with the world of appearances. These meditation techniques are known for their ability to lead to profound realization. In some Mahamudra schools, the word dharmakaya is used to describe the true universal Self.

The importation of Vedic teachings to America is a fascinating story and thoroughly chronicled in the book *American Veda: From Emerson and the Beatles to Yoga and Meditation. How Indian Spirituality Changed the West.*[56] Written by Philip Goldberg, the book shows how Vedantic teachings found their way into the libraries of John Adams and Ralph Waldo Emerson, as well as Henry David Thoreau and Walt Whitman. Later generations exposed to these teachings include Aldous Huxley and Joseph Campbell. These teachings have been so absorbed by America that meditation and yoga are now household words, and many Western academic medical centers routinely prescribe them as therapies.

The work of Paramahansa Yogananda and Swami Vivekananda should be especially recognized. Several of the scientists in this book noted how reading Paramahansa Yogananda's book, *Autobiography of a Yogi*, first published in 1946, helped set them on their spiritual path. For some, it was important because it helped validate their own spiritual experiences. For others, it presented a new way of seeing the world that they felt needed to be explored. Paramahansa Yogananda's organization, the Self-Realization Fellowship (SRF), has a particularly strong presence in Southern California, including in Encinitas, CA, where I used to live. Just off the SRF's grounds, down the cliff face from the hermitage where Paramahansa wrote *Autobiography of a Yogi*, is a world-class surf break named "Swami's." I spent endless enjoyable surf sessions there with my good friend and UC San Diego colleague Professor Michael G. Ziegler, M.D.

Some twenty-five years before Paramahansa Yogananda's arrival to America, Swami Vivekananda came from India to speak at the 1893 Parliament of the World's Religions in Chicago. From there, he toured the U.S. and Europe, teaching from the Vedantic perspective. As described in *American Veda: From Emerson and the Beatles to Yoga and Meditation— How Indian Spirituality Changed the West,* in addition to teaching the

Vedanta philosophy of meditation and yoga, he also worked to increase interfaith awareness.

Years ago, my wife Tiffany and I were visiting our dear friend Monica Behan in Thousand Islands, New York, on the St. Lawrence Seaway. She offered to take us on a surprise excursion from Murray Island, where we were staying, to nearby Wellesley Island. Upon arrival we promptly headed off into the forest and after some time came across a large stone and brass memorial dedicated to Swami Vivekananda. It turned out the Swami had visited Wellesley Island in 1885, and on the very spot where the memorial stands, he had a life-changing experience, a "transformation and epiphany in consciousness" that helped guide his subsequent work in the U.S. and when he returned to India.

NO SELF?

Bernadette Roberts was a former Carmelite nun and modern mystic in the Catholic tradition. I first encountered her work in the early 1990s through her book *The Path to No-Self: Life at the Center*, which describes her spiritual journey after entering the Monastery of Discalced Carmelites in Alhambra, California. She followed up this work with subsequent books, *The Experience of No-Self: A Contemplative Journey*, and then *What is Self?: A Study of the Spiritual Journey in Terms of Consciousness*.

One of the reasons I was intrigued by her books is because she found herself opening to transpersonal states of nonduality and eventually arrived at the experience of no-self. Her descriptions of the state are similar to those in the Vedantic traditions. The rub though was she didn't initially understand what she was experiencing as she hadn't heard of this state of consciousness before. In addition, she wasn't able to find any teachings in Christianity to guide her.

She stated that the purpose of her writing the books was to put into the Christian contemplative literature something that had not yet been there, i.e., "the no-self event," which she felt had not been accounted for in that tradition. She based that position on the fact that she had "read over one hundred of the spiritual classics" in the Christian literature and found only a few hidden suggestions of such no-self experiences, which to her was not sufficient.

For the most part, I agree with Bernadette in the sense that most of the esoteric and mystical Christian texts I am familiar with stop with the discovery of the "unitive state," the self experiencing its oneness with the divine, and do not progress to the no-self state. Eventually, Bernadette came to understand and integrate her experiences and put them into a broader perspective of the larger human spiritual journey. For decades afterward, she taught retreats on Christian mysticism. She passed away in 2017.

Beautifully so, Bernadette wrote that primary amongst her experiences was silence itself. In *The Experience of No-Self: A Contemplative Journey*,[57] she writes: "Through past experience I had become familiar with many different types and levels of silence. There is a silence within, a silence that descends from without; a silence that stills existence and a silence that engulfs the entire universe. There is a silence of the self and its faculties of will, thought, memory, and emotions. There is a silence in which there is nothing, a silence in which there is something, and finally, there is the silence of no-self and of God. If there was any path I could chart my contemplative experiences, it would be this ever-expanding and deepening path of silence."

Anyone who is familiar with contemplative and metaphysical literature will be familiar with such descriptions of the ubiquity of silence in its many permutations. Thomas Merton, an American Trappist monk and mystic, whose book *Seeds of Contemplation*[58] I found on my Aunt Nora's bookshelf when I was a boy, wrote, "Not all of us are called to be hermits, but all of us need enough silence and solitude in our lives to enable the deeper voice of our own self to be heard."

What is actually meant by "no-self"? All meditative and mystical traditions, in one way or another, conceptually refer to a "small self" and a "large Self," the latter being understood as our true identity.

Bernadette defines the small self, or ego-self, as "the totality of the self-experience; it is consciousness centered on and around itself." This self-center is experienced as the center of self-will. It is "the center of desire, clinging, fear, and much more." She goes on to describe that as long as this self-will or ego remains, all of consciousness must function according to that as our center. Should that center fall away, however, consciousness then has to learn to function around that egoless or empty center, which she calls the "divine center."

Some people vigorously oppose the term "no-self" because it implies there are people walking around who don't have a self, something that is taken as the very foundation of being a person. The truth is that when a person drops the experience of being the ego-self, the experience is very much that of no longer having a self. "Self" perception is radically changed because what was once the center of experience, through which all perception had been in reference to, i.e., the ego-self, is no longer there at the center. As Bernadette notes, it appears "empty." As a result, the prior sense of being an individual, a person, of being a self, ceases. Obviously, experience is still going on, but it's just not happening through the same lens of perception that it had been.

Because there is still memory, things being perceived look *familiar,* yet they look *foreign* at the same time because they are no longer being perceived through that filter of the ego-self, the self which had previously built the life being lived. Memory of the past provides to experience a sense of familiarity, but the sense of identity with those things of memory, be it people, places, or things, is no longer there. In the book *No Man Is an Island*,[59] Thomas Merton wrote, "The discovery of ourselves is always a losing of ourselves."

While the term "no self" is a reasonable description of the experience, it is a misnomer. There is a self, but we can say that it is now a universal impersonal self. Terms such as "waking up", "self-realization" and "self-transcendence", among many others, describe crossing the threshold of living a life from a personalized small self perspective to an impersonalized universal Self perspective. Being more impersonal, the experience is initially interpreted as no self, as having an empty center, but eventually that experience is transformed. For Bernadette Roberts, over time, her experience of the empty center became the divine center. The Maharishi spoke of the perceived disparity of the emptiness and separation found in Cosmic consciousness eventually being replaced by the totality and fulness of Brahman consciousness. In Esoteric Christianity, people speak of merging into the true Christ Consciousness. St. Paul said, "It is no longer I who live, but Christ lives in me." The Buddha's immeasurables of love, compassion, sympathetic joy, and equanimity characterize a life that can be lived from that state of universal Self.

As I described earlier when summarizing some of Maharishi Mahesh Yogi's description of the higher states of consciousness, the movement is from separation to unity. It is perception through the ego-self that supported the perception of separation. Once the ego-self is put aside, so to speak, the perception automatically becomes that of unity. All is seen as consciousness itself in endless myriad forms.

Many of the scientists interviewed in this book described this very phenomenon, where their sense of egoic-self was suspended. When that happens, there is then an inevitable consequent insight into the true Self.

I say inevitable because the only reason it isn't consciously experienced all of the time is because our awareness is overly focused on the egoic-self. The Maharishi spoke about over-identification with the small self and its objects of perception. Bernadette wrote how surprisingly self-absorptive the experience of ego is, of being "consciousness centered on and around itself." Bernadette calls the falling away of the ego-center the beginning of "the true transforming process." She also describes that new state as one of emptiness or negation of ego. At the same time, "it is the positive presence of the divine."

THE GIFTS OF THE WORLD AND BEING HUMAN

As I had touched on in my interview with Menas Kafatas, one of my complaints about the nondual Vedantic teachings is the leaving out of the importance of developing the human being, of our innate capacities as spiritual beings to be experienced fully while living a human life. What I mean is that the emphasis in the Vedantic system is to become "Self-realized," to drop the ego-self experience. There's typically no insight provided into the nature of the human being that merits parallel attention and development.

Several times, I heard the Maharishi tell stories about living in the caves in the Himalayas in India. He and his fellow sannyasis considered the world down there as "mud full of ignorance," and why go down into that mud? Eventually, however, he did, following an inspiration that he received to go down into the valley and give a talk about the meditation methods he had learned from his Guru. He intended to return to his cave after a couple of days but never did; he just kept teaching.

There is a nondual system that does see the world differently than the Vedantic system. Kashmir Shaivism, or Trika Shaivism, is a nondual tradition out of Kashmir. In Kashmir Shaivism, like Vedanta, all things are understood as a manifestation of one Consciousness arising from the true Self of awareness. In Kashmir Shaivism, however, the phenomenal world is also given importance. It is considered real and not to be misunderstood as solely illusion. The great Kashmir Shaivism teacher, Swami Lakshmanjoo, wrote that Kashmir Shaivism teaches us to realize that the physical world in which we exist is not separate from God-consciousness, that "God and the individual are one. To realize this is the essence of Shaivism."

Along these lines, there's an aphorism I learned years ago that I appreciate and is consistent with such a more vital and holistic view of the world. The saying starts out, "The world is illusion, only Brahman is real." Recall that the Sanskrit word Brahman refers to the eternal absolute reality, the totality of all existence. This part of the aphorism is a Vedantic point of view, the idea the world is illusion, maya, and of not much value. But the saying has another phrase and goes on to add, "The world is Brahman," that is, in the completed version. "The world is illusion, only Brahman is real, the world is Brahman." It thus comes around full circle to show that the totality of all existence, Brahman, is the world, too—not simply illusion, but the totality of all that is. If we discount the world, which so many philosophies and religions do, then we too are discounting the totality of Brahman itself.

Ken Wilber writes of Integral Spirituality as including Waking Up, Growing Up, Cleaning Up, and Showing Up, each being equally important dimensions to cultivate to help ensure a human being develops the fullest states of spiritual and human enlightenment. The Maharishi's way of saying this was to encourage people to develop "two-hundred percent of life," with one-hundred percent living full enlightenment and the other one-hundred percent fully enjoying the gifts of the relative existence as a human being.

I very much appreciate philosophies that do encourage the full development of the human being, not only developing enlightened consciousness per se, as do the teachings of Sri Aurobindo. Aurobindo was a brilliant Indian philosopher and guru with an adventurous early life. He was born in Calcutta and later educated at King's College in Cambridge, UK. Upon his return to India, he started working for the state service but soon took an interest in the Indian independence movement against British colonial rule. He eventually became one of the leaders of the Nationalist Movement. He ended up being arrested and spent a year in prison where he had mystical experiences that changed the direction of his life. Aurobindo said he was visited and instructed by Swami Vivekananda in those experiences. Upon his release from jail, he had a premonition that he was to be arrested again for more serious charges and so fled the British-occupied lands to the French colony of Pondichéry in Southeastern India. There he spent the rest of his life developing his philosophy of Integral Yoga.

As described in his 1133-page tome, *The Life Divine*, Integral Yoga has as its aim "a fulfilled and spiritually transformed life on earth." Aurobindo rejected the traditional Vedantic approach of striving for liberation only as a means of reaching happier, transcendental planes of existence. Like Kashmir Shaivism, he posited a world that is not illusion (maya) but one that can evolve further and support the human species spiritually evolving beyond its current state. He foresaw an eventual descent of the divine into the world. The world itself would be transformed into a divine existence characterized by love, unity, and knowledge.

I grew up in the Catholic church, which, like much of the various Christian religions, doesn't have this particular vision of a future world. Rather, the world is seen as "fallen" from a former glory of a Garden of Eden and is now a testing ground of sorts of the soul's merits, one that determines whether you ultimately make it to heaven or end up in purgatory. Purgatory is described as an intermediate state for the soul where it has the opportunity to make amends for its past sins and thus, eventually, become fit for heaven. For those not fortunate to make it to heaven or to purgatory, their stop is hell, an eternal endgame of suffering never to have the hope of seeing God. I totally reject such dogma.

There are Christians who have a future vision for the world like Aurobindo and others. Matthew Fox is a former Roman Catholic priest who was silenced and excommunicated for his book *Original Blessing* and its Creation Spirituality teachings. Creation Spirituality celebrates the awareness of the goodness of nature, of humanity, and of creation itself. Fox's Creation Spirituality recognizes the divinity that permeates all things and the deep potential that inherently resides in them. He presents a vision of the earth and humanity designed to manifest divinity. Imagine being silenced and excommunicated for teaching such a beautiful message.

Fox edited *Hildegard of Bingen's Book of Divine Works*.[60] Hildegard was a tenth-century Benedictine abbess and mystic whose mystical visions started in early childhood. In his introduction to the book, Fox considers Hildegard a "prophet to our day because she lays out the possibility of, and therefore hope for, a living cosmology." To create a living cosmology, Hildegard wrote that we need science, a healthy mysticism rooted in the heart, and art. Being grounded in these three domains, Hildegard experienced the "interconnectivity of all things... the psyche and the cosmos, the divinity and humanity, and humanity and nature".

This reminds me of one of the more modern thoughts on this subject, called Gaia Theory, as put forth in the 1960s by the British scientist James Lovelock. His work points to the earth as a massive interconnected self-regulating system with consciousness, sentience, and intention. Stephan Harding, head of Holistic Science at Schumacher College, worked closely with Lovelock and wrote the book *Animate Earth: Science, Intuition and Gaia.* Harding sees humans not as objective observers but as "participatory experiences radically embedded in the world." He writes that participatory holistic science is more than just an intellectual approach but involves a radical shift in our fundamental perception of nature, needing us to rediscover our felt sense of the living qualities of nature. He typifies a scientist who adopts the holistic science approach and its tools, then embarks on a transformative journey toward their own wholeness and a radically new understanding and intimate relationship with nature.

This approach to science creates a living cosmology. I think it's the kind of science Hildegard would advocate. In his book, *The Story of Our Time: From Duality to Interconnectedness to Oneness,*[61] Dr. Robert Atkinson shares with us a similar narrative of conscious evolution guided by love's unifying power—not only for humanity but for all life, all lifeforms—an "interconnectivity of all things" as Hildegard would put it, driven by the power of love.

For Aurobindo, the path to developing the new human being and earth is raising our awareness to include the higher planes of mental being. He describes several sequential states of mind consciousness available to human realization, including the Higher Mind, Illumined Mind, Intuition, Overmind, and, ultimately, the Supermind. He writes, "Our first decisive step out of our human intelligence, our normal mentality, is an ascent in a higher mind, a mind no longer mingled with obscurity or half-light, but a large clarity of Spirit."[62]

What is the ultimate ascent, the Supermind? He writes, "It is hardly possible to say what the Supermind is in the language of the mind, even the spiritualized Mind, for it is a different consciousness altogether. In the Supermind all is known self-luminously, there are no divisions, oppositions, or separated aspects as in Mind whose principle is division of Knowledge into parts and setting each part against another."

He writes that Supermind is not completely alien to us and can be realized within ourselves. It's an intermediary power between the unmanifested Brahman and the manifested world.[63] "We discover that this Spirit or Oversoul is our own highest deepest vastest Self." We need these states of elevated mind consciousness to experience the unity of all.

In the Introduction, I shared my experience of communicating with a vast consciousness that said it was keenly interested in humanity's development and that, one day, we would manifest something into the physical world that has not yet existed—love, as I understood it to be. Perhaps that vast consciousness I encountered was one of Aurobindo's higher minds and perhaps that vision of the earth's future is Aurobindo's "descent of the Divine into the world." Aurobindo wrote that realizing our higher evolutionary stages is the real goal of creation.

The concept of humanity evolving to higher mental planes is discussed at length in Richard Maurice Bucke's *Cosmic Consciousness: A Study in the Evolution of the Human Mind*.[64] Bucke, a psychiatrist who lived in the late nineteenth century, shared the Cosmic Consciousness experience he had when he was thirty-six years old. He was in a state of quiet, almost passive, enjoyment when without warning of any kind, he found himself "wrapped around as it were by a flame-coloured cloud… I knew that the light was within me." He adds, "Directly afterward there came upon me a sense of exultation, of immense joyousness accompanied by an intellectual illumination quite impossible to describe. Into my brain streamed a momentary lightning-flash of the Divine Splendor which has ever since lightened my life; upon my heart fell one drop of Divine Bliss, leaving thenceforward for always an aftertaste of heaven."

He wrote, "I saw and knew that the Cosmos is not dead matter but a living Presence, that the soul of man is immortal, that the universe is so built and ordered that without any peradventure all things work together for the good of each and all, that the foundation principle of the world is what we call love, and that the happiness of everyone in the long run is absolutely certain." What is often so typical of such experiences, of what we can call a Gnosis experience, he also wrote, "I learned more within the few seconds of that illumination than in all my previous years of study and I learned much that no study could ever have taught."

His experience led him to develop a theory of the development of human mind consciousness, going from the "simple consciousness of animals, to the self-consciousness of the mass of humanity, to cosmic consciousness, an emerging faculty which is the next stage of human development." Bucke hypothesized that cosmic consciousness is slowly beginning to appear in humans and will eventually spread widely throughout the human race. May it be so! His work became foundationally influential in the field of transpersonal psychology, which integrates the transcendent aspects of human experience within the framework of modern psychology.

POTENTIAL CHALLENGES ALONG THE TRANSPERSONAL JOURNEY

There are, of course, many teachings about the human spiritual journey, and it is important to speak about it responsibly. With that in mind, I want to include further consideration of the journey as a transpersonal journey, as the eventual rebirth of the individual as it is often understood across meditative traditions.

Gita Vaid , a psychiatrist, was asked in my interview about that aspect of our spiritual evolution which is to bring those dual aspects of our heritage, the human and the divine, into a full harmonious integration. I asked for her insights into the potential limits for a human being of what is commonly called ego dissolution, the falling away of the small self as a step in our human development. How far can a human go with that experience yet still function within the mindbody system? Recall the initial challenges that Dusana Dorjee described regarding her nondual experiences and how to function in such a state. She asked a Buddhist teacher about it and he said to go with it, that she had to make a choice to further explore the experiences or accept the conditioned reality.

"That is the crux of it," she said. "The journey towards wholeness is to have those two natures of ours coexist together, each fully expressed." Gita said that in her work, she focuses a lot on the person's "character structure." An adequate character structure is needed to support the person when they're out of their egoic separateness to have a healthy experience of ego dissolution and unity. Can the person get a sense of an "I" beyond the trappings of their usual self? "It's like discovering the world is not flat. It's a whole different perspective and incredibly helpful."

Gita said she's seen several so-called "psychonauts" who are interested in pushing the limits of ego dissolution, exploring the phenomenon as far as possible. For the majority of people, with proper guidance, Gita sees they can very much come back intact and often with new knowledge. There are some, though, for whom it's best not to get too ungrounded because they may not find their way back into a healthy state. She said, "It's not something you want to just assume because some of that is the cards we've been dealt in terms of our capacities."

Gita continued, "As an analyst, I'm also looking at how, with our stories, we keep ourselves in our prisons. More and more, I see character as a defense against the ecstasy and wonder of existence. The journey is to expand ourselves and get beyond our constraints." She explained that once a person opens up some of those character structures, more of what's beyond them can then be integrated into the person. This helps bring together the duality; it can come together more easily. "You can have access to the beyond, and the beyond can come more into your world. From that perspective, once you come back from ego dissolution, you get a chance to experience how you're constructed and what's your prison." Once a person's seen what freedom is and what the prison looks like, "that's a big deal to have a sense of all of that, the whole map." This work can build a stable foundation for eventually being able to permanently maintain the expansive state of ego dissolution.

What Gita shared reminded me of a saying: "You have to become someone before you can become no one." In this saying, the "someone" refers to a person who has become individuated, to use Carl Jung's language, or they have self-actualized to use Abraham Maslow's language, or perhaps Grown Up, to use Ken Wilber's language. Gita speaks of having an adequate character structure. These ideas point to the need that for a person to successfully tread the journey they need to have a reasonable degree of ego development in order to make it through the ego dissolution phase, to be able to carry it in a healthy way. The "no one" in the saying refers to the state of no-self that was described earlier, living from the position of the true Self. As it is highly relevant to these ideas, I'll share more of my interview with Jeffery Martin.

Jeffery Martin said there came a time along his journey of interviewing hundreds of people about their transpersonal persistent non-symbolic experiences that he felt he had learned enough. He understood that he too could move into one of those states of consciousness—that he could "transition," as he put it. He had a decision to make. He saw himself as a budding scientist, still junior along the way, and he wondered if he allowed himself to transition, would it adversely affect his research on the people he was studying? He didn't want to bias his research.

For a scientist, it's one of the holy grails to maintain impartiality about what one is studying. For a scientist to have a strong bias about his or her research is to lose objectivity and begin, however unconsciously, to adversely affect the direction and perceived outcomes of the research. So, he decided to wait.

He shared too that in the back of his mind there was also a worry—a question, really—whether he was ready for this. While he believed he could indeed transition himself into a persistent non-symbolic state, he wondered if he could properly manage it to maintain a healthy and balanced life. Jeffery wondered if he had developed his psychological life enough to manage an enlightened state. He said he had been observing what seemed like random good or bad outcomes in individuals, and he couldn't be sure what was producing it and therefore if it would affect him too.

I mentioned earlier the idea that "you have to be somebody before you can be nobody." If the "somebody" isn't well developed as a foundation, the (no-self) "nobody" won't do so well in life. Bernadette Roberts notes that "should that center fall away, consciousness then has to learn to function around that egoless or empty center." For this phenomenon to be adequately supported, a certain degree of inner development is needed.

This is such an important point for the spiritual aspirant. There are innumerable examples of so-called saints and gurus who have indeed achieved higher states of consciousness but who were not developed mentally and emotionally. This leads to the many examples of gurus overstepping personal boundaries with their adherents; we've all read stories of sex and other scandals in ashrams and meditation centers.

Jeffery had interviewed plenty of people who had made it into some form of enlightenment state but who were not doing well in terms of their day-to-day human living. This gave him pause. Jeffery worried whether he might have trouble with the transition. He knew he was highly intelligent and wondered if that would work against him. Would he get "stuck in some kind of never-ending loop of mental concepts?" He thought to himself, "What if I messed up and all of a sudden started spouting some nonsense about how this one type of approach to enlightenment is the greatest ever? People would probably wheel me into a corner and say, 'Oh, he was a really good scientist and helpful, but just ignore him now.'"

Approximately a year later, he decided he could proceed with a transition. He also realized that if something did go wrong, he actually had a number of people to call if he needed help. He decided he wanted a "gentle transition," not severe, that many people he interviewed had experienced. To accomplish this, he let himself slowly lower the barriers he had constructed over the years, barriers that had helped helped him remain in a normal human waking consciousness.

Upon making the transition, Jeffery made a very relevant observation. Although he could intellectually recognize his experience from descriptions from his research, they were different. "Once I experienced them, I realized there was a gap between my descriptions and what the actual experience is." Some seven years after starting his journey, he arrived at what he had been seeking, an experience he likes to call "fundamental well-being," a well-being that exists within oneself independent of anything going on in the outside world.

It's worth commenting here on Jeffery's experience where he describes a kind of mismatch between what he had understood from his research of higher states of consciousness and what his actual experience was. This makes much sense as the mind has a limited ability to conceptualize experiences of awareness itself, and it is beyond what the mind itself can know. That is, the mind can prepare us to have a sense of what is to come and be able to recognize it when it arrives, but it won't be an actual match because of the mind's inherent limitations of the experience.

To help prepare for such transitions, Dusana Dorjee addresses the need to build adequate self-regulation before getting too deep in self-transformative practices. How to best prepare ourselves for a possible awakening to Self? In her book *Neuroscience and Psychology of Meditation in Everyday Life: Searching for the Essence of Mind*,[65] Dusana writes about compassion, self-regulation, existential well-being, and the four immeasurables as vital ingredients to successfully traverse the journey of consciousness transformation. In her Spotlight feature in this book, she does a deeper dive into compassion as a catalyst and compass for self-transcending experiences along the journey of consciousness development. This is highly important.

Tiffany Barsotti is a therapist and, over the years, has had a number of clients come to her who, for one reason or another, experienced ego dissolution—some for just a few moments and others for prolonged periods of time. Unfortunately, they had no reference for it, no knowledge about the state, and were having a great difficulty managing their lives. They needed help. Recall my interview with the psychiatrist Rael Cahn, how as a boy he experienced ego dissolution and as a result identified more "as an unbounded consciousness with a oneness with all things." Rael didn't have the kind of internal development or external support he needed, but that is, fortunately, becoming more available with each year. There are more and more therapists and teachers who specialize in this type of work.

Tiffany's work includes what is called Personal Self Integration. C.G. Jung said, "Until you make the unconscious conscious, it will direct your life, and you will call it fate." Personal Self Integration supports the individuation process, helping to make the unconscious mind more conscious and the ego self more mature and individuated. The psychologist Dr. Judith Blackstone teaches in the field of nondual realization, having developed The Realization Process to support people through the nondual, no-self experience. Jac O'Keeffe is a longtime friend and master nondual teacher who supports people through the awakening process by helping them attain a level of spiritual maturity and autonomy to authentically live moment by moment. What sets her apart as a teacher is the precision with which she is able to see the exact next step required on each person's unique path. Personally, I'm grateful to the Maharishi for his teachings and the hundreds of hours of his lectures I listened to which prepared me for such experiences. I had an immediate understanding of them as they presented to me.

Gita said that much of her therapeutic work is about placing an emphasis on the mindbody to help a person uncover their mystical nature and to integrate this into their day-to-day life. It helps us to recognize and support the movement towards wholeness. "We have to unpack and discover all that—that's the big piece of the human journey." While there is a tendency to want to discard the self, Gita said, "We have the responsibility really to develop our self to support and allow for emergence of the true Self. Then you can have new growth, new potential, and unexpected possibilities."

IN CLOSING

How might science look when more scientists take up the journey to awaken to their true Self? What are the implications of a science operating from that perspective? We have some insight into how this might look from the interviews and Spotlights in this book as well as from the work of Hildegard, Johann von Goethe, Matthew Fox, Rupert Sheldrake, James Lovelock, and Stephen Harding, to name just a few.

In the same way that a human being must transcend their egoic small self perspective and embrace their universal Self, so too science must now transcend its materialist perspective and embrace ways of observation and knowing based on love and compassion. Imagine how science and the world might look as more scientists take up the journey to awaken to their true Self!

This is especially important today when there are increasing forces in the world averse to freedom and human sovereignty, which are characteristics of life that naturally arise when we are living with love. It is thus more vital than ever that more of us embark on the journey of consciousness awakening.

May science fulfill its promise and become an instrument to help support the spiritual awakening of humanity and bring to fruition the vision of love eventually descending upon and transforming life here on the earth.

COMPASSION AS A CATALYST AND COMPASS FOR SELF-TRANSCENDING EXPERIENCES ALONG THE JOURNEY OF CONSCIOUSNESS DEVELOPMENT

Dusana Dorjee, Ph.D.

C ompassion can be described as an emotion we experience when we are concerned for another's suffering, and, at the same time, we are motivated to help them.[66] It is a positive emotion associated with openness to others, rather than withdrawal. Because of these characteristics, compassion has been of therapeutic interest for over a decade. Similarly, there is growing interest in compassion in the educational context, not only because of its mental health benefits but also due to its potential for fostering a sense of connection and prosociality, qualities much needed in the face of the current societal challenges of political polarization, climate change, crisis, and war.

Despite this increasing scientific, therapeutic, and educational relevance of compassion, the reasons compassion has these positive impacts are poorly understood. In particular, the transformational effects of compassion in expanding our sense of self towards nondual states of awareness are rarely mentioned. Such effects are more obvious in conceptualizations of compassion in the Tibetan Buddhist tradition of Dzogchen, which focuses on realization of the nondual innermost essence of the mind referred to as pristine awareness (rigpa).

Dzogchen teachings distinguish three different gradients of compassion.[67] The first gradient is closest to the Western scientific and therapeutic applications of compassion. It describes compassion in terms of motivation to alleviate suffering in its obvious forms, such as illness or conflict. The second gradient aims to go deeper and see not only the obvious suffering but also links between suffering and afflictions of the mind—anger, attachment, and ignorance. Part of the compassionate response is helping with the obvious needs but also trying to address the afflictive sources of suffering in oneself and supporting others to do the same. The third gradient of compassion goes beyond analysis and concepts. It is a type of awareness that naturally responds to others with deep compassionate concern without effort. It is sometimes described as unconditional non-referential compassion.

From the perspective of Dzogchen, the unconditional non-referential gradient of compassion is inherent in pristine awareness. It arises naturally when we experience a shift in our awareness toward this innermost essence of our mind, but it can also serve as a bridge enabling the experience of pristine awareness. This is why the practice of compassion is strongly emphasized.

From a wider perspective, it's easy to see how the practice of compassion can break down our habits of subject-object duality. In the first gradient of compassion, with the experience of reaching out to others who are in distress with genuine concern, we expand our sense of self by connecting with other's pain and wanting to help, rather than slipping into protecting ourselves and ignoring other's distress. In this way, we gradually progress from a sense of fixed self-focus toward flexible extended sense of self that brings us closer to experiencing nonduality.

My experience of compassion practices certainly attests to this. Some of the most transformative experiences of nonduality in my meditation practice arose as a result of intense compassion practice. But there are further reasons why I find compassion practice particularly useful when exploring nondual states. A key reason is that experiences of unconditional non-referential compassion can be a useful distinguishing feature between different types of awareness states which are sometimes described as nondual but are not the experience of pristine awareness. For example, there are states of meditative absorptions that can be very peaceful and self-transcending but do not encompass the experience of unconditional non-referential compassion.

There are other experiences, sometimes described as experiences of emptiness of self, where the inherently non-existent nature of the self is realized. These experiences are accompanied by self-transcending feelings of peace and contentment, but again they would not necessarily involve inherent experience of unconditional compassion. Thus, such compassion experiences can serve as a compass when exploring nondual states of awareness.

Compassion practices can also offer useful emotional grounding during some types of self-transcending experiences. When working with different states of awareness in meditation, some of the meditative experiences can be unpleasant—for example, those involving a sense of an inherently empty nature of self. This can give rise to feelings of anxiety accompanied by worrying thoughts in some meditators, particularly if they have not built up initial stability of awareness through calm abiding practices and compassion practices. Such self-transcending experiences can also sometimes arise spontaneously, outside of a meditation context.

These anxious feelings in themselves often narrow the attention focus on oneself. Practices of compassion can help expand this self-focused state and bring about more ease, balance, and anchoring in the experience.

Aside from its grounding effects, the practice of compassion can also become a way of overcoming stagnation in some of the self-transcending experiences. These experiences are often accompanied by a sense of peace, joy, and contentment that may be associated with lack of overall desire to change anything or do anything else other than savoring this state. These are positive experiences in which the afflictions of anger, attachment, and, to some extent, ignorance are temporarily suppressed. Such states are valuable because they are experiences of non-harm, but the cognizant-knowing aspects of such states are limited, and the experience of unconditional non-referential compassion is either absent or present only to an extent.

Practicing compassion in this state can deepen the self-understanding and understanding of nature of reality further, strengthening unconditional compassion, particularly the motivation to help others. This brings one closer to the experience of pristine awareness and, in turn, results in more effort to help others based on the accumulated understanding of nonduality. Tibetan Buddhist teachings describe a progression of stages of realization and associated nondual states of awareness, called Bhumis, that involve ever-deepening understanding of both nonduality and compassionate ways of helping others.[68] Cultivation of unconditional compassion is instrumental to this progression.

Finally, the deepening of insight in self-transcending states with the practice of compassion points to the central role compassion plays in answering the question about the purpose of such states. In other words, compassion helps us answer the question, "What are nondual states of awareness for?" The answer to this question is closely linked to the notion of existential instinct—a fundamental drive of our mind to find meaning and purpose in our existence, to find out "who we are."[69]

The existential instinct drives us to look for experiences that help us fulfill the need for existential purpose and self-understanding. The ultimate fulfillment of this need comes from experiencing pristine awareness. This means that we have a natural drive to experience pristine awareness because it's the source of finding the ultimate purpose and self-understanding. As we mentioned earlier, this state is a combination of unconditional non-referential compassion and non-conceptual ontological knowing. It is a state of well-being and flourishing we are all striving for.

What happens if one does not experience pristine awareness? How does existential instinct express itself? As we have discussed earlier, the experience of compassion is the closest emotion to the experience of a nondual state of pristine awareness. Therefore, responding with compassion to others and seeking altruistic ways of engaging with the world on a daily basis is a healthy way of fulfilling the need for meaning, purpose, and self-understanding arising from the existential instinct. The experiences of compassion then gradually bring us closer to experiences of self-transcending states and, finally, nonduality. However, if the existential instinct does not have the opportunity for expression via compassion, this can lead to states of mind with narrow self-focus and poor well-being, such as anxiety and depression.

An unhealthy way of expressing the existential instinct is through addictions, which bring about facsimile experiences of self-transcendence followed by craving. Another unhealthy way of satisfying the need for purpose and meaning is through seeking out experiences of dominance and power over others that provide a false sense of self-transcendence. This shows how practice of compassion is fundamentally important for healthy expression of the existential instinct and can serve as a bridge to its ultimate expression via the nondual state of pristine awareness.

It seems clear that the practice of compassion is indispensable along the journey of exploring self-transcending states of awareness, understanding nonduality and its meaning in our lives. Current applications of compassion practices are mostly limited to their health and well-being benefits. Much of the potential of compassion in supporting our self-understanding is untapped. Unlocking this potential may prove essential not only to further investigations and applications of self-transcending states of awareness in healthcare and education, but also to answering fundamental questions about the nature of humanity and its future. Given the current existential crises humanity is facing, we need to do whatever we can to enable ourselves and others to express the existential instinct—not in destructive ways, but rather in healthy ways through compassion and self-transcendence.

ACKNOWLEDGEMENTS

I am very grateful to each of the scientists who agreed to be interviewed. Clearly, the book would not exist if it weren't for your courage and willingness to openly share your personal and transformational stories.

Additionally, my sincere gratitude to Drs. Eben Alexander, Robert Atkinson, Dusana Dorjee and Kyriacos C. Markides for their invaluable Spotlight contributions which help to further illuminate the more important themes of this book and to Ken Wilber for sharing his vast integral wisdom in the Special Commentary.

I am grateful to Drs. Kurt Johnson and Robert Atkinson and Nomi Naeem at Light on Light Press for their belief in this project and for their time and collective wisdom during the many conversations which brought the book to reality, and to Nomi Naeem for "downloading" the book title.

My thanks to "book doctor" Sara Stibitz for guiding me through the proposal process and for the book's final edits.

To my friend Ruth Westreich, thank you for your creativity mocking up the original book cover design. I couldn't be more pleased with how it turned out from your original inspiration.

My thanks to the many students over the years, including my dear daughter Mariah, for through our conversations about your desire to find ways to pursue the spiritual in the context of the academic mindset helped me to further understand the need for such a book. May this book help you to be true to your Self and steady at the helm.

My heartfelt gratitude to my many mentors—personal, academic, and spiritual—who provided me guidance with their knowledge and wisdom along my own monomyth journey these many years.

Finally, I am deeply grateful to my beloved wife Tiffany J. Barsotti for her encouragement to pursue this project and for her steady support to see it through.

INTERVIEWEES

Tiffany Barsotti, M.Th., Ph.D., Biofield Sciences, Heal and Thrive, Weaverville, NC; The University of California, San Diego, La Jolla, CA

Thomas Brophy, Ph.D., Physics and Planetary Sciences, Former NASA astrophysicist; President, California Institute for Human Science, Encinitas, CA

Mary Jo Bulbrook, R.N., Ed.D., President, Akamai University, Hilo, HI

William Bushell, Ph.D., Medical Anthropology, MIT, Boston, MA

Rael Cahn, M.D., Ph.D., Professor of Psychiatry and Behavioral Sciences, University of Southern California, Los Angeles, CA

Gaetan Chevalier, Ph.D., Physics, California Institute of Human Science, Encinitas, CA

Melinda Connor, D.D., Ph.D., Akamai University, Hilo, HI; Chair, National Alliance of Energy Practitioners, Redondo Beach, CA

Arnaud Delorme, Ph.D., Neuroscience, Swartz Center for Computational Neuroscience, The University of California, San Diego, La Jolla, CA; Institute of Noetic Sciences, Petaluma, CA

Dusana Dorjee, Ph.D., Neuroscientist, University of York, York UK

Michael Goldsby, Ph.D., University of Louisville, Kentucky

Tamara Goldsby, Ph.D., University of Louisville, Kentucky

Richard Hammerschlag, Ph.D., Director of Research & Innovation, The Consciousness Healing Initiative (CHI), La Jolla, CA

Shamini Jain, Ph.D., Clinical Psychology, Consciousness & Healing Initiative (CHI), La Jolla, CA; The University of California, San Diego, La Jolla, CA

Wayne Jonas, M.D., Integrative Medicine, Executive Director of Integrative Health Programs, Samueli Foundation, Corona del Mar, CA

Menas C. Kafatos, Ph.D., Physics, Chapman University, Orange, CA

Hollis H. King, DO, Ph.D., Integrative Medicine, Center for Integrative Medicine, The University of California, San Diego, La Jolla, CA

Rajnish Khanna, Ph.D., Plant Biology, Carnegie Institution for Science, Stanford, CA; Founder & CEO, i-Cultiver, Inc. and Global Food Scholar, Inc.

Thomas Liu, Ph.D., Neuroscience, Director, Center for Functional MRI, The University of California, San Diego, La Jolla, CA

Jeffery Martin, Ph.D., Psychology, Sofia University, Palo Alto, CA; California Institute for Human Science, Encinitas, CA

David Muehsam, Ph.D., Biophysics, Subtle Energy Collective, Stanford, CA

Julia Mossbridge, Ph.D.; Cognitive neuroscience, Experimental Psychology and Technology; Co-founder of TILT: The Institute for Love and Time, Sebastopol CA; Affiliate Professor at University of San Diego Dept. of Physics and Biophysics, San Diego, CA; Fellow at Institute of Noetic Sciences, Petaluma, CA

Christine Tara Peterson, Ph.D., Microbiology & Immunology, The University of California, San Diego, La Jolla, CA

Dean Radin, Ph.D., Electrical Engineering and Psychology, Institute of Noetic Sciences, Petaluma, CA; California Institute of Integral Studies, San Francisco, CA; Co-Editor-in-Chief of the journal *Explore*

Marilyn Schlitz, Ph.D., Social Anthropology, Institute of Noetic Sciences, Petaluma, CA; Sofia University, Palo Alto, CA

Joseph Tafur, M.D., Family Medicine, Ocotillo Center for Integrative Medicine, Phoenix, AZ; Host, Modern Spirit Podcast

Rudolph E. Tanzi, Ph.D., Joseph P. and Rose F. Kennedy Professor of Neurology, Harvard Medical School, Boston, MA; Director of the Genetics and Aging Research Unit at MGH and Director of the Alzheimer's Genome Project

Neil D. Theise, M.D., Professor of Pathology, NYU Grossman School of Medicine, New York, NY

Tawni Tidwell, T.M.D., Ph.D., Tibetan Medicine physician, doctorate in biocultural anthropology, Center for Healthy Minds, University of Wisconsin, Madison, Madison, WI

Gita Vaid, M.D., Psychiatry and Neurology, Center for Natural Living, New York, NY

Daniel Vicario, M.D., Integrative Oncology, San Diego Cancer Research Institute, San Diego, CA; The University of California, San Diego, La Jolla, CA; Medicine of the Soul Foundation, Del Mar, CA

Cassandra (Cassi) Vieten, Ph.D., Psychology, The University of California, San Diego, La Jolla, CA

NOTES

1. Ken Wilber (2000), *Integral Psychology: Consciousness, Spirit, Psychology, Therapy*. Boston: Shambhala Publications.

2. Maharishi Mahesh Yogi (1963), *The Science of Being and Art of Living*. New York: Age of Enlightenment Press.

3. R. K. Wallace (1970), "Physiological effects of transcendental meditation," *Science*. 167 (3926):1751-1754.

4. https://pubmed.ncbi.nlm.nih.gov/

5. R. K. Wallace, J. Silver, P. J. Mills, M. C. Dillbeck, D. E. Wagoner (1983), "Systolic blood pressure and long-term practice of the Transcendental Meditation and TM-Sidhi program: Effects of TM on systolic blood pressure," *Psychosomatic Medicine*. 45:41-46.

6. R. K. Wallace, P. J. Mills, D. W. Orme-Johnson, M.C. Dillbeck, E. Jacobe (1983), "Modification of the paired H reflex through the transcendental meditation and TM-Sidhi program," *Experimental Neurology*. 79(1):77-86.

7. R. K. Wallace, D. W. Orme-Johnson, D. E. Jacobs, P. J. Mills (1979), "Meditation and the Hoffman-Reflex," *Iowa Academy of Science.*

8. R. K. Wallace, Mills, Orme-Johnson, Dillbeck, Jacobe (1983), ibid.

9. https://cih.ucsd.edu/

10. https://imconsortium.org/

11. https://cih.ucsd.edu/coertih

18. P. J. Mills, C. T. Peterson, M. A. Pung, et al. (2018), "Change in Sense of Nondual Awareness and Spiritual Awakening in Response to a Multidimensional Well-Being Program," *Journal of Alternative and Complementary Medicine.* 24(4):343-351.

19. P. J. Mills, T. J. Barsotti, J. Blackstone, D. Chopra, Z. Josipovic (2020), "Nondual Awareness and the Whole Person," *Global Advances in Health and Medicine.* 9:2164956120914600.

20. Joseph Campbell (1949), *The Hero with a Thousand Faces.* New York: Pantheon Books.

21. David A Leeming (1981), *Mythology: The Voyage of the Hero.* New York: Harper & Row.

22. Phil Cousineau (2014), *The Hero's Journey: Joseph Campbell on His Life and Work.* Novato, CA: New World Library.

23. Joseph Campbell (1949), *The Hero with a Thousand Faces*, ibid.

24. Abraham H. Maslow (1943), *Hierarchy of Needs: A Theory of Human Motivation.*

25. Abraham H. Maslow (1971), *The Farther Reaches of Human Nature*. New York: Penguin.

26. Kyriacos C. Markides (2021) *The Accidental Immigrant: A Quest for Spirit in a Skeptical Age*. Lanham, MD: Hamilton Books.

27. Joseph Campbell (1994), *Pathways to Bliss: Mythology and Personal Transformation*. Novato, CA: New World Library.

28. Rupert Sheldrake (2012), *Science Set Free*. New York: Deepak Chopra Books.

29. Ken Wilber (2000), ibid.

30. Kurt Johnson and David Robert Ord (2013), *The Coming Interspiritual Age*. Vancouver, BC: Namaste Publishing.

31. Leon R. Kass (2007), "Science, Religion, and the Human Future," *Commentary*, April.

32. Rupert Sheldrake (2012), *Science Set Free*, ibid.

33. Arthur Koestler (1967), *The Ghost in the Machine*. London: Hutchinson Press.

34. Owen Barfield (1974), "Matter, Imagination, and Spirit," *Journal of the American Academy of Religion*. Vol. 42(4), pp. 621-629.

35. Owen Barfield (1947), *Saving the Appearances: A Study in Idolatry*. Middletown, CT: Wesleyan University Press.

36. Max Weber (1918), *Science as a Vocation*. Munich; reprinted by Unwin Hyman, 1989.

37. Joseph Campbell (1972), *Myths To Live By*, New York: Viking.

38. Wouter J. Hanegraaff (2012), *Esotericism and the Academy*. Cambridge: Cambridge University Press.

39. John Horgan (2015), *The End of Science*. New York: Basic Books.

40. Rupert Spira (2017), *The Nature of Consciousness: Essays on the Unity of Mind and Matter*. Oxford: Sahaja Publications.

41. Mattias Desmot (2022), *The Psychology of Totalitarianism*. White River Junction, VT: Chelsea Green Publishing.

42. Joel E. Dimsdale (2020), *Dark Persuasion*. New Haven: Yale University Press.

43. Deepak Chopra (2020), *Metahuman*. New York: Penguin Books.

44. Nisargadatta Maharaj (1973), *I Am That*. Durham, NC: The Acorn Press.

45. Joseph Tafur (2017), *The Fellowship of the River: A Medical Doctor's Exploration into Traditional Amazonian Plant Medicine*. Self-Published.

46. Ken Wilber (2001), *No Boundary: Eastern and Western Approaches to Personal Growth*. Boulder, CO: Shambala.

47. Ken Wilber (1998), *The Essential Ken Wilber: An Introductory Reader*. Boulder, CO: Shambhala, 117.

48. Elaine Howard Ecklund, David R. Johnson, Christopher P. Scheitle, Kirstin R. W. Matthews, Steven W. Lewis (2016), "Religion among Scientists in International Context: A New Study of Scientists in Eight Regions," *Socius*, Volume 2, January-December.

49. Henri Bortoft (1996), *The Wholeness of Nature: Goethe's Way Toward a Science of Conscious Participation in Nature*. Lindisfarne Books.

50. Connor MH, Connor CA, Eickhoff J, Schwartz GE (2021), "Prospective empirical test suite for energy practitioners," Explore (NY). Jan-Feb;17(1):60-69.

51. Joseph Tafur (2017), *The Fellowship of the River*, ibid.

52. https://med.virginia.edu/perceptual-studies/wp-content/uploads/sites/360/2018/09/Greyson_-Alexander-JNM.D.-2018.pdf

53. Ken Wilber (2007), *Integral Spirituality: A Startling New Role for Religion in the Modern and Postmodern World*. Boulder, CO: Shambhala.

54. Ken Wilber (1977), *The Spectrum of Consciousness*. Theosophical Publishing House.

55. Ken Wilber (1995), *The Atman Project: A Transpersonal View of Human Development*. Theosophical Publishing House.

56. Philip Goldberg *(2010), American Veda: From Emerson and the Beatles to Yoga and Meditation. How Indian Spirituality Changed the West*. New York: Harmony Books.

57. Bernadette Roberts (1993), *The Experience of No-Self: A Contemplative Journey*. Albany: SUNY Press.

58. Thomas Merton (1979), *Seeds of Contemplation*. Westport, CT: Greenwood Publications Group.

59. Thomas Merton (1955), *No Man is an Island*. New York: Harcourt.

60. Matthew Fox (1987), *Hildegard of Bingen's Book of Divine Works*. Rochester, VT: Bear & Company.

61. Robert Atkinson (2017), *The Story of Our Time: From Duality to Interconnectedness to Oneness*. Ft. Lauderdale, FL: Sacred Stories Publishing.

62. A. S. Dalal, ed. (2001), *A Greater Psychology*. Pondicherry: Sri Aurobindo Ashram.

63. Sri Aurobindo (1990), *The Life Divine*. Lotus Press.

64. Richard Maurice Bucke (1901), *Cosmic Consciousness: A Study in the Evolution of the Human Mind*. Reprinted by White Crow Productions Ltd, 2011.

65. Dusana Dorjee (2017), *Neuroscience and Psychology of Meditation in Everyday Life: Searching for the Essence of Mind*. Oxfordshire: Routledge.

66. T. Singer, & O. M. Klimecki (2014), "Empathy and compassion," *Current Biology*, *24*(18), R875-R878.

67. Dzogchen Ponlop Rinpoche (2004), *Trainings in Compassion*. Ithaca, NY: Snow Lion Publications.

68. Gampopa (1998), *Jewel Ornament of Liberation*. Ithaca, NY: Snow Lion Publications.

69. D. Dorjee (2017), *Neuroscience and Psychology of Meditation in Everyday Life: Searching for the Essence of Mind*. London, UK: Routledge.

ABOUT THE SPOTLIGHT AUTHORS

Eben Alexander, M.D., an American neurosurgeon and author. His book, *Proof of Heaven: A Neurosurgeon's Journey into the Afterlife*, describes his near-death experience that happened in 2008 under a medically-induced coma when treated for meningitis.

Robert Atkinson, Ph.D., Professor Emeritus, University of Southern Maine, Director of One Planet Peace Forum and StoryCommons. Dr. Atkinson is an author and speaker on interfaith, interspiritual, and interdisciplinary approaches to individual spiritual development and the processes, patterns, and promises of humanity's conscious evolution.

Dusana Dorjee, Ph.D., Psychology in Education Research Centre, Department of Education, University of York, York, United Kingdom. Dr. Dorjee investigates neurocognitive mechanisms of well-being from a developmental perspective, with a particular interest in neurodevelopmental trajectories of the self-regulatory capacity and the self-world capacity linked to existential awareness as possible key sources of well-being for the individual.

Kyriacos C. Markides, Ph.D., Professor Emeritus of Sociology, University of Maine. Dr. Markides' areas of interest in teaching and research include the sociology of religion, non-medical healing, the lives and teachings of Christian mystics, healers, miracle workers, and monastics around the world.

ABOUT THE AUTHOR

Paul J. Mills, Ph.D. is Professor of Public Health and Family Medicine, Director of the Center of Excellence for Research and Training in Integrative Health, and Former Chief of Behavioral Medicine at the University of California San Diego. He has over 400 scientific publications in the fields of pharmacology, oncology, cardiology, psychoneuroimmunology, behavioral medicine, and integrative health. He published some of the earliest scientific research on meditation. His work has been featured in *Time* magazine, *The New York Times*, National Public Radio, *US News and World Report*, *Consumer Reports*, *The Huffington Post*, Gaia TV, and WebM.D., among others. He's presented his work at hundreds of conferences and workshops around the world, including at the United Nations.

MESSAGE FROM
THE PUBLISHER

Light on Light Press produces enhanced content books spotlighting the sacred ground upon which all religious and wisdom traditions intersect; it aims to stimulate and perpetuate engaged interspiritual dialogue for the purpose of assisting the dawning of a unitive consciousness that will inspire compassionate action toward a just and peaceful world.

We would like to acknowledge the Science and Spirituality Synergy Circle of the Evolutionary Leaders Circle coordinated by Dr. Kurt Johnson of Light on Light Press.

Managing Editors—
Kurt Johnson, Ph.D.
Robert Atkinson, Ph.D.
Nomi Naeem, M.A.

www.ingramcontent.com/pod-product-compliance
Lightning Source LLC
Chambersburg PA
CBHW021615120626
46545CB00001B/232

MY
AUTO-
BIOGRAPHY

BY JOE BIDEN

BASED ON A TRUE STORY, MOSTLY

J. GALT, ESCORT

ISBN
978-1-958690-97-0 (Hardcover)
978-1-958690-98-7 (Paperback)
978-1-958690-99-4 (eBook)

Warning! Warning! Warning!

Hi, I just found out that some buddy or not so buddy buddy wrote stuff about me and made pictures of me and stuck it all in this book without my permission. That stuff isn't true, even the parts that are. Kamala said she didn't do it and I believe her. And Nancy — well Nancy liked me, I think, before the Chinese thing happened. Anyway, I just don't know who did it but I for sure did not.

Joe for 2024 and 26!

(You don't think Hunter...Nah.)

P.S. I will autograph this book if you bring it to the White House. It's the "Big" one which is mostly white, if you get my drift.

P.P.S. Remember, if you want your kids to be successful like mine, let 'em watch the radio and play things on the pornograph.

When you're dead,

you don't know

you're dead.

All of the pain

Is felt by others.

The same thing

happens when

you're stupid.

– Philippe Geluck

AN IMPORTANT NOTE FROM THE PUBLISHER AND THE EDITOR

We are too proud to present to America and to the entire world the inner thoughts, the recollections and the non-recollections of America's greatest living, currently alive president sleeping in the White House and in his home in New Jersey and elsewhere. When the American economy goes bust, he goes boom, when the drums of war are playing, he dances, and when everyone is crying, Joe is spying. We here at the Delphic Oracles Publishing House cannot help but salute the man, a man of inexhaustible urges and profound profligacy. He is a paragon of urges and psoriatic arts. The melody and beauty of his voice is both profound and base like much else about the man we so admire. A man of the utmost personal integrity, my friend Joe never shoots from the hip as he engages in great research to validate everything he says in public and private. He instead engages in a kind of source-e-ry that makes everyone cry with delight. How could one man know so much? Is he possessed or just truly a genius? His DNA is inhuman, just like the man. His recollection of things both past and past, as well as future, makes him a fitting presence in today's White House. As Joe himself says, "Don't do now what you could have done then, like in the future, or the past." Such profundity even now eludes most great minds. Like Einstein, Joe knows his own relatives, for he is one of them. And I ask that you the reader remember that, for Joe, life is truly a re-rendering of all things past even if they are not or have never been present. He is a man beyond his times. He is a visionary who has seen beyond the final frontiers of the heavens and embraced its empty spaces, bringing them into his own brain. And like all great genies, Joe thinks outside the box and destroys its rigid confines. As he once said to me, "Who needs outer space when I got inner space right here," as he brought his pointer finger to his temple. Yes, his temple, how symbolic. Now join with me as we harken back to those glorious days of Joe's youth and early childhood in Scranton, Chihuahua, where he first learned to samba and to covort with his Mexican playmates. As Joe likes to say, "O Cisco, O Pancho." Not only is he the man, he is the wanderer of the world about whom we all wonder.

One final thought. Some say that Joe exemplifies the very essence of concentration with his laser-like focus upon any task he ignores to undertake. I say Joe is beyond mere concentrated power of the mind. Joe is instead positively dense. Like the universe itself, Joe's mind can bend time and space thereby bringing darkness to light. Truly, Joe is the black hole of humanity.

Ali Bin Ben
Allahu Akbar

CONTENTS

The Mean New Deal

PART I

THE EARLY YEARS

Better to remain silent and be
thought a fool than to speak
and to remove all doubt.

Abraham Lincoln

MOTORS, VOTERS, AND JOE

JOE TALKS

When I first told my voters I wanted to write my auto-biography, they said, "No, no, Joe, it's too hard. Please don't, you'll hurt yourself." Well son of a bitch, I wrote it anyway and proved, proved... something, like when the people are for ya, I'll just ignore ya, 'cause I'm Joe. Believe me, cause I ought to know. So these are some of the thoughts I was thinkin', as I was, about why I have never been selected for jury duty and other strange things in my wonderful life filled with Ho downs and other Chinese women and mischief makin' mamas, if you know what I mean.

So one of my greatest thoughts was everyone needs this book about my auto-biography because without it you'll never learn to drive a car. And that is how it all started. But first I was thinkin' and thinkin' rhymes with Lincoln, that's the guy who started the War of Northern Aggression against us. Cause he didn't like Democrats much. So we use to say, "Lincoln, Lincoln, you been drinkin', smells like wine, oh my gosh, it's turpentine." And that ain't good to drink, Hunter told me about it. And I didn't like Lincoln much cause he didn't like us but I do like his car. I got one for my rides around the parking lot at the White House. It's smooth as Jill's behind and she ain't even a cat.

Some say, "Joe, you got no Lincoln, cause you were too poor growin' up." That's true, I was poor but I worked in the coal mines in Detroit to bring home stuff achin' like bones and stuff. So when I was workin' all day in the coal mines I turned black. 'Cause of that, I was liked by all the other black kids in my neighborhood. They use to call me, "Little Black... Little Black Little Black White Boy who ain't." But I didn't like it much cause it was too long and I couldn't remember it all. So I just walked around the neighborhood all day long turnin black more and more each day. I was like a tranny, you know. I was transitioning from one color to another like one of those lizards that climb the walls and drapes in your house, especially the curtain. Then you don't know where they are until they fall on you at night when you're sleepin' and you think it's Giraffish Park and you're like bein' eatin. But you ain't cause they're too small and they can't eat ya if you can't see 'em. I think they're called Kamaleons.

But sometime there were problems like when I was walkin' around tryin' to grow my face more black so I could fit in more. This guy came up to me and was sayin' I shouldn't keep walkin' around the neighborhood with a black face when I was cause it was evil. He was named Justin ya know. But he wore pajamas a lot and use to climb up on the school during recess – oh, oh, oh, and I forgot to tell you this—he had a black face just like me. Oh, hey, that reminds me of a book I wrote about my whole experience when I was with Justin the white black boy. I called it *Black like Me*. And I wasn't afraid of him and I told him, "You ain't black like me if you ain't white." And he was a chain smoker too so he use to cry a lot from skin cancer cause he said black people get it more and he thought he had it and a kind of sickle from sales in Armenia --I think it was called. I never understood that. I like Armenians but he didn't cause he was black. But I'm not sure cause he knew he didn't cause I could see his white skin kept poppin' up especially when it rained. He had leprosy! When we played basketball, he kept his shirt on even when we was skins. He was kinda strange too. He ate a lot of frogs and said they were good for ya. But he still wouldn't take his shirt off 'cause people would see his whiteness and start yelling" "Leper, leper! Ring around the roses. You got no noses. Ashen, ashen, you all fall down."

Oh, oh, oh, an' I forgot to tell you about his peeing. When he was on the roof of the school, he would pee on the little kids down below and tell them they were stupid and some day he would grow up and be a king and they would be mothers of truckers an' he would reign over them. But I knew that wasn't true cause he was peein on the boys too and they couldn't be mamas unless they became lesbians first.

MY FIRST OTTO

**Don't be
Ashamed to fart
While you pee...
There is no rain
Without thunder.**

-- Somebody

My first car was a self-made car and I made it mostly myself, especially after I got some help from a friend of mine named Johnny. First I bought a Revel Kit Car and put it on the floor after I opened the box, but it weren't no use. It just sat there and never moved or made anything of itself. It was born dead so I quit and just poured all sorts of glue on it and lit it on fire to watch the black smoke comin' out of it remindin' me once again about my bein' a black kid and all. He said some day we would have a whole car if we stole it just one piece at a time from the GM plant where it was bein' made. He even made a funny song about it.

But I had a German friend who I played football with me all the time who was really good at throwin' long passes. He became a famous football player after he taught me how good it was to learn to throw long, long passes. He was always' sayin' "The pass is the gas of the game and the air in room." In fact that's how I met Jill. We met at a party once and I had been eatin' a lot of beans with my Mexican playmates named Cisco, Pancho, and Goya. And what happened next was just the funniest thing. Ya see, after all them beans, I wasn't feelin' so good and I was getting' kind of crampy and all. Well anyway that Jill was kind of flirtin' with me and one time she said, "Don't you like me? Tell me how you really feel, Joe, just let it out." And so I did, Whew! What a relief! And as I did that I remembered Otto tellin' me, "If you pass the gas, you'll always leave air in the room," and boy did I! That may have been the longest gas pass I ever made. Jill was impressed, I could tell when she expressed her admiration for me when she wrinkled up that cute little nose of hers and said, "Sniff, sniff, was that you, Joe?" Because I'm kind of the manly type when I was a kid, I didn't say a word. I just turned around with my back to her and did it again. "This one's for you, Otto, let's build that car." Jill giggled and said, "Joe, you had me at BRRRRRRRRRRRRRRRRRRRRRRRRRRRRRRRRRRRRRR RRRRRRRRRRRRRRRRRRRRRRT." Yep, she said it, truly. "You may be smelly but you are the king of farts, no one else like you." Two days later I saw her again and didn't say a thing, I just let my body do the talkin, BBBBBBBBBBBBBBBBBBBBBBBBBBBBBBBBBBBBBBBBBBBBBB BBBBBBBBBBBBBT. "Joe, Joe, Joe, you are so special. Maybe some day you can do it when you grow up and do it front of a queen or maybe even the pope. What do you say, big guy?" And, if I can re-collect, I think to myself, "Boy did it! "

Every party needs a pooper...

So my friend Otto and Johnny and I, after I was workin' all day in the Detroit coal mines, began constructed a real, live, self-made car. But I had learned that just because somethin'says its self-made doesn't mean it is. You still have to make it even if the instructions say it is self-made. So we would go to the trash every day outside the Ford Plant and pick out the GM parts we wanted. Some fit together, some didn't. Then one day I found a pumpkin field near the Chrysler Plant and it had pumpkins growin' next door. So I know what you are thinkin', how can a body make a car out of pumpkins? It was actually easier than you can imagine. Remember, I am Joe and I ought to know.

So by and by, we started collectin the parts and draggin' them out of the trash all the way to my secret hidin' place which was a deserted barn near the coal mine where I worked. It took us months to put it together. Then finally one day, we had put most of it together but part of it didn't quite seem right and by and by I contemplated my design. We had forgotten the one thing we had forgotten that was absolutely necessary. Yep, sompin' was amiss and now I knew what it was. Wheels, we had forgotten the wheels. I screamed out loud in the barn, "Da Wheels! Da Wheels! " Well, I plump dat a doo masawaki hollobaloooooyeee! Remember I talked like that back then cause I was transitioning into a black person from workin' in the coal mines all day. But we didn't have any wheels and we noticed that GM just didn't seem to want to throw any of them away. We found some but they were all flat on the bottom and no matter which way we rolled 'em they were always flat on the bottom. And we were just about to give up when a miracle happened!

There she was!

Oh, she was sooooo pretty. She had long golden-like hair and beautiful black face just like me. I had seen her before at the coal mine givin' out canaries and cages to the other miners so we had somethin' to eat when we went into the coal mines. But I guess I hadn't noticed so much her bein' pretty and all and I thought she was too nice for someone so manly and all like me. But there she was standin' in the doorway of our deserted barn. When she smiled, her black face and against it them thar pearly whites. Whew! Goodbye Jill! Yep, after hangin' around the coal mine all day long she had begun to transition from white to black just like me. So I asked her her name, she said, "Black Cinder." "Ah, that's beautiful," I said. But then Otto said, "Her name's re-dundant." "No, it's not," I said, "It's Black Cinder." "That's what I'm sayin'," he said. "No, you just said her name's Redundant." "Exactly, it's her name. I agree, she's Black Cinder, and she's redundant."

Otto was getting' kind mad and I didn't want to embarrass her and so I invited her into our barn. To humor Otto, I just started callin' her Redundant and Otto just called her Black Cinder, I guess to humor me. And that is how I first learned to negotiate and compromise, by understandin' the situation. And, Otto, he just ate some Graham crackers and din't say much the rest of the evening. But later as we were still refin' our thoughts about how to get the wheels to our auto-mobile, I caught him lookin' at her kinda subtle like. So I winked at him and he says, "You're dumb." "Yeah, dumb like an ox, I think." He just rolled his eyes and looked away. I think he was scared at my wit. But Johnny didn't say much, even to Redundant. When I asked him why he didn't seem to take to her, he just said, "Cause I walk the line and I keep a close call on this heart of mine." I thought about it and all but I couldn't make sense of it. So I just stopped thinkin' and that has been one of the great lessons I have carried with me all my life." If you're not sure what to do, just stop thinkin' and I been doin' that ever since. And once when I became an alcoholic so I could see what it was like, I just started sayin' to myself, "Don't think, just drink." Been doin' it ever since.

JOE SAYS TAKE BREAK BECAUSE READING IS HARD. DID YOU EVER READ A TELEPROMPTER?

JOE DOES THINKIN'

So when we was thinkin' and workin' on a way to get tires on our auto that weren't flat on the bottom, Redundant says to us, "Why not turn the pumpkins into the wheels?" "Of course, I said, that's it! That's it! Eureka! Eureka! " That was an expression I learned from people in California and they said every time they struck gold but which was mainly only if they went to a dentist and had their teeth fixed, I guess. An' then later on I went to a Greek bathing party where they baptized a new baby and poured stuff all over their bodies and then left olive oil streaks in the tubs and kept yellin' Eureka every time the tub overflowed. I never did see the fun in that, especially when they had to clean up the grease streaks they left in the tubs. I tried it once by pourin' olive oil all over myself but I kept slippin' in the tub and spent two days in there until a plumber came to my room and pulled me out 'cause the neighbors were complainin' about the water and the olive oil that kept flowin' into their apartments. An' I was mighty hungry. A body can only drink so much bathwater.

So after that I kept to myself kinda and showed 'em how tough I was by spittin' when they walked by and poured more olive oil on myself to be kinda greasy even though I didn't really like grease. But that was the only way I could pretend I was Aye-talian which I wasn't completely because I never had a greaser black leather jacket to wear when I was poor and black. But I did have an Aye-talian girl friend once who was named Sophia and She was beautiful and use to dance a lot to make her blouse real sweaty-like so that her nipples would punch holes in 'em. Man she was a real hot tamale but not completely 'cause she was still not Mexican and came to

9

America legally, they tell me. But she was really beautiful and so I never understood why she had a fat old bald guy who made her do tricks and stuff in the movies and then he would just go around tellin' the guys she was doin' things with in the film, "Basta!, Basta! " and then clap his hands like a seal. So I didn't get it, if you want her to do things in sheets and stuff with other men and take pictures of it why do you keep tell her to stop after you make her start in the first place. I guess it's just the way Aye-talian men are. I like spaghetti and maybe I'll go there to Aye-taly and find out what "Basta! " really means but not until I find out if Sophia was born in Rome or Nipples. I think it's more likely Nipples.

So like I was sayin' until I was so rudely interruptin' myself, we had Redundant tellin us about putin' (somethin' about that word bothers me) on the pumpkins which made me think of another song I was workin' on while Johnny kept playin' a banjo he made out of an old hubcap and some strings. I still don't get it and told him so, "Johnny," I says, 'Why don't you help us get the pumpkins by the coal mine instead of just singin' a song about that line you keep wantin' to walk. It's stupid, nobody walks a line even when you're fishin'. You might cast your line into the water but it won't catch no fish unless you got bait on it and big old carp come up and bite it. Even then it won't do you no good, no how. After all, you can't just throw the line in the water without you tie it to somethin' like your big toe or a fishin' pole. But that Johnny, he was real stubborn and kept ignorin' Redundant and pretended he didn't see her. But then, real funny like he started smiling and said, "I know, let's give her a real name." "Like what?" I said? "Well, let's ask Otto," he said. Otto was pretendin' to work hard but the car was almost finished except for the wheels and the batteries and stuff and he looks up and says, "Give her a nick name that makes sense. Since Black Cinder is redundant, make her name somethin' that says that to someone who is even stupider than someone else we both know." I knew he was talkin' about Johnny and I didn't want to hurt his feelins so I say, real wise like, "Yes, what would you suggest, Otto?"

"How about 'Ree-Ree, then her redundancy makes sense." Maybe it did, but not to me, so I just says, "OK, but if we do that I'm gonna give you a nick name too." "Like what?" he says. "From now on I'm gonna call you 'Otto Otto." "That makes no sense," he says, "My name isn't redundant."

"I guess, but I can't let you steal her name now you gave it to her. So, if she is Redundant, you are Otto Otto." He rolled his eyes and shook his head and slapped his own forehead like a retard. Boy he could be dumb and I knew I was right and he was almost spazzin' out.

But while he was spazzin' I noticed that Redundant was startin' to cry for the meanness of changin' her name to Ree-Ree and I was manly but kinda sensitive too so I walked right up to her and says, "Ree-Ree, don't you cry, you'll always be Redundant to me. No one can take that away from us" and I was about to hug her and make her feel good when I saw the tears was washin' away her blackness. Oh my! That's when I realized, she was just like Justin, her whiteness was comin' through. She had leprosy! " I left her alone for a while to get over that and promised myself to never fall in love with a leper ever again even if she had a beautiful face like Ree-Ree, who Johnny still didn't like 'cause later on he told me he thought she was two faced and that made her Redundant all over again. I was so confused when all of sudden there he was! And I'm pretty sure I don't need to tell you who it was. But if you're not real bright, I'll give you a hint. He likes to eat frogs, he thinks he's a girl and he looks like one and he has leprosy.... I'll wait...Still don't know? Well I ain't a gonna tell ya then 'cause your so dumb and I don't want to appear redundant because I just learned what the word means, and I ain't sayin' the same thing twice. So bye bye for now. Get it? I said what I wasn't gonna do but I just did. Smarter than you thought, huh?

A NOTE FROM THE EDITOR

Joe's manner of speaking and thinking is some time nearly incomprehensible to mere mortals like you or me. That, of course, is Joe's gift. He is an enigma wrapped into a cocoon of willful indifference to the urges and concerns of normal people. In short, Joe is the forerunner of the modern enigmatic trend of crypto currency. Before there was Bitcoin, there was Joe. Before there was Ethereum there was Joe, and before there was modern welfare, Joe was already understanding proof of work and proof of stake. Joe thought that neither was necessary to get a government grant, stipend, or to qualify for anything. Joe's genius is to make even the most comprehensible beyond the understanding of even the most brilliant. Joe has long said, "government is a tail held by an idiot schooled by the dumb and indifferent connected to nothing." Be patient, Joe knows even more, and like the Almighty, he will soon bring to you peace and understanding beyond your comprehension. After all, he is Joe. Joe the indifferent, Joe the inscrutable, and Joe the undefinable. Joe is Genius.

"Basta! Basta!, I like Pasta."

PUMPKINS

So there he was just tryin' real hard to act tough, chewin' on a frog which he said was poisonous but it didn't matter because he was black like from Africa and those were his people and he ate poisonous frogs all the time 'cause they were use to it. To show him he wasn't no tougher than me, I picked up a couple of snails and stuffed one in each of my nostrils and then ate two more just to show him. I told him, "You dough thnare me, you thnob. I eat snogs like you for breffas." I admit it was difficult to breathe and like talk with the snails in my nostrils, especially when they started crawlin' down my nose into my throat. It was snot pleasant, if you know what I mean. And then when I sneezed, Oh Boy! The snails got stuck back up in my nose and my ears about blew off my head and I about choked on the ones I was chewin' on. Justin was impressed. I could tell. But the blackness was still pealin' off him from his sweat and so he was lookin' strange still. I knew then the leprosy just wasn't so easy to get rid of even if a body catched a body a comin through the pumpkin patch. And, cause I felt sorry for him, I told him he could help us fetch the pumpkins for the wheels so long as he addressed Miss Redundant as Ree-Ree for no apparent reason. I just wanted him to have to obey me even if he didn't want to or know why I was tellin' him to do something'.

Beware the Jabberwocky, my son!
The jaws that bite, the claws that catch!

Lewis Carroll

JABBER NO. 1

TRUTH INJECTION: All of you reading this story of Joe need to make certain you do not skip any of these injections. In fact, if you ignore them, you will be required to pay a fine and you may even lose your job and your eye sight. Without these injections you cannot hope to comprehend the total and uncompromising truth and wisdom of Joe Biden. Like Joe says, "Let no one buy or sell unless he has the mark…" That means if you don't read this, Joe will make you take the injection jab so you can know the truth. So here is more information about Joe from one of his childhood friends when he was a black Mexican living in Scranton, Chihuahua.

Secretly recorded interview by a whistleblower from an organization named Veritas. That means truth in Latin because Joe was living among Latinos when all this happened. Now for the truth.

NAUJ (A secret name): When we first met Joe, or, as we liked to called him, Jose, he was kind of strange. He was living with his mama in a hacienda on the outskirts of our village. We don't know where he got the money to buy such a beautiful house but he had money all the time even when he wasn't working for it. He said he had a friend from a place called China where they met with him under a table. We never understood that. He had another friend, he said, who gave him money every time he said "son of a bitch." That was from a country we never heard of that grew a lot of sunflowers and the girls there were beautiful like his friend Redundant. He would walk around the neighborhood singing about them, like… "they leave the West behind and Georgia's always on my mind." We knew about Georgia and didn't know if he had lived there or not because it was not Detroit where he once worked as a coal miner. He confused us. Later on when he got older we found out that he owned a ranchero in that country. It was near a town by a river he said that was hidden by a bunch of fronts. We never understood why anyone would hide his ranchero behind a front. That just seemed stupid. But we knew it was true but Jose said he did it so no one would know he owned the ranchero. He was showin' off his humility, I guess.

So he was only six years old and had a lot of money, which was a mystery to us. Why would someone give you money just for sayin' "Son of a bitch" all the time. But then we understood it better after we got to know him more. When he was tryin' to impress his mama or his father, he would call us 'beaners." We didn't like that because we knew what it meant. Later on when he became a honcho in Washington D. C., the capitol of America, he pretended he didn't say those things about us. But my father had a camera and at one of the parties he filmed us having fun. But while we were dancing and laughing, Jose kept calling us "stupid beaners." Back then we didn't know much English but we knew enough and my father posted the pictures years later on a thing called "YouTube," but they took it down after they saw it because they said it wasn't true. Sometime when he got really angry and he hit us he would call us "niggers." But we said, "Jose, how can you call us that when you are the only black person in the room?" So he started to cry and said he would some day go back to Detroit and become a king in America and he would let us all in even if we were all niggers and spics and other stuff he called us. But I think all that has been taken off "YouTube" now because they like Jose. But we still have the original film.

But one thing about Jose, he liked to go to church a lot or pretend to. But Joe didn't always get along with the village priest. Some time in church Jose would fart during communion which we thought was a sin. But he would fart real loud and long even right in front of the priest. Because of that people were afraid to get close to him when he was having communion so they stopped going. But, I have to admit, no other boys could fart like him. His farts, they were something else. Like Jose use to say to us after mass, "Fart is an art." Maybe, but all I could think was "who's the beaner now, Mr. Leper?" And here is something else about Jose you didn't know. Sometime during Lent, Jose said the priest's sermons were stupid and he didn't like them, so he just went to church for a minute and then went outside and put some dirt on his forehead. Then when the cameras came, Jose walked around the village like he had been blessed and he would mock the priest by saying, "I ain't dust and to dust I won't return." But still he did have good farts and we thought some day he would become famous for them.

One more thing you should know. Jose would play marbles a lot with us, but when it was over, he would steal the marbles he couldn't win in a fair game. He liked to steal cat eyes and puries, which were the most beautiful of all the marbles. We all wanted them but Jose was always stealing what he couldn't get fair. But one day, even though he was mean and stupid at the same time and did a lot of mysterious things we didn't understand, God got even with him for farting in church. You see, Jose lost all his marbles in a card game called Poke Her. Jose misunderstood what it was about and so Jose thought he could sniff and touch the other guy's girl friend and maybe do other things to her. Jose was wrong. That's how Jose lost all his marbles and never got them back to this day.

PUMPKINS GALORE

So by a while Justin and Johnny and Ree-Ree and Otto Otto and someone else we didn't know – well we went to the pumpkins patch to get some magic pumpkins to put the wheels on for our auto-mobile. It's important to know I am telling you this so you know how I came to write this here auto-biography, in case you missed the point. And somepin' else you need to know. Remember when I told you about the pumpkin patch bein' next to the coal mine and all. That's not true. I lied to conceal its real location so you can't find it even after all these years. It's our secret and a lot happened there. You'll c (this secret way of my writin' will be clear down below --stop thinkin' dirty it ain't that, not yet, anyway). But what I'm sayin' is this sentence is a kind of 4 shadows. One would be good enough for most people but I got 4 of 'm. When the sun is right above me, I'm pointin' 4 different ways at once. It's a kind of miracle.

But when we got to the place where the pumpkin patch was situated, we found it had a brand new fence around it and we couldn't get in. That is, we couldn't just walk in because there was a brand new barn right in front of the one and only gate. You'd have to go through the barn to get to the pumpkin patch. But we was scared and a little bit stumped because we didn't know who would built a barn to protect a pumpkin patch unless it was GM or maybe Ford tryin' to prevent us from stealin' wheels for our auto-mobile.

So the barn was big and red. I think there's a law that says barns have to be red. But on each side there was a picture of a guy with a big head of yellowish hair and under his picture it said, "This is HaHa country." And he was pointin' his finger at a picture of another black guy with a rope around his neck who was cryin' and sayin' "I'm not suicidal." We didn't understand it at all. But we look before the barn. We look behind it. We didn't see nobody. So we walked in da doe. I was talkin' a little mo differen now on account I was really, really black that day and had lived south of Detroit for a while where there were a lot of poor southern white folks, just like me. Yup it was true, for though I was black, I still had a whiteness behind my head and my neck was bright red and I still loved to eat them crackers.

Now by and by (I know I was sayin' I wouldn't be redundant, but I couldn't help myself. Sayin' things twice or more just made me think of Ree-Ree who I still loved even though she was a leper, just like me and Justin). So, by and by, (I know, I know, I did it again) we start traipsin' through the pumpkin patch lookin' for the biggest most round on the top and bottom pumpkins we could find. But the truth was there were so many, we just kept goin' and goin' just like in that battery commercial. But we weren't rabbits, just mostly black kids, lepers, and a banjo beatin' Johnny and Otto Otto who passed a lot and maybe was smarter than the rest of us, except maybe me.

Then finally after a real long day we came to another barn with a sign with just a great big circle in white paint with an X in it right on the door.

"What does it mean?" Ree-Ree asked. Then I says, "You know, *I* before *e* except after *c*, except as in **nay** bor and **way**, and sometime *y*." She smiled and said I get it now, Joe, and she made a motion with her hand and points at her eye and says, "I" and then she made a half circle with her hand again and said "c" and then she made another one with her hand shaped like "u." And I guess she did 'cause she was smilin' right at me. But Otto just rolled his eyes again and says, "No sense at all." So I couldn't help myself, I was sweatin and all from our pumpkin patch exertion, but I hugged her right off, even though she was a leper and we smeared ourselves on each other and exchanged bodily fluids. Some of her black was now on me and mine on hers. Otto Otto didn't like it much and threw up. Johnny just stood there with his homemade banjo and began pluckin and singin', "If I were a horticulturalist and you were an agronomist, would you marry me anyway, would you plant my seed?" He used more big words in one second than any of us heard in a life time but we was curious so I says, "Johnny, keep singin'" and he did. By the time he was through, I was so excited that my seed was on the ground and no one would get near me except Johnny and I told him how to improve the song and I changed the words a little to make 'em clear for ordinary folk and years later it became a big hit song and he cartered away a lot of cash from it. Even today when I hear the words to that song about me and Redundant I have an organism. That's the way it is, I guess, when you're in love.

REE-REE FINDS HER UNCLE

Well we all stood there lookin' up at that big red barn, just as big as the other with its funny lookin' circle on the door with an X. So we stood there and weren't sure what to do, so just pretended like someone was talkin' in my ear and givin' me secret instructions about what to do next. And I kept sayin' just loud enough for most anybody close by could hear me sayin', "That's right, that's right, and that's exactly what I was thinkin' of doin' too. Glad to know you agree." I was talkin' like this to make certain everyone else thought I knew what I was doin' and to make them not afraid to follow me into the barn past that big ol' X.

Even though I was leadin' us all into the barn I thought it might be better if I led from behind, which, I just admit makes me think of a certain someone named Jill. She's so soft and her behi-- Well I ain't sayin' no more except to say that she likes it when *I* play the doctor. But, Basta, Basta, I'm thinkin' of Pasta. Back to my story. So we all stood there and thought someone has to go in first and it might as well be the new guy who kept followin' us and who we didn't care about. So, I says, "Hey, new guy, with the funny face, curly hair and a guitar, I am your leader. You go in first and we'll be right behind you with me leadin' from behind." All nasally like he started complain' sayin'" I'm not going in lessin' I get to go in last with you leadin'." "Why?" I says. "After all I'm the leader, do like I say." "I'll tell you why," he says. And he started singin' "The answer is blowin' in the wind." Right then and there, I had the answer to everything and I started thinkin' about a guy who fought with *windmills* and lost but I was gonna *wind*." I'm bein' funny, that ain't no error. I did that on purpose, twice. HaHa. Smarter than ya thought, huh? That's when I first said to myself "you wind with wind." Great slogan.

So anyway, I did what I needed to do and pushed my lovely Ree-Ree through the doors. They swung open and the rest followed while I stood aside to let 'em pass while I guarded the rear. Even Mr. Nasally passed me, still singing' about wind blowin' and stuff like gettin' stoned and stuff. I didn't quite understand it but it seemed kinda funny. In the barn it was all dark except in one corner where we heard like an old man's voice singin' "Zippity do da, zippity ay, my oh my what I wonderful day..." Well we all rushed to see who it was when all of sudden, he stopped singin' with the most wonderful smile on his face and Ree-Ree run right up to him and hugs him straight away.

He says to her, "Oh, my Cinder! Cinder!, oh, my, my, my! You are such a sight to behold. Behold! Behold! Behold! But there is somethin' different about ya, now. I cain't put my finger on it just yet gimme a minute. Hair still blonde, teeth still bright white, and you're still pretty as a popsicle on a hot summer day. What oh what could it be?" I think he was playin' 'cause I think he knew all along what was different but he was just beside hisself with delight and fun. So she hugged him again real hard, so hard like she smeared her face against his so that he grew blacker as she grew whiter. "Oh, Uncle Remus, seein' you is my most wonderful surprise! " But as she was squeezin' him again and again, she looked down 'cause there was a bare rabbit with hardly any fur at all that was sweatin' and who started playin' the symbols (That's right this kind not *those cymbals*). He was holdin' two brass phallic like symbols which he clashed against one another to make a sound. It was somethin' special so much so that I almost had another organism. He did it over and over again and started marchin' around the room. Why he didn't care who was in the way. He just stepped over each and every foot just to have his way. But good ol' Uncle Remus he was happier than the sun trapped in a solar panel. He was positively beamin' even in the brightnessof his gas lamp. Yessir, it was a room all aglow with so much happiness that I was exhausted from the sheerness of it all.

So that evenin' we was so exhausted that we decided to spend the night in the barn and to go back to the coal mine later, like in a day or two. That evenin' was the most tumultuous time ever as everyone was listenin' to Uncle Remus tell us his tales of wonder growin' up in the deepest part of the south, like South Dakota and Utah where they was a heap of stuff goin' on about blackness and whiteness a goin' on all the time. And don't forget South Detroit and especially South America which really was far south but not so far south as Mexico, I think. So we listened and listened until our ears hurt. But by and by we began to drift off and to fall asleep what with bein' tired and all and our ears hurtin' from his joyous tales. And then it happened...

BEARS, BUCKS, AND BEAVERS

Sooner or later, mostly later, we all fell asleep and we were breathin' easy in the night. I was thinkin' some kinda thoughts I don't think I woulda been allowed by my mama. Durin' the night when I noticed everyone snorin' and squirmin' real gentle and all, I reached out and touched her behind for the very first time. She turned toward me and smiled them pearly whites and says, "Let's do some exploring." So real quiet, Ree-Ree and I rose real quiet from the floor by Uncle Remus' feet and tip toed out of that room hand in hand and began to explore the rest of the barn when we come across some stairs.

At first we hesitated because stairs is always kind a creaky and dangerous in the dark. Still we felt our way along a railing and made our way in the total darkness up the stairs to the next floor. Then we stopped and I felt a door knob and pushed it. As quiet as could be the door opened and still no noise because the barn was new and I suppose everything worked real fine because of it. Sos we steps through da doe. In our quietude, we shuffled our feet upon the floor and found that it was covered with real smooth and soft straw. It smelled real nice and kind of barn like in a good way, just comforting and outdoorsy but with no cow or chicken poop which can spoil a romantic evenin' without even tryin'.

Once we shuffled all the way to the right just to find a light switch. There mighta been one but we couldn't find it. So we just drew our hands along the wall to our right until we touch another wall goin' the other way. Easy like, we turned to our left and bumped into something hard. I felt it, and it seemed like it mighta been a bed and it was really hard and I said to myself, "No, I ain't sleepin' in that bed."

So we shuffled on down a little more when we bumped into another hardness. This time, I felt somethin' a little softer. A bed, maybe, still too hard for us. We shuffled on in the darkness until we came to somethin' else. A little bit hard and a whole lot more too soft but too small for us. Finally we came upon somethin' firm but real soft and spread out in what seemed like an entire square of straw. It seemed big enough to hold a whole bunch of people and at last we were where we wanted to be, alone in the dark in what seemed like several bales of hey and somethin' more. On top of what I thought might be bales of the stuff was enough loose straw we could get real comfy in. And in the darkness we committed the first great sin of my life and I was thinkin' this might be the time to do more of it.

And, Oh Boy! Was that ever a sin. Before I knew it we was both naked and sweatin and I was on top and she was a doin' things which was crazy to me. I was so desperate for her full affection I couldn't even talk. Whew! I was bein' scrumptious and she was bein' delicious and together we were makin a banana split with a cherry on top! And for the first time in my life I was harmonious with another body and I was in heaven, at least for a while, when all of a sudden, it happened. That's right. For the first time ever I had an organism with another person and not just by myself. I was beyond harmonious. I was downright flabber and ghasted all at the same time. Especially when somethin' entered the room completely by surprise and boy was it ever.

While we was a covortin' a light suddenly emerged in the darkness right next to the straw. It was a little girl lookin at us with a flash light in her hand. She was kinda short and when I saw her, I will never forget. She was wearin' knee pads and smilin' and she said, "Are you finished? Me next! " "What do you want?" I asked without knowin' what else to say since Ree-Ree was in shock and could say nothin' at all. The little girl, all feminine like said, "Fweeedom! " and waited next to the bed for us to finish. It was humiliatin' but it only got worse. But I always remember what she said and then wrote another song about that moment, "Fweedom's just another word for someone who's real loose." I had to change it a bit over the years before it became another big hit.

In the darkness, someone else came up and stood next to the little girl. She had nothin' on but a stocking cap and wrinkly old skin and eyebrows that looked like McDonald's arches. She didn't say nothin' except with her hands and they was movin' around wild like. She just made gestures without sound and I thought she must be rememberin' movies from when they had no sound. Years later I would meet her again and she learned to talk by then but I think it was too late. She was Aye-talian and they tell me a Mafia Queen. Sho' enough. But then while the little girl was waitin' for me to finish, the Aye-talian one ran around the bed and I heard her jumpin into the bed next to mine. And man, that was just the beginning, because of what happened next.

In the complete darkness except for the flashlight that the little girl had in her hand, I heard a growl from one of the beds and then a snort from another and finally a loud "smack." I was putrified with fear but still more was a comin'. No matter what I could not move. And the little girl with the knee pads and the flashlight, she ran all the way to the other end of the room and shined it on the big hard bed. And there, was sitting up just then, a great big ornery lookin' bear who didn't seem pleased at bein made to wake up. But what was even more amazin' was the fact that he had no fur. He was a bare bear (I know, I know, redundant, I get it but it was mostly true except for some hair on his chinny-chin-chin.) But then the little girl ran over to the next bed and shined it on the snortin' critter. Lo and behold it was a naked Buck with no fur, hence, the term, Buck Naked. Never before had I understood it. Never before had I ever thought about it.

Finally, she ran to the next bed and shined it on two little beady eyes that shined right back at her. Somewhere under its blanket was the little gnarly old lookin' woman with wrinkly skin and big arched eyebrows and I didn't have to imagine what was goin' on under the covers, especially when the beaver smacked its tail so hard against the old woman it threw her out of bed. I know what you're thinkin' so I'm not sayin' any more about that because of your gutter mind. And just when you thought it couldn't get any worse...

SURPRISE!

There was now noise enough to wake the whole barn when boom, the door swung open and someone did what Ree-Ree and I couldn't do. They found the lights and switched them on. Oh, the humiliation. I was still putrified and unable to do anything more than move my neck and my eyes. Uncle Remus! Justin! Otto Otto! Johnny! Mr. Nasal Voice! They froze where they stood, their eyes bulging. Beaver sat straight up in bed and the gnarly lady, still naked, looked in her compact and adjusted her hair and said, "My nephew is coming to pick me up in a bit so we can go to a nice French restaurant." She then got somethin' from I don't know where and tied it around her head like a bandana, like a Mexican bandido. She smiled at me and said, "Hi ya, Joe, I'll see you in the future." With that the naked lady walked out past Uncle Remus and the rest, singing "I left my heart in San Francisco... high upon a hill..." What a voice I thought, but the body – well not so much. Maybe once upon a time but not any more. And the little girl with the flash light, she was makin' eyes at the bear, until Uncle Remus came in and she tottled over to him and smiled broadly, as only a girly girl or full grown woman can do. Speakin' of groan. That's what was comin' out of Ree-Ree. She was in-ca-pac-itated. There I said it right. My weight was full upon her and she wanted me to get off but I kind of already had and couldn't do much else. Then everybody walked over to where me and Ree-Ree were a covortin, but not so much any more under the circumstances.

Well, by and by, they got over their shock, includin' the buck who was still naked. But then I found out why he was a naked deer. Ya see, he took off his antlers which was just a head piece. Underneath the antlers and long snout was an actual man with a southern accent. He looked at me with a smile and said, "I feel your pain,." He wiggled his hips I think a bit as he walked past as I was now startin' to recover from my state of putrification. But was still havin trouble movin' about when Ree-Ree turned her neck to look at me and she said, "Relax, Joe, and just go with the flow." I couldn't believe my luck. I was almost hallucinatin' when she said that. I was under such pressure and I erupted. It was loud and long, I had done it again. They were all stunned, but no one dare to move and the stench was God awful. I felt mighty proud. Uncle Remus, he looked like he was about to pass out. But just then, Mr. Nasally Voice walked out and sat on a bail of hay and started singin', Hey lady lay across my big bail of hay. Why wait any longer for the... "I couldn't help it, but I had to interrupt." Listen here, Mr. Nasal, I don't think she is waitin' and I got news for you. When you use "hey" and "hay" in the same line, that's confusin. Them two words form what they call a homosexual phone (I never understood that either because most homosexuals I know have the same kind of phone I do). So as well as I could I straightened him out on how to make the song better but I was in a bit of a predicament as I was still naked in front of everyone and so was my Redundant.

Even though Uncle Remus was shocked, he was kind of receptive to the little knee pad girl with the flash light and she invited him and the bear outside for a "skirmish", she called it, in the briar patch. But both of them said they didn't really like prickly things but she said she did and so to humor her they both left the room and went out in the night time darkness and nobody came til mornin' when Uncle Remus came back all by his lonesome. But I heard later on that the bear was afraid to come back for fear she would just a keep pesterin' him for more playin'. So he just hid in another briar patch tryin' to escape her. But it wasn't no use, she was relentless and for a bear there ain't nothin' for annoyin for a fella who's getting' ready for his winter time nap.

Justin was still there and he started cryin' and speakin a lot of French because he said he was a BoneyPart who conquered most of Europe and she shoulda **asked him** to go to the briar patch. Well, I suspect that if she had known about his bein' a BoneyPart she woulda, but she didn't so she didn't. Sometime the French are just too late for the fun.

But just when I thought everying was settlin' down and Otto Otto and Johnny was already outside some where lookin' for more pumpkin wheels, and the beaver had left with Ms. San Francisco, and Mr. Nasally was still perfectin' his song about women in hay stacks, I thought I could finally get off of Redundant without being entirely exposed and she and I still bein' naked and all. But as I started to get up my head hit the shelf above the bed and knocked down a can of somethin'. It fell right on me and covered my chest and legs and other parts. Oh my! I was now tarred and strawed together and Ree-Ree, well she was screamin' and cryin' Ever since that moment I hated anything' made out of oil and I heard tell that that was where tar came from. But since we was mostly black anyway, I wasn't so hurt by it and it did cover our nakedness long enough to get out of that room and into an outside pond to scrub stuff off. And it wasn't even light out yet. I hate tar and I hate oil and I talk about it a lot. Guess that makes me what they call a fossil-fool-hater. Smarter than you thought, huh?

JABBER NO. 2

TRUTH INJECTION: JOE DOES FOREIGN POLICY AND STUFF.

Hello, you don't know me unless you are what we call a foreign policy wonk and because I am a very secretive fellow who has a **TIPPY TOP SECRET** clearance for intelligence type stuff. There's a name for that and it is called a **Q** clearance. No, no, no, I am not a conspiracy type guy who thinks that Joe Biden is actually one of the lizard people. No, Joe's kinda smart but he is not smart enough to be one of them. Maybe Mark Zuckerberg is and that's evident in the way he looks and smiles (he really is a robot and I know). By the way, do you know that I helped set up Facebook. Back in the 1990's the CIA had a program to spy on Americans called Lifelog. It was an illegal program which Congress found out about, much to our regret and ordered us to shut it down. Well, we dragged our feet until our toes bled and we transferred a lot of information and a good deal of money to a young lizard type guy named Zuckerberg. Once we had it all set up, we shut down Lifelog and several days later a number of our people began working for Facebook. And, guess what, what we did illegally, i. e., gather information as we spied upon Americans -- Facebook got you to volunteer the very same information. We had been trying surreptitiously to collect that information but now we don't need to. Now you are doing all that work for us for free and we couldn't be happier. Also, just thought you should know that J. Galt, the fellow who is helping Joe write his autobiography, has a clearance just like me and that is why we keep kicking him off Facebook and then frustrating his efforts at time to publish things. Not all Q types are bad, but they are rare, very rare. But I will tell the name of one Q guy you already know and he is famous, really famous and you will be surprised. His name is Charlton Heston. Not even the Vice President had a clearance as high as his. Ben Hur was really just like Moses in that he heard things most mortals never even dream of. Well, enough about us, let's see if you can solve the next riddle made especially for you who are reading the true life adventures of Joe Biden as told by himself. Just for a minute there I bet you were reconsidering who you thought was the guy writing all that **Q** stuff, huh?

So now for all you Q type wannabees and gurus of government gossip let's play a game. I will drop a hint as to my real identity and you will with your magic cryptographic cipher rings determine who I am. It's really not that hard. Hint number one: I was once a Secretary of Defense and my first name is a palindrome. Hint number two: GAy Turtles Eat Snails. Still don't have it? Then put this book down and ask for your money back. You aren't smart enough.

(MORE HINTS) TO BEGUILE THE BIGIDEN

I once famously said that Joe has been wrong on every foreign policy decision for the last four decades. I said that before he started the fifth. How, you might ask, can one man be so wrong about so many. The answer, PRACTICE, PRACTICE, and MORE PRACTICE. Or in the words of that currently famous late night metaphysical guru of Yogi and Kundalini Yogism, "Joe does not know what he does not know and is proud of it." The man who so aptly captures the essence of sweet, old Joe is none other than Swami Suwannee from Georgia who said, "Joe did not know that the Russians who took over the Republic of Georgia in Asia never set foot in the state of Georgia. Thinking that they had, Joe concluded that Georgians talk funny because they have a Russian accent. Nyet, they do not. *By the way, Joe, Joe, over here! I assume you'll read this so I'd like to bring your attention to another matter regarding Georgia. Remember when you commented on the Swaney River way down in the Republic of Georgia? I bet you don't but we almost bust a gut laughing about it. In the Republic of Georgia, people do not sing about being "way down south upon the 'Swaney Ribber'," or even the more contemporary "Suwannee River." They might, however, sing something like, "How I love ya, how I love ya, my Suramula." So, Joe, you were kind of close, but no cigar. 'Cigar' brings up another matter but, of course, we won't talk about that in polite company. Still, I'm sure your Mammy loves you no matter what ever you're up to.*

And for other foreign policy extravaganzas, Joe, what exactly did you do when Vlad the Inhaler sucked up Crimea? Did you wave back to him when he laughed at you? He was looking to see what you would do. And when the government of Ukraine complained on the world stage, thinking surely the United States under the Barack and the Biden will do something to help them make their case, what was your response? Humm? Instead you just smiled, took your money under the table from Moscow as well as Kiev and sang Crimea River. From the Volga to Dnieper, Joe, they're either laughing or crying, but no one's respecting. By the way, when the Russians close in on eastern Ukraine, will you suddenly care about your ranchero and maybe send Migs and Javelin missiles to defend your own private turf or will you just take it as another tax write off? Gee, Joe, say it ain't so.

Remember that poem by the Ilhan the Omar? How do I love, thee, Joe? Let me count the ways. She smiled when you left thousands of Americans behind in Afghanistan while taking in her distant cousins by the tens of thousands therefore affirming her home state as a safe haven for 3M corporation: Minnesotans, Muslims, and Mayhem. But those Americans in Afghanistan, do you even know how many thousands you left? Do you know their names? And, by the way, why did your own Secretary of State try to thwart U. S. retired Special Forces and private contractors travelling at their own expense to rescue some of them? I heard Messrs. Blinken, Winken and Nod even tried to take credit for the operation after they found out that the American rescuers outsmarted our own State Department as well as the Taliban and smuggled a number of their brethren out of the country. Guess they didn't notice since they were on the phone warning the Taliban not to let the Americans land. I could go on but you know the old saying, "Brevity is the soul of wit." Since I wish to be deemed witty if not brief, let me count the ways in which you have so sorely tried the patience of your countrymen. Feel free to chant or sing along to the tune of the Beach Boys, "Barabara Ann" Ben, Ben, Ben, Ben, Ben, Benghazi, just me and Hillary, Just a bombin' and abandon, bombin' and abandon, just the squeal of the SEALs Went to Kadafi, left him for Assad, took all of his guns and left behind his bod. Yeah, yeah, Joooooooooe, that is your name. You don't love but you leave 'em, d'love 'em and you leave 'em all the same. "By the way, did ISIS or the Taliban send you a Valentine's day card this year? They should.

Oh, and remember this hit from your days as VP? "Bin, Bin, Bin, Bin, Bin, Bin laden." Fortunately, Joe, without you, the hits just kept on comin' even when you were tryin to thwart them. Even Barack and Hillary thought you were wrong.

MEANWHILE BACK AT THE BARN

Whilst Ree-Ree and I were a washin' ourselves off, somethin' really strange happened in the pumpkin patch. It was dark as my face and twice as scary. It was a moonless night so that I couldn't even see the tiniest bit of my reflection despite there bein' some light throwed from the house in the upstairs bedroom where we had been sleepin and doin' things in the straw, along with sleepin' with the buck, the beaver and the bear. I always liked the sound of that, buck, beaver, Biden, bear. Of course, there was everybody else, too.

Well, after a while, I was out lyin' in the grass which was still kinda warm for that time of the year. Ree-Ree was right next to me with her hands behind her head. The two of us was gazin' straight up at the beautiful stars which we could see so spectacular on the night on account of the moon not brightenin' them out against its own reflection. That moment with Ree-Ree was maybe the most happiest one in my entire life. Though we was still naked, we weren't embarrassed, lessin' my body started to respond to her presence. Then I would just go back into the water and cool off. Some time I would even bob for an apple or two what had fallen in from a nearby tree. But I could never catch one until she walked up to me and handed me one – the apple that is. I took of it and did eat. Whew! Boy! Big Mistake!

At precisely that moment when I did eat of the apple whereby it was given unto me by the woman, my little Ree-Ree, lo and behold a headless rider did suddenly appear with the sound of thunder comin' from the hooves of its horse. Whilst it didst whinny and neigh, I didst hide while about chokin' on that bite of the apple. By the way, up until that point I didn't have an Adam's apple, but after that I did even though I only took the one bite. Strangely, she did not have an Adam's apple even after that. Up until that point, even though she had leprosy and all, I had forgotten and forgiven. But then, I knew she had brought the wrath of orange headed pumpkin creature upon us whose body was headless except for the one he didst carry under his arm. How a body couldst navigate like that whilst holdin' his head under his arm was beyond me. I was scared, especially when the horse started walkin right into the pond, knee high, and the pumpkin with its glowing eyes and jack-a-latern grin was spittin fire and fumes like a volcanoe. I was so scared and Ree-Ree was too, but I think I was the worst for it. I was trembling when the pumpkin head looked right at me and said, "Didst thou eatest of the apple which I have forbidden?" I tried to speak but a bite of the apple was stuck in my throat when all of a sudden the rider dismounted with his head under his arm and stood smack in front of me. "Turn around, you jackass," saith the pumpkin head creature, and after I did so and was about to die from near choking, he didst kicketh me in the ass and then cracked me on the back. I spit that apple right out so fast it hit poor Ree-Ree before she get out of the way. Right in the eye (I know, I know, she had two but I'm only talkin' here about the one that got in the apple bite's way).

As I turned, there was a tear comin' down out of each of my eyes, not just one. Ree-Ree, she was cryin' again, out of both eyes even though only one got hit. It is peculiar how nature works against the commonist of sense. He was a lookin' real stern and he says, "What is your name, you fuckin' son of a bitch?" When he cussed at me like that, at first I was relieved until I looked real close at the pumpkin man's face. Not only did he have that God awful grin but he also was the onliest pumpkin I ever saw that didst haveth orange hair, just like his pumpkin skin. I knew, I knew, I knew for some reason that that grin and that orange hair would cometh to haunt me some day in some way I couldn't not yet figure. But then I was still relieved because now I knew he wasn't the one they call Annointed and even I knew what that meant. Still, he didst curse me by sayin' "Because thou hast eaten of the one damned tree in the entire pumpkin patch I have forbidden, thou shalt worketh all the days of your life as a politician and be slimed and slithered by the corruption that greases thy palms. Hence forth because thou hast acted as a jackass, thou shalt be eternally associated with that symbol. And, you, woman, or girly girl, thou shalt be forever accursed with acne as thou gets older and so shall your beautiful hair have unbearable knots in it when thou goest to comb it or even brush it." I guess from that moment on, I could not elude my fate, try as I might as you shall soon see. And to make matters worse, he just didn't seem to want to leave. I was thinkin' what else is in store?

Well his flamin' eyes look right at me as he asketh this question: "Boy, what is your name?" Scared as I was, I still was manly and spoke right up, Joe Biden." "What kind of name is that?" He just stood there with his flamin' eyes and orange hair flowin' in self-propelled audacity, until I saith, "My name is Joe Biden." "Say it again," he saith unto me. "Joe Biden," I against didst say. "That is what I thought you saith, Joe. And I don't mean to be prophetic, but, boy, since you are Joe B, I shall just call you Job, and, son, that ain't good." At that point I was just as scared as ever and immediately I became afflicted with the leprosy again with the whiteness tinglin' all over my skin and my feelin' kind of wouzy from that creature callin' me Job Iden. I didn't like it much, but then again, I guess Ree-Ree didn't like it much bein' called by a name she never invented or even said she liked. But maybe Black Cinder wasn't the best name for a girl either. So there we stood with the chewed apple floatin' in the dark water of the pond, the headless orangey headed creature man splashin' out on the his horse and me and my onliest real friend. Yep, it was just Job Iden and Ree-Ree from then on out. Scared though we were, we again lay down in the still kinda warmy grass and fell into a deep sleep and I went into a fitful dream...

JOEB THE JOB KILLER

First thing I noticed in my dream is that I knew I was dreamin' and I wanted to wake up from the fright that was envelopin' me all over but I couldn't 'cause I was paralyzed. In my dream, all around me, was great big signs that said, "Joeb the Job Killer." Even though I was dreamin' I knew this was not good and not likely to end well. Then, while I was standin' paralyzed a fellow with bushy longish white hair and cloven feet and a funny woman with big eyes whom he called the Ass of Congress, came upon me and touched me with a long pitch fork. "Hey, Joeb," he said, "You, you wanna be a good man and be poor all your life or a rich dude with young hotties surrounding you and constantly trying to pull your pants down?" Before I could say a word, "I heard the sound of thunder off in the distance and I saw "The Big Fella," the Real One, and I was once again, plenty scared. The Real One was leanin' against the hughest oil derrick I had ever seen and he was a vulminatin' like a creature in the Good Book, because I had the sneakin' suspicion he may have actually wrote it.

The Big Fella was vulminatin' like a volcanoe and walkin to and fro like he was in a huff about somethin' when he all of sudden stops and sees me. He up and spoke directly to me and he smiles and says, "Joeb, I have told everyone about your goodness and your manly religiosity whereby you go around once a year with ashes on your forehead for all the people and the cameras to see." I did not know whereof he spoke at the moment but I had the feelin' he was prophesying an projectin' what I would someday do.

Then the white haired old guy with the Ass of Congress by his side and some other woman dressed in a long table cloth or somethin' and a brown face said, "Don't listen to him Joeb, he's the devil, he's just like that other devil…" He did not mention who the other devil was but I suspected I was soon enough gonna find out, or maybe not.

Then the Mighty Big Fella spoke again and he was vulminatin' right at the old guy with his crazy hair all white like a smoke from a dyin' fire. "How could I be the devil, you maple suckin' huckster of fake frugality and socialistic impotence? Huh? Huh?" Right then and there I might have sided with the Big Fella had I not been distracted. Perhaps seein' I was wooblin' a bit and thinkin' about goin' over to the Big Fella, the Ass of Congress, sidled on over to me and whispered in my ear (Yes, I know, I have two but she only whispered in the one, hence, the term ear and not ears). She says, "Joeb, would you like to mate me, I'll give you a treat, you know, your very own Joeb Job." I was smitten then and there until I looked at her friend with the long table cloth and like a man wobbly and horney, I said, "No offense, but I like her. She got my eye." "Well, you can't have her because she's already married to her brother and she comes from the land of 3M." "3M?" I asked. "Yeah, Minnesota, Muslims, and Mayhem. But we call her the Somali Walli, which is short for Walleye." "Why's that?" says I. "Well cause she was trying to get elected in a state with a lot of lakes and fish. So if you say you like fishing even if you don't, you get elected there because they have a lot of Swedes who don't read and stuff." I can't pretend that I understood everything she said, but I was hooked, and like a fish out of water, I just was hopin' to flop all over her. And the next thing you know, that's exactly what we were doin'. Ooooooeee! Goodbye, Jill, all over again and goodbye… I can't remember even now, what's the name of my leper girl?

At that point, Mr. White Head, as I took to callin' him, comes over to me and says, "Because you have defeated temptation, you are now one of us." I felt really happy, even though I was kind of embarrassed again 'cause I was once more naked and lookin' like a stump with a trunk growin' out of me after all the pleasuring the Ass Of Congress provided. He must have known what I was feelin' cause he says to me, "Don't worry, Joeb, I got one myself. What she gives, I receive and I went from bein' a Meanie to a Greenie. And that, as you shall see, is how our government works."

I thought about it some more and while I was still thinkin' Mr. White Head came back to me and said, "Oh by the way, because you scared off the Almighty, from now on you'll be known as the new Big Guy and you'll never have leprosy again. And with that he waves his magic wand which looked like a large wooden table spoon and he hit me on the head with it three times. When I came to, my leprosy was gone. And get this --I was now completely black. Even my hair was kinkafied. I was an all-out purified, rarified, Afro-American oreo, white on the inside but black on the outside. I was so proud. I swear, even my muscles were bigger and, of course, I was happy and simultaneously oppressed. From that point on the people all around me called me Bro Jo, which was lot better than JoeB the Job Killer. I never liked that name because it made it sound like I was contemplatin' suicide. 'Cause, you know, why would a JoeB like me want to kill a Job just for mouthin' off the Big Guy, which I now was. I was not, not, I tell you, contemplatin' suicide because, even then I knew from bein' a coal miner and a sometime farm boy that "sui" (pronounced sue we) was a term people used when callin' pigs. I liked pigs and lard and all, but I certainly didn't want to kill one. Still, I must admit, they did taste good.

But that night, because I was a new black man, I slept like a baby all by my lonesome back in the straw all by myself on the second floor of the barn. No one else was there. Even Ree-Ree was gone and I knew my experience was not a dream because my real identity had been affirmed. I was now black and I was as proud as a crow or a raven or top soil. And though I was sleepified most of the night after that, I did have a dream, a real honest to goodness dream which almost scared the tar outta me and given my relationship to it and my embarrassment just a short while ago, I wasn't apt to forget.

I once was a slave who dropped the "e."
So now I'm a Slav
Who works for me.
The Russki they love me
For the way I play.
But Joe is a wuss
Who played with power,
Now he's the slave
Of my golden shower.

Vlad the Bad

JABBER NO. 3

A MUSICAL INTERLUDE

TRUTH INJECTION: In the course of assembling Joe's notes for this book, we thought we would invite certain celebrities and well known political figures who have worked along side Joe to comment on the virtues of the man. One of them offered his commentary, unabridged, and uncensored so long as we promised to print it no matter what. We agreed. It may surprise you, as he surprised us. It not only speaks to who Joe is but also to the man who takes his measure from the virtues and talents of Joe. To compare himself to Joe might be downright foolhearty. However, since his commentary is in another language, you will need to press right **HERE** in order to get the translation. Unfortunately, I myself am not able to read the language. Incidentally, some may think it risky to allow foreigners to speak plainly about Joe's influence on them. We here at Delphi Oracle Press do not. After all, the man told us that he wished to provide an original musical composition that would enlighten the whole world as to those characteristics that Joe so consistently exhibits. Since we know that the man recently learned to play the piano himself and since his favorite tune is Fats Domino's rendition of *Blue Berry Hill,* we knew then and know now that it will be a work of art that graces us all with creativity, beauty, and respect for the man. And now, for our special guest writer and musical interlude entertainer, Mr. President of Russia, the one, the only, Vladimir Putin. And, as they say in the caged ring, "Let's get ready to rumble!"

Vladimir Vladimirovich Putin: Thank you, thank you, or as we like to say in my country, "Eat dirt from a dead man's grave." Here is my totally original song, originally composed by me from handwritten notes from beloved Russian composer and choreographer – George Balanchine, who is not only a great Soviet composer and musical type teacher, but also a dirty, rotten traitor because he fled mother Russia, like the lousy people of the Republic of Georgia where they do not sing about the Suwannee River no matter what Joe thinks. I hope you like my song and remember how much I like the music from mother Russia and not the American state of Georgia where an American Jew named Irving who was named after the capital of that country of Germany, stole this. His parents must have been stupid and that may be why he was always stealing Russian music like White Christmas, Moon Over Moscow, and Kiev is My Kind of Town, and, also, Yankee Doodle Dandy. (our most patriotic Russian song about noble Belorussian Mother who wave her hanky at son leaving for war which she make out of knitting needles and she carry in her hand. Originally called Hanky Doodled Handy. You see, you steal everything from us, even Alaska.) For you and all America I sing my new song.

PUTIN ON THE BLITZ

I am Vlad// the big inhaler
I suck in air// and you grow paler,
If you cross me// with your rancor,
I'll invade your country// with my tank Corp.

The Ukraine girls// they knock me out,
But you dear Joe// you got no clout,
Wack a do, wack a do and I'll
Wack wack you.

You know you got your trillion dollar navy,
Navy.
What good is that when you're the captain
All you'll do is make it happen,
Happen.
Hmmmm,
I am Putin, Putin on the Blitz.

Well you got billion dollar sanctions,
Ummm Humm
But we got land with lots and lots of tanks in
Uh huh.

We got cannons,
You're the fodder,
Bring your troops
And we got slaughter.
Here and there,
Bodies everywhere.

You may be thinking about your NATO,
Way Ho!
What good is it if you're hung low,
You, Joe.

You are fool and we see it,
Face it Joe, tit on a bull.

Now you may be thinkin'
You're such a trooper.
But we know
You're just a pooper.
Soil every where.

We got film
And lots left over,
Called you names,
Even Rover.
Poopin' Everywhere.

So that's the tally,
Now let's rally,
Migs and Mayhem,
Murder everywhere.

So in conclusion,
If you want to,
I can pester
Then come haunt you,

'Cause -- I'm Putin on the Blitz

From Russia, Your Pal Val the Impaler XXOO

Remember what we say, "If you go Russian you always come back – dead."
Have a nice day. Tanks a lot.

We want to offer free cosmonaut training to your Elon guy in our space
station. We will make him an ***astronyet! We make him go KaPutin!*** HaHa

EDITOR'S NOTE: Musical interlude is over. Keep reading.

I DREAMED A DREAM

IN THREE PARTS

PART 4

First I want to apologize for bein' redundant. See, I told you I knew what I was doin', I did learn it's true meanin' 'cause I'm Joe and, yes, I oughta know. In the midst of dreamin' I was lookin at a tall, tall building (I know, I did it again), a skyscraper (that's called threedundancy). It was so tall it was blottin' out the sun when I looked up at it. And then, by and by, I discerned that I was in a city called New York and I was standin' on a busy street on great big letters that said BLM, which I thought meant "Biden's Lies Matter." How I came to lookin' so disparagin' of myself, I do not know. But as I was lookin' up a city worker who said he was from the department of sanitation walked up to me with one of those electric sniffers that follow a smell. He was standin' right next to me and he said, "Hey, buddy, get outta the way, we got a complaint about an awful smell somewhere around here and you seem to be standin' right on the spot. So I moved, but everywhere I moved to, he kept a followin' me with that electronic sniffer until finally I stop again and so does he. He wrinkled up his nose and stares at me and says, "Bowels Loose Maybe?" He then gagged and put a towel over his nose and walked away. I found out that I was no longer a kid but was prophesyin' just like a full grown man.

And in the dream, I was no longer Mean Joeb Green, but a white guy who mumbles (WGM).

Then while I was standin' there, I saw the pumpkin headed man got out of a limousine right by me. He was no longer carryin' his head but was instead wearin' it on top his shoulders just like a real person but his orange hair was still flowin' from his own audacity even though there was no wind. As he is walkin' past me into the big building, he stops, looks at me, sniffs a bit and says, "Hey Joeb, you're full of crap and get off my street. Hey, and when you do, stop by Danny's Diaper Depot on the way home. You need to, here's twenty bucks." For a scary pumpkin' headed orange guy, he was all right. I watched as he walked into the building like he owned it or something.

Just then, a black guy who was also kinda white, who was yellin' about not bein' suicidal, came up to me, and handed me a rope and check for $5000. He sniffed but didn't run away, but instead he says, "I need a black guy like you, to beat me up. First, hold my subway and then hit me." I didn't know what he was up to so I just ate his sandwich while he just stared at me and shook his head. "Gimme my check back and I at least want the other half." He grabbed both and walked right into an open sewer near by and started yellin' but I didn't much care about that. But I did not understand how he could see me as a black guy when I thought I was again completely white and no longer Mean Joeb Green. But then it hit me. They were both wearin' really expensive lookin' sun glasses. I thought to myself, maybe the glasses make a body see what they wanna see. I knew then I was onto somethin'. The glasses. By the way, it wasn't long after that I went to see the doctor about the smell that kept following me around. You know what he said? "Joe, you're incontinent!" Well if that don't beat all. "Everybody's incontinent unless they're from a big island like Guam which some really stupid people in Congress think can flip over. I'm American," I said proudly, "like in North America, of course I'm incompet – incontinent. Sheesh."

14

Then while I was lookin' up a wrinkly old lady with those funny McDonald's arches stood next to me with a hammer and her nakedness and hit me right in my knee cap and then bit right into with her voraciousness, sayin', 'I love you, Joeb, you Job Killer you.' Suddenly, I went blank. I guess I was havin' a white out, or a black out. These days I'm not sure.

When I awoke, I was standing in an orchard entirely naked with my son Hunter. He was naked too and surrounded by little girls wearing the Ukrainian flag, all blue and yellow. They was hopin' around and saying stuff in Russian which I didn't understand until a beautiful woman walked up to me and say, "You luke like something da cat druug in. I like cats, I one." I didn't know the meaning of her words exactly or what she might be hinting at as I was contemplatin' her attire. She was wearin a negligee with holes all over so that I could see everything and she was wiping honey all over her body underneath that negligee thing. In her other hand she was carryin' a potted plant. I thought there might be some cymbalism there and I am almost certain I was right. Because she grabbed my now enlarged magic wand and says, "Take me to your Ukrainian ranchero and pay me lots of money for your energy which you will give to me. Hunter too. Maybe you adopt me and let me ride on your ranchero and I give you sex massage. What you say, Big Guy? If you say no, I think maybe you wussy man." She paused real long and looked right at me and smiled sly like. She stopped rubbin' the honey all over her and put the pot down. Then she grabbed a jar from a table which miraculously appeared next to her. But she dropped the jar and the glass shattered and there were beans everywhere.

I read a book once by a Swiss type guy named Karl who said that people in dreams are cymbals. But at that moment, all of them were clashin' in my brain and I couldn't make heads or tails about anything, so I fell asleep right in my dream about dreamin'. Life is so complicated and Hunter wasn't sleepin' though and when I saw him next in my dream he was covered in honey and had some flowers in his hair and beans stickin' to his behind and back. I knew what beans could do to a man, so I told

Hunter, "You better leave this place where we are before you relax too much and we find ourselves snortin' somethin' more foul than parmesan from a dirty old carpet." But he just took a path right in front of me and disappeared below the water in a bath tub until the tub overflowed and magic rubles and a thing called creepy cryptos started over flowin' which said, "His bod we hushed" on one side, and had a picture of a guy named Epstein on the other. It was confusin but she was still there holdin' on to me like can only happen in a dream. Like I say, I believe in cymbals but this time they were really clashin'. I don't get it.

11

Then as the dream about dreamin was fadin' she put on her sun glasses and says, "I never been with black man before, you must be Mean Joeb Green and your kid here is Cocaine Kid, I hear of him when he come to Ukraine to get special massage." Then she said somethin' real strange. "Joeb, you are Dyslexic, I am told by your country men, which explain a lot. So read my special message for you and good-bye or you give me what I want. Joeb, remember we have cameras everywhere even at your ranchero, Mr. Beaner guy." The words was strangely powerful but I didn't understand a one and Hunter wouldn't help. He said, "Come on, Big Guy, you know you don't know about my business dealings, that would be wrong if you did. Hey, pass me my pipe." I did as he asked and read her message which I had to descramble because she said she wrote for someone named Dyslexic. It said, "AM is for mother, SI is after was, and RAB is for drunks." I knew there was a meaning there but even a Mr. Dyslexic might have trouble understandin' this until suddenly it hit me! The cymbals were really clashin' now and I knew that Swiss guy was right, cymbals in dreams are all about the dreamer and I was that dreamer along with Hunter. Whooo Weee! It was tough but then I got it and I was exhilarated. I looked down at what she was still holdin and said, "I think I'll take that massage now." Afterwards she gave me her business card, it said, "For a neurotic massage, contact Titiana, and I will take your energy and pump your oil. For this you will pay me in American dollars and much love. Just say Da, Da, Da, Da, Da, and smile, you on KGB Kamera. XXXXXXrated." I love havin' my picture taken even in dreams.

When I was writin' this part of my dream it was a long time after my dream and was havin' trouble remberin' it all and so I started takin' large, large doses of vitamin B after someone told me it would help me recollect what I had been dreamin. But then I took so much of it I had to see a doctor because I was havin' headaches, nausea, diarrhea, clyptomaniacus, pregnancy, menopause, feelings of abandonment and the worst of all was Nero-opthamy, or somethin' like that. The last one was the meanest of the lot because it caused all sorts of tinglin' stuff that made me just want to have sex and eat ice cream at the same time and to pour whip cream into my ears to block out the painfulness of hearin' my own screams. And I was growin' weak in my legs and arms and most especially my hands. It was so bad I had to even sit down when I was standin' up. And the worst of the worst was the dream came back all jumbled in the last part.

The worst part is I was sittin' down behind this big desk in a big white house which had lamps outside that kept showin' different color lights against it and a little girl in braids and stuff from Kansas, I think it was, showed up with a dog and two ruby slippers which she stole from a lesbian witch who she killed when she came in 'cause she wanted to take her rainbow. Or maybe somethin' like that. Well, I was gonna arrest her or somethin' in my dream but when she told me she was from a place called Guatemala or Kansas, I get 'em mixed up sometime because they're so much alike—well I couldn't bring myself to do it. She said she he had a present for me called fen-ten-all. I wasn't hungry, so in the dream I give it to my little friend the Kamalean but it didn't do her no good. I think it wasn't havin' no effect on her cause she said she was a woman of color and she liked Chinese stuff like great walls, chop suey, fireworks, and Uyghur meat balls which came directly from a place where they sold 'em only to rich Americans. The guy who brought them to her was named Fart Face and he had a girl friend name Fen Fen or somethin' like that who sold the meatballs in America and then left in a hurry, I think, to get some more Uyghur meat. She said it only could be found in her country of China in certain concentrated places. That's all I remember, but I think that might be enough. I still have the diarrhea and a general looseness. And that person that I was callin' Kamalean was being called VP as well. I was told it stood for Virgin Person, kinda like Jesus' mom.

I never made a mistake in
grammar but one in my life and
as soon as I done it I seen it.

Carl Sandburg

JABBER NO. 4

TRUTH INJECTION: JOE DOES GRAMMAR.

Perhaps it is best as we get older and more stubborn and set in our ways that those who remember us most fondly are those individuals who first became acquainted with us in our primal innocence. Certainly, Joe's grade school teacher qualifies as one such individual. Fortunately for us, it is she who reached out to us to give a glimpse of the great man's immersion into the world of childhood academia and pedagogy. She seemed breathless in her anticipation of relaying to us her take on the man who now governs the country if not the entire world. As Saint Thomas Aquinas once said, "Give me a child until he is 4, or 5, or 6 or more and that child is mine forever." Well, because of Joe's uncanny ability to see the world in slow motion and thereby comprehend its intricacies in fascinating and downright surrealistic ways, we could not ignore her offer. She extended a helping hand in our attempts to understand the legend that is Joe Biden. In the letter to us she referenced a common adage among those of the teaching profession. She said, "As a teacher, I can honestly say as the hundreds and eventually the thousands pass by my desk over the years, you only remember the really bad ones and the really good ones. I leave it to you to recall which of those two categories Joe fit into. I will have my say no matter what! " Knowing full well what that category will be, we responded unflinchingly with the cry of "Thank you. Fire away!" Below you have her commentary word for word as it was eventually relayed to us through a zoom call just a short while ago. Like you, my readers, I will react to it with you now as I wish to provide an unvarnished assessment of no doubt flattering remarks. So, shall we? Together in love and affection for the Joseph of the variegated achievements. We started it all with one simple question: What do you recall most about Joe, my friend? Incidentally, Joe tells me you are the teacher he remembers best and the one who had the greatest influence on his command of the English language.

REMARKS OF JOE'S TEACHER, MRS. GRAVESTONE: "Son of a bitch, if there is any such thing as a place between a rock and hard place, it's got to be Joe's head. It's denser than pencil lead, thrice as brittle, and as erasable as a blackboard. My God, My God! Why didst thou desert me? The agony of his presence supersedes that of any and all my other students combined. If he had joined the Hitler Youth as a toddler, he could have destroyed the Third Reich. If he had marched with Mao on the thousand mile campaign, Mao would have died of a coronary. If he had worked for Stalin, he would have called off the purges over Joe's urges to diminish all knowledge to one formula best summarized in the expression, 'I don't know and neither do I.' Joe could confabulate anything and all things into an eternal quest of nothingness.'"She paused for a moment, and took a deep breath as she asked, "He's not still alive, is he?" Upon receiving a response from a member of our publishing staff at the Delphic Oracle, she sighed, and said, "I guess God sees everything and I couldn't help myself when I kissed that stranger who entered my bedroom that one night and my husband was away on business. Maybe there was some heavy petting but only for a minute or two before I came to out of a deep sleep. Besides, how was I to know it was Joe. 999 times out of a thousand, anything naked in my bed would have been my husband. He was only eight at the time. Joe, not my husband. Who would ever suspect? I mean what kind of little boy goes around sniffin' full grown women in their nighties. Well, I been flagillatin' myself about that ever since he appeared to have gone to the wrong house in the middle of night while sleep walkin' which Joe could do even in daylight when the sun was out. But, by damn, no one woman should be punished for eternity for such a sleep-deprived indiscretion. I am punished, O Lord, and I wish you would take me now to escape the wrath of stupidity that has plagued my life since Joe came into it." Here Mrs. Gravestone paused to catch her breath.

"Well in reference to this child thing, the Joe, I can think of a line in a play I once read about an honest woman about whom her good husband says, 'There are them that cannot sing, and them that cannot weep—my wife cannot lie.' "Well, it was true of me. I could not lie and I confessed to my husband when he came back from his business trip. He fell on the floor laughin' so hard I could hardly believe it. He had tears comin' down his face and then, and then -- he cramped up all of a sudden and keeled right over from a heart attack. That little son of a bitch killed my husband by sleep walkin' right into our happy marriage and destroyed it. I've mused upon that line from that play quite a bit over the years. It's true, some, they cannot weep, but not me, I've been weeping since *it* came into my classroom. And singing? What for? Until the angels call me home, there'll be no singing in my life and I am going on 97 and I hope it doesn't last much longer. He's really not still alive, is he? I mean all that stuff about him on television and all, they're all the results of Hollywood graphics and special effects, right? If nothing else, after all, it is fair to say that Joe is the prima facie evidence that Father Flanagan was wrong when he said 'There's no such thing as a bad boy.' Yes, there is. Forgive me for a moment while I gather myself and take a nip (She snapped back her head as she downed a shot of something). "Ahh," she sighed, "Johnny Walker Red, it can sometime kill the torment, the pain."

A NOTE FROM THE PUBLISHER:

At this point in this most vivacious recollection of Joe's exploits as a boy and young man, we paused to prevent Mrs. Gravestone from exorcising her own personal demons that might later come back to haunt her. We were concerned for her well-being and feared that she was hallucinating. However, a day later, she called us again and we made arrangements for the second part of her commentary which we include later in this our most cherished publication, the life and the legend of Joe Biden.

JOE C'S THE I DOCTOR

More of me by me in my own way.

- Joe

The next mornin,' I awoke from my startlin' dream about a dream to find the pumpkin patch empty, and completely bare. The pumpkins were all gone. And, the barn, as I was soon to discover, both at the one gate and the back gate, both were deserted. Maybe, I thought, I'm seein' things, but then I thought about it and realized I wasn't seein' **nothin'** and that was a bigger problem than actually seein' **somethin'**. But I wasn't scared or nothin' cause it was a nice day and I thought I would just go to south Detroit. Yep, I could wander about kinda poetically and just – just avoid trouble and find me an eye doctor to make me more perspectivous so that I could see more of what wasn't there where I was or where I was movin' to. So, you know, I wandered lonely as a cloud until I upped upon a fine eye doctor, or so his picture window suggested, two big eyes on an owl and no glasses. This eye doctor, I thought, must be fine, so I walked in.

No one was there exceptin' two sets of twins who were sittin' kinda quiet, waitin' their turn. No mamas or papas to help them. The peculiar thing was both of them had bright blue eyes like the sky but they was much like me when I was black, like right then. I waited a spell and finally the doctor came out and waved the two children into another room which was partially open and I saw a nurse in there who had a familiar accent. She may have been from Alabama or far, far South Detroit. I say that mostly because I saw she was white and she was carryin' a not so concealed pistol on her hip. It was a lady's pistol, I could tell on account of the frilly pink handle which shootin' people call a grip. So, like when people, say, "Get a grip on it," I don't like it 'cause I think what they are really sayin' is "Shoot somebody." I don't much care for that unless it's someone I don't know or don't see often. Then I don't miss 'em when they're dead.

"What's your name?" he asked.

"JoeB," I replied.

He looked me over and real serious he asked, "JoeB, are you a Jew?"

I didn't know, so I said, "Maybe, is that a kind of sneeze or maybe a chewin' gum for black people?"

"Humm," you might just be one because you are engaging in the kind of trickery and off putting finesse that answers a question with a question. And by your coloration, the only real question might be, Sephardic or Ashkenazi. Which is it?"

I was stumped. To be black or a Joeb Mean Green was one thing but now to be called a Sephardic which as near as I cold tell was a kind of angel, well that was goin' too far. It might be what Mrs. Grave--, Gravely, Pavestone, nope, that woman with soft skin and lavender dresses with rosewater on herself—it might be what she called a blasphemy. I was no angel and I certainly didn't want to be a Ashkenazi, 'cause I grew up knowin' that the only good "nazi" was Deadenazi. And that is all I'm sayin' about because I didn't exactly like the way this conversation was goin'. So I just stood there and didn't say a thing.

He stared right at me and squintin' as if to pull somethin' outta me, when he smiled a little and says, "That's strange, you have Teutonic features, black skin, you're free of leprosy, and blue eyes." "You, my friend, are an anomaly."

I was relieved as Anomaly was some one I liked because he was my favorite author who had written so many books that were by a secret guy who only had one name. He was like Zorro or Soros. They both did a lot and left their initials wherever they went. I didn't know Zorro but I sure knew Soros. As I always said to myself, I knew about him because I read he put an "S" on things and then took all the people's money and pretended not to be what the eye doctor said was a Jew. During the Great War of the by gone era, he must have been what they called an Ashke*nazi* because he liked them so much he use to work for them. That was the day I learned a secret. You see when the eye doctor was through givin' me the exam, he went into a back room where he had a bunch of skulls and I heard him talkin' to himself. I peaked in and he was lookin' directly into the eye socket of a skeleton and he was sayin', "Alas poor Soros, I know him well." Well I didn't but I guess he did and, well, I did learn something very valuable. If you ever need to learn somethin' just remember this wonderful phrase, "Ashkenazi."

As I was about to leave, the eye doctor called out and said, "Wait, you came for good reason. Don't you want the special glasses I make for you?

"Oh, sure," I said, "I about forgot."

He sits me down again in his office and he says, "These are special glasses I myself have made. They remind us all that we only see the world as we wish to see it. For example, I do not believe you are a true Negro, but you have suffered some trauma. As a result, you will be black only for a certain period of time and then you will revert to your true Teutonic form. So tell me, what color am I?"

"Well, I think you're white. At least you seem that way."

"You may be right but you also may be wrong. For I think with my special glasses, you see me for who others think I am. They say I am a monster who does bad things to people even children."

I was quiet as I contemplated what he was sayin' and now I was wonderin' about the two little sets of twins he had sent into one of the other rooms with the nurse who had the not so concealed pistol on her hip. And at that precise moment, they both walked out and looked my way, and, glory be, if it weren't peculiar. Both sets of twins now had one green eye and just one blue. Never had I see such as that before. All of a sudden I didn't like the eye doctor so much any more and before he could say anything to me, I thrust the glasses onto my face and looked at him. Now I saw him as a snake, a snakified eye doctor who I suspect was doin' things he oughtn't. They twins weren't cryin' but the area around their eyes was bruised blue and puffy. So as I was lookin' back at the snakified doctor, he smiled and said, "See they really work. Never lose them. You look sharp in them, aviator glasses with the magic lenses."

I was havin' trouble bein' polite at that point but I thought he had given me the eye test and the glasses for free so I should at least be thankful and forgivin'. So I reached out real manly and shook his hand and said, "So, sir, what is your name?"

"Joseph, just like you, JoeB. But my friends call me JM." He was silent for a while and then said, with a peculiar accent which I suspected may have been that Teutonic thing he was talkin' about —he said, "Remember, vee hav vays to make you see. HaHa." It was too strange for me and I got up to walk out when he said again, "Remember JoeB, never lose these glasses so that you can see the world both the way you wish it to be and the way it truly is. *Auf wiedersehen, mein kind.*" After that, I ran all the way back to where we had been buildin' the auto-mobile.

I loved teaching except for my students.

Retired teacher

JABBER NO. 5

TRUTH INJECTION: MRS. GRAVESTONE RETURNS OR THE
UNBEARABLE BRIGHTNESS OF BIDEN.

EDITOR'S NOTE:

**After Mrs. Gravestone's peculiar hallucinations in her first
zoom meeting with my staff and me, we thought it best to
allow Mrs. Gravestone a period of grace. In this way she could
better gather her thoughts concerning her encounters with this
man of inestimable talent and generosity, her former pupil, a
godsend for any teacher. She's back with more truth and more
wisdom than we have a right to expect of anyone. These again
are her uncensored remarks.**

"Let me begin by saying, he was just a boy when I met him and generally
the kids at his age can be sweet and kind. Let me stop right there. I want
to be kind. I really do. I really, really, really do. (Here she paused for
a whole minute during the Zoom meeting and she clutched her hands
together in a sort of prayerful gesture even as she looked upward and
whispered something to herself and, I assume, to some divinity). Then
she looked straight at the group of us gathered on the Zoom and said,
"Well, well, I … I try but I can't…He, he…broke me, he broke me…He's
the devil! I hate him! I truly, truly do. My God! How can a student, and

in his case I use that term most loosely. He's a pupil at best but never a student. He may have been trying but he was trying beyond anything or any student I ever had. And when I say "trying" I mean "annoying" and "frustrating," not like some one making an actual effort. Until I met him I use to teach my science class that the densest metal on earth was **osmium**. Then I met Joe. I have an enlarged photograph of him made to fit my dart board with his nose right over the bull's eye. Right under that I wrote the word the word ***jomium***. No, no, I didn't even capitalize it. I want him to (she uttered through clenched teeth)..." She couldn't finish. So she started afresh.

"Who forgets his teacher's name over the weekend? Week after week after week after week. And he's not embarrassed. And after that incident in which he killed my husband he use to come up to my desk pretending that he wanted help for some assignment. While I was trying to explain something to him he would nonchalantly begin to hump the corner of my desk. Hell, all the girls in class called him "Big Dog" because he sniffed everyone of them and then, if he were standing behind one, he'd start sniffing their hair and I know it didn't stop when he got to the White House. When the girls would play on the monkey bars outside, instead of joining in, Joe would stand straight underneath the parallel bars to look up the girls' dresses. I hate him I tell you. I absolutely hate.

When he was here, I use to teach a whole section on word etymology. The kids in class thought it was fun. They liked learning Latin phrases and things like the origin of their names. But not Joe, he never understood what we were doing. When I said we need to keep in mind the influence of Latin on our language, Joe thought we were talking about Mexicans, or as he called them. "Beaners." He would make fun of the boys calling Jose and Miguel "Beaners with wieners." We had a girl in class from El Salvador named Chiquita. He wouldn't leave her alone and he kept trying to get her to eat his banana. I knew what he was doing but his mother denied there was a problem and his father was a drunkard. So, oh my, oh my, oh me."

"So one day just for fun I had my students and my one pupil watch part of the movie *Mary Poppins*. It was to have been a fun day, a reward for all the boys and girls, except one, who worked so hard to pass the state exam for their grade level. At that moment, of all the moments in the year, he wanted to talk about Latin and etymology. At first I was just stunned. Could Joe really be trying to learn something. Well maybe, but instead of understanding the appropriate application of Latin and Greek prefixes, suffixes and a few base words we still use today, Joe insisted that we parse the word – ready? Ready for this? He said, "Supercalifragilisticexpialidocious." "'Joe," I replied, that is not a real word. "Yes, it is," he insisted. "I saw it and I heard it in the movie." Since Joe was, as always, disrupting our class, I relented and said, "Ok, Joe, explain it to us." He did and I will never forget the moment. He took the word apart syllable by syllable. Never had I seen Joe so engaged. Syllable by syllable he explained the word's true meaning, or at least he tried. He said, the meaning of the word is "Extremely soft milk comes from hostile cows."' Unfortunately, Joe was kind of right and he made, at least for the day, a real impression on the class, especially the girls. Oh, young girls can be such fools. The result was that the sniffing got worse until the older brother of one of the girls told her big brother about it. He broke Joe's nose and I was so happy! Even now the moment brings back such happy thoughts. Then, when I had him in the 6h grade, he ran for class president and he won. How, I do not know. Something, to this day, tells me it was not a fair nor honest election."

"Joe was constantly getting one subject mixed with another. When I taught him and the others about equations, how hard can it be? For example, 2X + 6= 12. How hard can it be? As always, I insisted that they solve for X. Isolate it I said. With every problem he would begin throwing out the letter entirely. It didn't matter if we were solving for Y, X. A, B and more. He would just toss the letter out and so such a simple problem, 2X+6=12 became for Joe 2+6=12. So, of course, he forever got it wrong."

"And when we did geography it was just as bad. One day Joe was standing in front of the class spinning the globe around and looking at all the different countries of the world. For a moment I was thinking that maybe, just maybe, Joe would take an interest in other cultures, other climates, anything. I was wrong, for Joe so misinterpreted the world he causes me to this day to marvel at his near ingenuity at misreading any and everything that attempts to break through into his cranium, which I assume is at most a cavern with rocks in it. While he was spinning the globe, he suddenly stops and points to the line that traverses the earth. 'What's this?' he asks me. I said, "Joe, that is the equator, it divides the world in two."

"You mean like in math class?"

"Joe, I don't know what you mean."

"Where's X? Is it isolated and hiding some place?"

"Joe, X has nothing to do with it."

"Well if there is no X, how can the world divide itself in two?" he asked me smugly.

'Joe, it's just an imaginary line, just forget the X, it's gone, caput, it's...' Before I could finish, Joe got this gleam in his eye and pointed at another line and read the words slowly out loud, "Tropic of Cancer." I waited, I knew another question was forming. 'Is that where people go to die from cancer.'

"No, Joe, it's just a place on the globe where the sun shines directly overhead "

"So they die from getting too much sun?"

"No, Joe, they don't die there, it's just a..."

"Well, then can they live there if they have cancer?

"Joe, it's just a name, a Latin name, it's a word for crab. The Romans thought cancer made people's bodies have things in it that looked like crabs."

"So people and crabs went there to die from cancer."

I concluded that logic was not a discipline Joe would ever master. Finally, "Yes, Joe, people and crabs go there to die of cancer. Most of them do."

Joe was satisfied until he discovered the Tropic of Capricorn. Two weeks later, each student had to give a report in front of the class on a different country. Joe chose Mexico. I was afraid, very afraid. Joe began his speech by saying, 'Mexico is a country between the pillars of Hercules and the pillars of the Tropic of Cancer and the Pillars of the Tropic of Capricorn. Most people who go there die of cancer and the crabs which I discovered is a disease of the pee pee and the girl thing. If you live near the other place like Capricorn you die mostly from eating too many goats and sometime the horn gets stuck in your throat. Then when you die, especially in the ocean, the sharks and the crabs eat you.'

"I tried to be patient. I really did. But Joe was converting the class to his world view. When he made that comment about the crabs and sharks several hands went up, wanting to know if the sharks bit the heads off the people and if they died on land, did the crabs eat them or were there land sharks. One young prince of knowledge went so far as to explain that he thought his own father might have eaten some people. He said his mother told him that daddy was in Mexico to buy some land there because daddy was a land shark. So the boy wanted Joe to tell him and the rest of the class whether or not his father had eaten people on land when he was in Mexico or did he have to go in the water to grab the body first and drag it on the beach."

"Joe admitted he wasn't sure but he thought his father probably was already dead or would soon die from eating cancerous crabs and goats which came from the land of Capricorn which lay somewhere past the pillars of Hercules. He explained that Hercules never died of that kind of stuff because he didn't need to eat anything, he just went around smashing things like crabs because their shells were so much fun to crack. THE END."

"As Joe finished his presentation, he turned to me and said, 'Geography is so much fun, learning about dead people, and cancer and crabs and sharks and more dead people with goat horns in their throats. That's the best.' Joe's powers of persuasion exceeded all expectations and to think that Joe was the only black child in the school. I few days later, as we approached spring break, I received a nasty letter from the parents of one little girl in the class. They explained that their darling had a fit when they told her they were going to Mexico for vacation. She apparently screamed that she didn't want to go for fear of being eaten by crabs or having to eat dead people with goat horns stuck in their throats. After that, there was no going back, Joe truly was the devil. I knew it for sure. And now he lives in the White House. I've lived too long.'

Laughter is the best medicine.
-- Book of Proverbs, 17.22
King James Bible

Really?

The Author

JABBER NO. 6

TRUTH INJECTION: MRS. GRAVESTONE RETURNS OR THE UNBEARABLE BRIGHTNESS OF BIDEN –– PART II

Well, I suspect this may be my final commentary or transmission about Joe. He did something beyond the killing of my husband that has forever damaged even the most consoling of quiet activities in life – reading and writing poetry. At times, I swear, it brings me closer to nature and to God. That is, it did, until Joe touched it.

As had been my custom for years, I enjoyed many evenings, especially after Joe killed my husband, reading my favorite poems and, of course, the Bible, especially the psalms, most particularly, the 23rd one. It has always provided me with some comfort amidst the travails of life. So much so that I keep a copy of the Bible both at home and on my desk at school. Next to the Bible at home sits a copy of my beloved Robert Frost, his complete works. At school I have the same. Robert Frost consoles me with his beauty, his love of nature and his sometime haunting portrayal of the vagaries and vicissitudes of life. As few others can, that white haired man with his command of the complexity and nuanced texture of the English language conceals genius with his disarmingly simple and yet varied rhythms. Were he alive today, I should drive a thousand miles or more just to spend a moment in his presence. The aesthetics of his works combine a kind of linguistic virtuosity with a searcher's spirituality He was my guru, my avatar until that day when Joe…(**HERE SHE PAUSED AT GREAT LENGTH**).

The evening before, I was reading the poem "West Running Brook." The description of the well, the water, and the mystery of a certain whiteness beneath the water. The intrigue as he almost non-chalantly tosses aside the fleeting glimpse of something that seems to be there and then disappears to which he responds with merely, "For once then something." As I sat at my desk reflecting upon that poem before class had started, before a one of them had entered the classroom, that something walked into my room, smiling, and came right up to my desk and said, "I want ta ask you a question, OK?"

For a moment I looked at him with fresh eyes and I was able to bury my embarrassment at my own chagrin and general dislike for this innocent albeit cruel and sometime manipulative spoiled child of God. If Jesus can forgive me for all of my sins, then surely I could look upon this innocent and uncommonly muscular white miscreant with a truly burnt exterior that seemed a contrivance of Satan himself. Surely, he was indeed, as the great philosopher Elvis Presley once sang, "He's just the devil in disguise." Maybe I have indulged in a little bit of poetic license here, but it seemed then and even now somewhat justified. Perhaps I was guilty of a certain naivete when I responded to Joe's seemingly genuine interest in poetry with a whimsical admonition that he learn the greatest poem that I could bequeath upon this wayward child.

It is said that when the child is ready the teacher shall appear. It seemed that this was the moment and I was that avatar. I was the one whom God himself had selected to bring this child from the darkness of his self-indulgent recklessness to the light of the Lord's grace. This I would do for the good of this child and for God. All humanity would benefit for as John Donne once wrote, "No man is an island." For a moment I forgot that this, this, this Joe had been the thing that had killed my poor husband. For a moment at least he was just simple Joe and not the Mean Joe Green he tried to pretend he was. Or so I thought, for at the very least it was clear to me and all others that he was truly "simple Joe" whether he was being good or even when he was bad. For whatever had formed Joe had left out a piece of the jigsaw puzzle that was his DNA. He was indeed simple, like bread without butter, or a rock of ages that never quite evolved. If man is dust, at least he is animated dust, but with simple Joe, well, I was always wondering where the impetus of locomotion came from for it seemed separated from any semblance of an intelligent design that the Almighty could bequeath to humanity. As a good God-fearing woman, I felt it was my responsibility to support the mission that God has ordained be at he very center of my being. And Joe – well I saw him as my cross to bear. Frankly, ever since he came into my life I could not help but consider that Joe may have been the greatest indictment of the theory of Intelligent Design. Still, I was hoping for redemption for all my sins if I could touch his life. Little did I know that that Mr. Touchy Feely would be touching in a most delicate and well, you understand. But I digress. Joe was expressing some interest in poetry so I tried to go with the flow and play into his excitement over an art I deeply love.

"So, Joe," I said, "Whatever is your question?"

"Do you like poetry. It seems kinda stupid to me. So what poet do you think I should make my report on so I won't have to read someone who writes stupid stuff?"

"Well, Joe, since you put it that way, why don't we start with one of the greatest poems ever written. Some how I think it was written just for someone like you who doesn't want to read stupid stuff. God, I think, has other plans for you, my fine young fellow."

"So what's the poem? Who wrote it? And it's not stupid, right?"

"No, Joe, it's not stupid and that is why I am going to share it with you." At that point Joe put on his aviator glasses, which I had seen him wear before without understanding why he always put them on in the classroom but never outside when he was playing in the sunshine. But the moment he put them on, he smiled broadly, very broadly and seemed to be fixated directly on me. As he was watching me, I stood up and reached for my copy of the Bible that sat on the very edge of my desk right next to Joe. In reaching for it, I had to bend over and instead of his watching me grasp the Bible, I – well, well – I had the distinct impression he was trying to look down my blouse. Though I could not see through the lenses, there was a discomfiting seeming misdirection about his gaze. It should have been on the Bible, not my bosom.

So, as I stood erect, I noticed that his gaze remained where it oughtn't. I was about to open the Bible to the 23rd Psalm and instead I dropped the Bible right on my desk and said, "Joseph Mean Green, what exactly are you lookin' at?"

He smiled and answered with either the sweetest and most innocent lie a body could create or he was indeed a pervert who seemed bent on the personal invasion of my earthly being. My flesh felt a certain warmth and tingling.' Mrs. Gravely Grit, I can't see anything with these glasses so I don't know exactly where you are but I think they make me look manly and a good person you would be pleased to call your pupil, which I know you don't call no one else on account of you think I must be special.' I was defeated. If he was telling the truth, how could I scold him for seeing what he was seeing if he really was not seeing it because the glasses were just a toy of sort? So, I asked, "Joe what's so special about those glasses, there must be something?"

"Yes, Mrs. Gravulator, they're special. They were made special for me by a German type doctor who told me the lenses were extra special because they were made from crystallized mushrooms which was pounded and thinified so that a boy could see through them in a special way. They're the kind Green Arrow wears and Captain Marvel. Some say that Hopalong Cassidy wears 'em too when he isn't ridin' his horse or shootin' people like in the movies. So they're special to me and I never wear 'em outside so I don't break 'em, just inside where it's a bit dark and I can't see. But I think they make me look like a great hero which I want to be some day."

I was out and out charmed. When he used the word "thinified" and spoke of his heroes, I couldn't hold it against him even if I still thought he had indeed tried to look down my dress to my bosom. So, I smiled back at his grinning face and opened my Bible to the 23rd Psalm. I began. "

"The Lord is my shepherd, I shall not want –"

"Whoa, whoa, whoa," he interrupted, "What's he want?"

"What do you mean, Joseph?"

"I just want to know what does he want?"

"Who?"

"The guy in the poem when he was writin' it, what does he want?"

"Joseph, he *doesn't* want *anything*. That's the point of the poem. He's got everything he wants already because God has already given it to him.

"Who is this guy?"

"What guy, Mr. Mean Joseph Green."

"What guy who doesn't want anything? Is he dead?"

"Joseph, how can he be dead if he doesn't want anything?"

"Exactly! If he doesn't want, he must be dead. 'Cause I want things' cause I'm alive. If I was dead, I don't want nothin'."

"Oh, Mr. Joseph Interruptus," I sighed. "Perhaps we should try a different poem."

"Yeah, I wasn't figurin' on listenin' about a poem about a guy who was and writes about not wantin' anything 'cause he's already dead."

"Joseph, when you die, I hope they put those words above your tombstone, 'I shall not want.'"

"Yeah, that would be the best. Read me another poem, only this one about a livin' guy who wants things. Another one of your favorites, I like doin' this. Teach me."

I put my Bible down and reached for my most precious Robert Frost in hopes that we could get past the first line of "For Once, Then, Something." Such a lovely piece, the mystery, the innocence. The lack of guile, just a boy trying to understand the mystery of the enigmatic whiteness that stands deep within the water and temporarily appears and disappears. The boy sees it but then it is gone. The mystery is like life itself, we grasp at it's elusiveness and then, amidst the black and impenetrable darkness of our own ignorance, the white flashes before us like some illusory creature that may be evil or perhaps...I began to read, "Others taunt me with having knelt at well-curbs/Always wrong—"

"Wait, wait, what does *taunt* mean?"

"Ok, Joseph, taunt means to make fun of someone."

"So the other kids in the story are laughin' at him because he was lookin' at a dead whale next to the curb on the sidewalk? Right?"

"No, Joseph, there are no other kids in the story, they—"

"Are they dead?"

"Who, Joseph!, Who?"

"The other kids, did the whale eat them?"

"Joseph, there is no whale! " And in a terrible moment of weakness, I gave in to Joseph's remarkable ability to destroy the essence of poetry by misreading any and everything and said, "No, Joseph, the whale is dead. He ate the children and died of kid-ney disease and that's why the other kids who weren't even there were making fun of them. Just think how much more fun everyone would have had if the whale had not eaten them. But he did because they died when they tried to push the whale off the curb into the street where it was hit by a big truck that damaged its kid-neys"

"Now that makes sense," he responded, "Read another, please, I am startin to like poetry with all the dead kids and whales and stuff. And I remember the time you told the class about another whale that was white and had a big dick and he would hit people with it. That was the best, but I like this one too. So I'm gonna go play and I want to do more of this tomorrow…"

"Joseph, I have never told you about a white whale with a, with a – Joseph, the whale was evil –"

"'Cause he had big dick, right?"

At that moment, I recalled Sampson's utterance to Almighty God, when he stood before the two great pillars and uttered, "Oh, Lord, let me die with my enemies." He destroyed the building and the Philistines with it. I envisioned myself standing in the doorway of the school building with

Joseph seated triumphantly as I blocked the doorway and pushed against the frame and forced the building down upon him, taking myself with him to eternal damnation or bliss. Surely, I had already suffered greatly at the stupidity that walked about disguising his malicious whiteness beneath his burnt black exterior. Someday, somehow, I would have my revenge…And on that day, I would utterly devour him and take him into the depths of his own Satanic whiteness 20,000 fathoms below the surface.

"Good night, Mrs. Palestone."

"Good night, my little Philistine," I whispered.

A short while later, as I ruminated on the unbearable darkness of Joseph's being, I noticed that on the way out into the twindling sunlight that my torment had left his aviator glasses on my desk. I took them and was about to place them in my desk drawer for safe keeping. Then, just to see about the miraculous playfulness of his boyhood naivete, and perhaps wishing that I too had superpowers like his comic boy heroes, I put them on. The coloration was brilliant and almost surreal. The flowers on my desk were more red, more blue, and yellow than ever before. Everything about me had an enhanced quality. I understood why the torment of my life had guarded his glasses so closely and worn them whenever he was in a room like mine with its flowers, colorful maps, and a variety of posters and pictures. Smiling to myself, I then looked down at my dress with its flower prints. Then I saw it.

THINGS WERE GETTIN' WORSER AND WORSER

PUMPKINS SHMUNKIN

So when I got back to the barn where we were buildin' the new special car with one piece at a time like Johnny sang about and Otto Otto made fun of it all, well, there was nobody there. I felt some dang awful I about cried. In fact I wrote a song about it and sang it out loud 'cause I knew winter was a comin' and I didn't see no point in tryin' to finish my auto-mobile. The pumpkins – they had rolled some mighty fine big ones all the way here. But instead of puttin' 'em on the auto-mobile, they just left 'em leanin' up on all sides of the chassis, that's like a skeleton for a car. I knew I could not put the pumpkins on all by myself since I couldn't hold up the skeleton and shuck the pumpkins underneath it at the same time. At one point, I tried attachin' one pumpkin to the axel, thinkin' I could pump it up with air pump we had. But it blew up and there was seeds and pumpkin pie-like stuff all over and I was hungry but not so much that I wanted to eat it specially since I was so lonesome – well, here's my song anyway.

Pumpkins are orange
Just like the bees
Even if they Bumble
And they got no knees.
So I was so lonesome
That I could sigh
Smashed all the pumpkins
And I don't know why.

So anyway, years later as I contemplated that the song wasn't quite right as I could not find any word to rhyme with orange to complete my rhyme. That was important 'cause without a rhyme you can't make a dime, if you know what I mean. So, I sold the song to a nice white skinny fella who played a guitar and ate a lot of jambolaya or somethin' like that. He played baseball too some time later and hit a lot of homeruns. That was 'cause when he played the guitar people say he just hammered it with delight. Well, I didn't know how hammerin' a guitar could make it a "delight," but I could see it might work with hittin' a baseball. So they called him Hammerin' Hank until one day he had a total eclipse of the heart and transitioned into a black fella 'cause I think he wanted to hit more homeruns. Once he became a transi because he didn' also like his whiteness cause he was afraid of getting' eatin by a dead whale or something,' then the white people all pretended he was dead and so he went and started a another group about plants and stuff called "Crashed Pumpkins" after he himself supposedly died in a automobile wreck. I knew it wasn't true but I didn't say nothing 'cause I like to watch him hit homeruns and thwart his whiteness like I had done. Later I found out he had a sister who also became a transi but she was still white and a lesbian so I just called her my fella transister which gave me a really good idea about makin' tiny radios. Yep I invented that too although some guy from Tennessee who liked to get massages late at night 'cause he was feelin' stiff and sore – he said he invented that along with the internet. Sometime he would also play songs he said on the skin harmonica with a girl he met in Iceland who he was savin' from climax change. I guess it didn't work out so well because he said he was against her climaxes changin' so she told everyone the truth: she was his twisted sister who fled Tennessee 'cause he was pallutin' everything by takin' too much zinc out of his mines and given it to people who didn't need it in their diets. But he had a campaign slogan to get more zinc and make people like it: No Zinc, No Think. Well, I guess so, but I've never had an ounce of zinc and I think I think just fine (there it is again, I know, I know…) One day I found out he had a sister named Leslie who sang a lot until she died. But mostly I heard she went to a college named Vassar. And even though she ate a lot of zinc, they say you can lead a girl to Vassar but you can't make her think. True enough I hear.

Later on I wrote a song for her too. She had one about a sex relationship in which the guy was havin' sex but she wasn't. I still don't get that but OK. One of the lines in it was, "You don't bone me, I'm not just one of your little boys…" Well, by then, I had enough experience in song writin' to know that that just wasn't gone to work for sellin' a lot. Dogs might buy songs about bones and stuff like that but not most people. And dogs also need little boys to play with but not full grown men. So I changed the words a bit for her and it became a big hit – yep, another one.

You might be wondering' how old I was then. The truth is I don't know. Even now, I just don't know. As that guy once said from Germany or somewhere, "If you're not late, that doesn't mean you had to be on time." Now that's wisdom, I think. So I don't know if I was born late or if I was late when I was born. That kind of thought is what they called a sim-ant-ich division, or distinction. I forget which. Kind of like the story here I'm tellin'. I can't tell it quite right 'cause I forgot who it's about.

I giggled like a pancake in heat
(Russian spy quote with secret meaning)

JABBER NO. 7

JOE'S OLD ANTAGONISTS RESURFACE: BORIS AND NATASHA BADANOV TELL US ABOUT THE REAL MR. BIGG AND CONFRONTATION WITH MOOSE AND SQUIRREL.

EDITOR'S COMMENT:

THE FOLLOWING REPRESENTS A TRANSCRIPT OF THE ZOOM CALL BETWEEN OUR OFFICES IN AN UNDISCLOSED LOCATION FOR SECURITY REASONS AND THAT OF AN EQUALLY SECURE LOCATION AT ANOTHER UNDISCLOSED BUILDING TO PROTECT OUR GUESTS. I MUST SAY THAT THE ASSERTIONS ARE ASTOUNDING AND PERHAPS UNFOUNDED. BUT HERE AT DELPHI ORACLE PUBLISHING WE NEVER SHY AWAY FROM THE TRUTH EVEN IF IT HURTS:

GOOD EVENING, BORIS, NATASHA: We certainly are a long, long way from those freezing temperatures of the Cold, Cold War. Aren't we?

BORIS: Eeees good evening for you, maybe, Mr. Ali B., but eees good morning for us. But don't worry, sometime later today, you will catch up wit us. Eeees like Cold War, no?

ALI B: Da, it is like the Cold War but now it seems to be returning with much fervor and heat and true passion and accompanying death and destruction.

BORIS: Why are you, Muslim man, living in America with all those Christians and X-rated movies? You should live here where we are, surrounded by Muslims all the time who say if we even think X movies they will cut off Natasha's head.

ALI: Why Natasha? You are the man, why not you? You tell the people in our audience maybe where we are, huh?

BORIS: Da, Da, Da, we are like you but only different. We are Christians in Afghanistan whom they tolerate because we can still make things happen for the Taliban and for mother Russia. So, we are in a different world where men make the rules and the women submit or they kill them. So, as I say, not so good for Natasha.

NATASHA: Boris, you are so kind. That must be why I married you, so I could live in Afghanistan one step above the cattle and sheep. Oh, thank you, my love.

ALI B: But you still say the Taliban treats you well?

BORIS: I say for sure the Taliban treats us. How is another matter, but they need us, we need them.

ALI B: So explain why you are in Afghanistan.

NATASHA: Let me explain, darling. We are in Afghanistan because, I am sorry to say, because we are bored. When we worked for KGB in Cold War, it was excited, to always chase around Moose and Squirrel, Bulwinkle and Rocky. In fact, we miss them. But we still work for mother Russia and so we come here to pick up American weapons you have left behind after you leave Afghanistan.

BORIS: Wow! What you left behind! We couldn't believe it, enough there for entire army. That's why we are here. Your generals wanted to send to Ukraine before war start and your president said he did not want to pay for shipment so he left behind. We love him and it all shows that those meetings we had with him at embassy parties really paid off.

NATASH: And the night vision goggles! They are so stunning. You can see better at night with them than people can see in the day. Eighty-five billion dollars worth of equipment. We thought it was a trap like in the ancient story of that horse the Greeks snuck into that big old city. We thought maybe everything boobie trapped and then you blow up. But eeez great gift. On behalf of Vladimir, we say, "thank you." Please send more, please. We like the discount. Now our soldiers have best equipment in the world.

BORIS: I agree, Natasha, except for one thing. In Afghanistan, if woman create boobie trap, they cut off her head. No boobie trap, except for one – Ha Ha, is still Natasha for me. Her boobies trap me long time ago.

ALI B: So is it fair to say that you are now trafficking in American arms to different countries?

NATASHA: No, is not fair to say because Joe Biden don't want to pay shipping. He said so, we believe him. And, funny thing about what you just said. You say trafficking in arms.

BORIS: Yeah, like Natasha, she have beautiful long arms and she is always stuck in traffick. I think they call it pro—

NATASHA: Shut up, Boris or you get Molotov cocktail for drink and won't let you drive my new American tank. It's cool, really cool. Yay, we love America!

ALI B: Who sent you? Mr. Bigg?

BORIS: Ah, no. Mr. Bigg is dead. He died from overdose of Smirnoff's. Too much, too often. Like Natasha. Someday I can't walk she so aggressive. She is why I love Siberian women. Like Midwest farmers daughters they make their boyfriends warm at night. You know that song?

ALI B: So then, who did send you. FSB?

BORIS: Ha! You mean Fucking Stupid Bastards. Ha! I can still make joke like American only Natasha must be careful. She is woman, I am man. I run faster if they come for us. Natasha not so much.

ALI B: So, then who did send you? Vladimir Putin?

BORIS: Putin eeez the Great Man of History. He tells us that many times himself. And he have secret handshake and—

NATASHA: Boris, you talking too much. Put down Smirnoff.

BORIS: He is Vlad the Glad when he with me. We have secret handshake which says I am his friend and anything I say he would say. So who sent me? I give you one guess. And this is Pravda! Pravda! Pravda! " Big Guy" sent me with letter from Vlad telling Taliban and Isis to work with us for good reward.

NATASHA: Vlad and vodka do not belong in same room or conversation. Secrets are secrets. Stop drinking.

BORIS: So I am asking myself, what is question? Oh, da, da, da. Who sent us? Is not Mr. Bigg. He dead. But "Big Guy" send us and he eeeez…. I tell you secret and you tell me. OK?

ALI: OK.

NATASHA: Boris, I don't want you to –

BORIS: Natasha, please prepare, I am feeling romantic, and want you to execute me with your boobie attack. Is so much fun.

NATASHA: Boris, keep your pants on, you are not American like that man with Tobin. Keep pants on, stop drinking, and maybe we play late, my boobies will be yours.

BORIS: So "Big Guy" he come to Russia and then he go to Ukraine. We give him lots of money. So much, he had to have his son carry most of it home. Ukraine try to pay him too but not enough. So, we talk it over with Vlad and he say, I love Ukraine and we will help them too. So we give "Big Guy." And then –

ALI: But who is Big Guy!

BORIS: Ha! You funny. You pretend you do not know. But here is clue. I make rhyme like poet.

NATASHA: Oh, yes, Boris – he eeez great poet when he drink. He is romantic, I still remember poem you wrote first time in English. I still blush. Please, Boris, show your love for me and recite.

BORIS: OK, here goes. "Why I Love Your Boobies" by Boris Rachmanovich Aloisius Ivanovna Chicago Goodanov."

NATASHA: No, Boris, Goodanov eeeeze dancer, you are Badanov the poet. Recite for me. Start again. No name this time. We see you, we know. Please keep pants on and let me hold your hands as you recite. We do not wish a Tobin.

BORIS: "Why I Love Your Boobies" by me.

Why I love your boobies,
Eeeez because they are not Scooby's.
You say you like the doggie style
And that it eeeez that make me smile
But you're not so furry
And we never in a hurry
So when we in bed
You know I said, "roll over."
So I see your hot hot boobies!
Just not one but toobies.
So my poem it has an end,
Just like your big bottom,
So I love your boobies,
So happy I have got'em,
If you give not free,
You know I would have bought 'em.
The End, Your End. I Love You.
Your boobie are like Scooby,
Bow Wow! Wow! Wow! Baby!

NATASHA: Oh, Boris, you are so incredible, so romantic.

ALI: So, again I must ask, who is the "Big Guy"?

BORIS: Ok, so here is last hint I give you. "Big Guy" is someone you *know* and his name rhyme with *Joe*. Get it? I give you clue, "Big Guy" clue like Bolshoi. Maybe you get answer and I dance for you.

ALI: Are you saying Joe Biden is the "Big Guy"?

NATASHA: Boris did not say that, you did. But since you guess right, Boris will maybe later when he eeeze sober, dance for you.

ALI: If what you say is true, this – well I am stunned.

BORIS: Like my gun, it stun to.

NATASHA: Boris, remember, never point your loaded gun at anyone except me. He's not your type. Now put away and hold my hands again before the gun go off accidentally.

BORIS: So I see you are stunned like fish from grenade I throw in water. But is all Pravda, Pravda, Pravda. We meet under the table many times in Washington, Moscow and Kiev with Joe Biden and his son Hunter. He take money only when he meet with us under the table. American custom I think. Chinese do it to him first but we learn from them. Some time so many under table we have to meet in Kazakstan. They have really big table for Uranium. They buy so much from you they can call it Ouranium. Eeeez theirs now until they ship to Iran. Free delivery.

ALI: But what was the purpose of the money under the table?

BORIS: Eeeez real simple. We give money to JB and Hunter, we give Hunter honeypot girl, and JB agree to leave equipment in Afghanistan. In return we agree to ship for free to Ukraine for him. You see, everybody happy. What you Americans call "win win." We agree, you get free shipping from Afghanistan to Moscow, and we then send American equipment to Kiev. In fact eeeze already there. Your "Big Guy" is genius. So smart, we love him.

NATASHA: Da, new special delivery program, we call Afghan Prime. You don't have to pay extra for it. By the way, what do you think? (Natasha puts on American night vision goggles). Over $30,000 for each pair and we deliver to Russian troops helping with other deliveries of American equipment even at night in Kiev.

BORIS: Oh, Natasha, you are looking like Scooby. Like furry dog, you are also fetching! Grrrr! Bow Wow! Baby! Bow Wow!

NATASHA: Stop talking, stop drinking. Scratch my belly, Boris.

BORIS: Let go my hands, baby. I itchin' for scratchin' like American baseball player what do in public.

ALI: Well, I think we had better go before we have a Tobin or something more on our show. Before we do, Boris, Natasha, thanks so much, any last words?

BORIS: Yeah, you see my beautiful wife is 80 years old and still have figure like MM.

ALI: You mean Marilyn Monroe?

BORIS: O Boy! No, Minnie Mouse! HaHa! You Americans fall for that every time.

NATASHA: O Boris, you *funny* and romantic. You make me laugh but still I think I sad, vey sad. Yes, I still miss Moose and Squirrel. Tell them to come visit. We even eat boil acorns together like American squirrels. Yeah, like what they eat in Ukraine now! HaHa!

MRS. FLAGSTONE'S REVENGE

EDITOR:

As one of our more frequent guests on our show, Mrs. Flagstone has perhaps done more to shine a light into the early years of Joe Biden than anyone else. She brings a greater wealth of personal experience in dealing with "the Joe" than anyone else, including his parents who seem to be a conspicuous absence.

MRS. GRAVESTONE: Gravestown.

EDITOR: Pardon?

MRS. GRAVESTONE: Grave and stone, Gravestown. It's not that hard unless your name is Joe.

EDITOR: Oh, yes, yes, of course. I was just looking here at my notes from Joe. He referred to you by that name – and several others, I'm afraid.

MRS. GRAVESTONE: No need to explain. I have seen the enemy and I know how terrifyingly opaque he can be. But I will try to explain what happened next. For, however much you might think me to be too delicate in my femininity to confront Mean Joseph Green in all his majesty and mighty stupidity, I am not without my own godly endowed resources of persistence and outrage. In Joe I had occasion to bring both to the fore.

My turning the page on my erstwhile belief in the goodness of this child came the moment I put on Joe's glasses with what he called the mushroom crystalized lenses of magic. I saw things I had never seen before and I knew then why that little pervert was always smiling. You see, at first it just seemed that the glasses offered an intensity of color as I had never seen before. Every blue was bluer, every yellow yellower, and every red, much redder. It was surreal. But then, I understood why as I glanced downward at my feet. Between my feet and my head was the, shall we say for the sake of discretion, the immensity of my bosom. The endowment of my creator who had used it as instrument of masculine chastisement when my husband may have offended or slighted me in a hurtful way. For hours if not days I had in fact, on occasion, forbad their exposure to his lustful gaze. Like the moon at its ebb, they were eclipsed until he was dutifully reprimanded for his thoughtlessness and for sins he did not even know he had committed as he was, after all, just a man.

In short, with these glasses, I could now see the full unobscured immensity and contours of my most womanly endowment. I now saw without my having to see myself in my own domicile in a kind of nakedness. What I could now gaze upon was the result of a device as penetrating, I suppose, as anything man had yet invented to spy on that feminine animating force of sin: breasts, nipples, and areola. Joe had seen it all and I was hellbent on keeping the glasses and getting him expelled. I took small comfort in the fact that I often sat behind my desk during class so that it was less likely that my nether regions would have been subject to his lustful palpitations.

Upon Joe's arrival the next day at school, I marched him right into the principal's office where he sat down in a perplexed look even as glanced around wondering where he might have lain his glasses. For he seemed unbearably naked *without* them just as I knew then I had been before him when he *had* them. I once saw such glasses in a James Bond movie but until that moment I had no idea that Joe's eye glasses were a real thing. His and those used in the movie, I later learned, were invented by the same man, a fellow whose initials were JM. Joe refused to tell us where JM lived or worked and there were no identifiable source materials on the glasses themselves. But at the school's main office, after several minutes of waiting, we were escorted into the principal's private room. "Good morning, Mrs. Gravestone. Good morning to you as well, Joe. How's everybody?"

Despite his goodly nature and affability, I could detect a certain awkwardness in his demeanor as if he may have already read the tension on my own face and the suspicious look that Joe was giving him.

"Dr. Remmick, I will get right to the point, yesterday Joe left a pair of glasses…. " After a two minute diatribe on Joe's latest incursion into infamy, I paused to catch my breath and to give the principal an opportunity to respond. Instead, he just looked at Joe for a full discomfiting minute. Joe, in what may have been the only minute of humiliation and embarrassment I had ever seen in his time with me, he cringed and buried his head in his chest. Finally, the principal asked, "Do you have the glasses? May I have them?"

For a moment I hesitated, not wanting to take the chance that he would in a moment of misguided leniency return them to Joe. Instead, he asked, "Joe, do you have anything more to say?" Strangely, he was perfectly still and he said nothing. The principal then picked up his phone and dialed a number. The phone rang repeatedly and yet no one answered.

"Joe, Mrs. Gravestone and I had a long conversation last night about what you did. She has recommended that you be expelled and I concur. Joe, we've also made several attempts to contact your parents. We've not been able to contact them. Is there something we should know?"

Joe, with a woeful expression on his face, said, "My dad never comes home much because mom says he drinks too much and mom is workin' a lot to help out with the bills is what she says. But if you're expellin' me, then I know you have a good reason and I'm sure not right in what I did."

"Will your mother be home once you get there? I don't know, she works a lot at night now because she says she gets paid more for that." Then he looked directly at me and said, "Honest, Mr. Gravertone, I didn't mean any harm. It's just that you are the prettiest and most sweet smellin' teacher I ever had. An' from now on I will be smellin' every girl I meet to see if I can find someone who smells like the peachy smell you have." I felt positively awful but I couldn't think of anything to do, he had, with his conniving and all, invaded my very person. He continued lookin' up at me with those dear sad eyes and with tears staining his black cheeks. For a long time, he just sat there and then without a word, he rose to leave when he stopped and turned to me and asked, "Can I keep that book you said we were gonna read next about a guy named Injun Joe and Miss Becky Thatcher?"

"Certainly you can, Joe." With that he left the principal's office and the school itself. A minute later, I watched from the window of the principal's office as Joe shuffled across the school playground with nothing more than one book, *Tom Sawyer.* When I turned back there was Dr. Remmick looking right at me with Joe's magic classes. I gasped and picked up my purse and headed right for the lady's room. "Men! " It would be years before I would hear of Joe again.

MY LIFE ON THE RIVER OR HOW I CAME TO WISDOM AND KNOWLEDGE

On account of all my sadness and what with my not havin' any friends left, I took to wanderin' every where. And I was sometimes stealin' and stuff just so I could keep my belly full. I was still black then but I didn't any longer have my glasses and I was afraid to ask for 'em back so I was sad, bad boy.

So after wanderin' for nigh on two yeas or more, I came to a river, a mighty big river. The river didn't have any signs on it so I didn't know which one it was. Since I was hungry and there weren't no poor people or others I could beg from, I thought of learnin' to fish. So, I thought, I might as well try it. So I found a long piece of string and then I went lookin' for somethin' to make a hook. Well, I found just what I needed, a funny bent hook thing attached to a plastic drape of some sort that was hanging over the side a bath tub that looked like it had just floated down the river near to where I was standin'. Since it had rained hard just a couple days earlier I thought this might be somethin' that came ashore from it. Well, I took one of them hooks and bent it a little and saw to it that the end part was tied real tight to my new found fishin' line. But I still didn't have any bait. At first I didn't find anything I could use. But then as I was staggerin' through some weeds that waded right into the river itself. I smelt somethin' awful and I held my nose. Well, low and behold, I found what must have been the stinkiest, slimiest, most fly draped thing ever. Wooeeeeeee! Even with my nose pinched tight and my breathin' through my mouth, I could still smell it. I ain't sayin' I knew a lot about fishin' back then, but I bet you even God coulda smelt it. I pushed aside some reeds where the smell was comin' from where I had seen a bit of whiteness all bloated and flies all about. When I pushed the reeds completely aside, I saw what must have been the biggest carp I had ever seen. Now I have to admit that wasn't sayin' much since I had not really fished before but it wasn't hard to figure because of stuff on television and Boy Scout magazines.

Like it stunk to high heaven but I thought if I could smell it, then maybe a fish could smell it even under water and then it would bite and I could get somethin' to eat. Well, I took the biggest stinkenist part of that ugly carp as I could find and it slid it onto my new made fish hook. Whooee, I about gaged it was so stinky. Then, like I had seen in the cowboy movies, I swirled that string around over my head again and again until I built up some steam and then I let it rip right into the flow of that river. I knew better than to think the fish would bite right away and I knew enough to be quiet. If you make a lot of noise, the fish will swim away. That was the last thing I wanted. So I pulled up on a rock and just sat there awaitin'.

While I was waitin' with my fishin' line tied to my big toe like I had seen in some story books, I suddenly heard a sound like no other. I was a deep, deep human voice singin' like it was comin' out of the throat of the river itself. It came closer and closer. I looked about and then I saw a man, a big black man was singin' and sing' all about the river itself. He sang, "Dat ol' man river, dat old man river, he ain't seen nothin' he just keep rollin' along…" I really, really like his voice but I was concerned 'cause I was still hungry and I was afraid he would scare off the fish. So as he got closer and he was singin' about totin' that barge and a liftin' that bale, I stood up and said, "Who you to be singin' like that and makin' all that noise. You're scarin' the fish and I—" And just as said what I says here, I felt a strong tug on my big toe and wallooooo, I was toed right off my rock right into the reeds and I felt myself bein' pulled into the water. Here I am, I'm bein' pulled in but I can't swim. "Help! Help!" I screamed. I tried to reach for my foot to untie the fishin' line but the fish or whatever kept pullin' harder and harder which was makin' the knot on my big toe tighter and tighter (Now is not the time to question my redundancy). The big fish or whatever was pullin' me parallel right along the bank and I was gaspin' for air when all of sudden I felt another pull right on my arm. Two great big hands grabbed me up so I could breathe. He cut the line and the fish just about went off, when I saw it surface, the biggest cat fish known to man. He was movin' right into the reeds again and would a gone back deep into the river when the ol' man river guy jumped down hard right on the fish and stabs it again and again until it just floated dead and started to sink. Oh my, I was so thankful.

When he was through killin' the fish, he dragged it upon onto a dry space and then without sayin' a word or even lookin' any more at me, he started cuttin' off some mighty big pieces and he lays em' on a rock and then next thing, he's buildin'a fire right there. How he did it, I don't know. He didn't even have matches, least I could tell. And all the time, he doesn't say a word to me. Then he put those chunks of cat fish onto some sticks and starts cookin' 'em right in the fire. The smell was goodness itself. But he still ain't said nothin' to me and I was beginnin' to think he wasn't gone to give me any fish but what he intended to keep it for himself. But just as was I thinkin' that and startin' to shiver as a bit of wind was touchin' me what with the water still drippin' off me, he looks up and says, "Come here and eat." I came to him and for maybe the first time in my life I was really thankful and I said, "Mister, you saved my life," and without even think' it over I threw my arms around his neck and said, "I love you." He just smiled a big ol' smile and let me hug him for near a minute and then he gently pries my arms from around him and says, "Eat, I can see you's hungry."

So we ate for a while until I was feelin' full and happy even though I was shakin' a bit even though it was the high of summer and the sky was blue and sun was shinin'. Funny thing was I kept shakin' even after the coldness left me. After we was finished eatin' the fella reaches out to shake my hand and says, "What's your name?"

"Mean Joe Green," I said.

"Well yo' don't look so mean ta me. But yo' are a bit green. Yo' don't swim dose yo'?"

"Nope," was all I could say.

He held out his hand and said, "Name's Jim, Mr. Joe. How old 'er yo? 'Cause yo got a baby like face but yo got some mean muscles. Look mighty strong for a youngin."

"I don't know how old I am on account I got no parents any more and my dad he drank a lot and I don't know what ever happened to my mom. So I came to the river."

"Well, ol' Jim is gonna give yo some wisdom 'bout the river. Rule number one, on sunshiny days like now, never sit on a rock lessin' yo first look ta see 'iffin someone else is sittin' their first."

"Like rattle snakes. They love them warm rocks to sun theyselves on an' iffin' yo ain't watchin' yo just as like as not to sit on one and get a bite in the behind. Now 'ol Jim he will do a heap for any man what treats him right, but iffin yo bit on the bottom, ain't no way Jim's suckin' that poison outta there."

I learned that then and I ain't sat on a rattle snake my entire life. But as I was then lookin' around, I saw Jim lookin out over the river at a speck comin' our way with a sail on it. It drew closer and closer until Jim started wavin' and someone else on it wavin' back. Unlike us, he was white and was wearin' a straw hit and had red hair. Jim seemed mighty glad to see him. Before the boy's raft was all the way up pushin' into the reed beds, Jim was already standin' in the water knee high a smilin' like he was meetin' Jesus. The boy threw a rope to Jim who grabbed it and pulled the raft up on the shore crushin' down the reeds as it went.

The boy who was maybe my age, whatever that was, was just as glad to meet Jim. He threw his arms around him like I had, and they hugged. Jim says, "Huck, I want yo' to meet yo' new travelin' friend. Still goin' all the way south to Cairo, Illinois?"

"Yes, I am, and what about you, you're comin' to, right, Jim?

'Ol Jim, he hesitated and did somethin' I never saw a grown man do before. He got teary eyed and said, "No, Huck, I got things to do. I got to get my family out of Chicago. Theys a heap of murderin' and mayhem there and it ain't safe for no human bein' on account all the corruption. I got to save my family." Huck looked kinda sad like he felt for Jim and for hisself too 'cause you just knew he was wishin' for Jim to go down river with him. Suddenly, Huck jumps up and says, "All right then, I am Huck and you are?"

"I am Mean Joe Green."

Huck looked kinda quizzical at me and then smiles and says, you don't look mean, and you're black and not green, and you look more like a Daniel than Joe to me. So from now on I'm just gonna call ya, Nice Daniel Black. And that is how I got my new name. This was how I met the boy who become my friend. So, after a while, we said good-bye to Jim who left for Chicago so that he could buy his family out of bondage to the street gang that had been terrorizin' his family. They were called the El Rukins which is mostly like in the Bible about Egyptians.

So, as we were floatin' down the river on this beautiful raft Huck had made with his sail and all and even a tent that we could sleep in and not get hit by the rain and sun—well then it occurred to about somethin' my teacher Mrs. Grayslick had taught me after I had killed her husband. She made us each pick a country to talk to the class about stuff. I picked a country called Egypt which had smart rulers for thousands of years, what I think they called pharaohs or rukins which means a kind of smart ass or a wise guy but lots of real wisdom to build pyramids and stuff. And one day, I remember a Sunday sermon when the priest was talkin' about Moses and how he was like me and Huck floatin' down the Mississippi on his way to Cairo just like in the Bible. And just like that baby, I had got tar poured all over me once and maybe that was the real reason I didn't drown that day when Jim pulled me out. So I thought maybe God had made me special and that was why I had spilled lots of tar on me. It was a miracle and a symbol at the same time. God was chosen me to be a deliverer of some kind like Moses. He didn't work for the post office but he could deliver. And just like in the post office I heard God made a fella wait a long time to get delivered but I knew then that Huck and I was on our way.

And so, when we was floatin' down the river I was thinkin' how we were to go through a thing called locks which Huck told me blocked the river. He agreed with me that was a lot like in the Bible when there was a guy named Remus, Remus the Great, he said who was blockin' the Reed Sea which we was in already all along the river banks. But Huck said he didn't know how we could get past the locks unless there was some kind of miracles. I know though that God had chosen me to be a deliverer like at the post office and so I started thinkin' how I could be like Moses and lead me and Huck to Cairo where they had a bunch of piled high rocks with points on top and a Spinks called the Leon's which had a creature's body and a man's face. I heard it said that the Leon was actually livin' in Illinois at the time of our river ride. But that didn't make any sense to me until later when I found it was true.

Well, finally, then it happened one day there we were bein' blocked from goin' any further down the river on account of a lock. I knew that I had a purpose and I picked up a large piece of drift wood and held it up and pointed it right at the lock and a sign next to it that said, "danger." And right next to the lock was a man dressed like Smokey the Bear who had a sign on his shoulder that said, "Ranger." All of a sudden it came back to me – what I heard as a baby when people told me to watch out for Ranger Danger. Well, here it was starin' right at me and tried to get me and Huck to pull our raft over to one side to avoid the lock. But we was headin' right to it and I had no desire to avoid it. In fact we bumped right into it just as a light was flashin' red next to it and the Ranger was shoutin', saying things like, "Get back or you could get killed." I figured that this was one of the tricksters who worked for Remus the Great. I could not have helped but wonder if Uncle Remus might have escaped this Remus with Ree-Ree so that she wouldn't be in no bondage. And just as we was about to be blocked and caught by the Ranger Danger, the red light flashed again and the lock suddenly opened and a bunch of water flowed down before us. It swept us clean past the Ranger Danger and swooped us on the other side of the lock what I heard was like a Disney ride. I knew that we had been saved by the Rod of God which I held in my very hands. I was a pro-verb come to life and that was somethin' that I always remembered even years later when I was Joe the Egyptian Candidate.

Three days and nights later, during a deep bad fog, like the death cloud floatin' through Egypt killin' all those babies, Huck and I floated right past Cairo. We was now beyond redemption in the land of Elk and Money, Sometime called Caintucky which was named as near as I can figure after a guy who liked to kill his brother because he had better fruit than he did. I knew that was a shame. Well what happened next was a mystery and another miracle. God had me and Huck by the scruff of the neck and He wasn't about to let go. Nope, He wasn't so we went by just floatin', fishin' and eatin'. Until that day--In the mean time Huck and I became fast friends and I even wrote him a song about us. It went like this –

> I had a good friend
> An' his name was Huck
> All day long
> He wanted to suck
> On real good things
> That brought him luck
> Like real cool
> fish all wrapped in muck.
> He had a green back dollar
> He called a buck,
> Put it in his pocket
> Where it got stuck.
> Such is the story
> Of my friend Huck
> Played on his banjo
> Pluck, Pluck, Pluck.

I am lookin' at you who are readin' this here manuscript with a dirty mind. I know what you were thinkin' and I ain't that way and ain't gonna do that no how. Shame on you!

HUCK TEACHES ME THE TRUTH ABOUT EGYPTIAN RELIGION

One of the best things about travellin' down the river with Huck is he just knew so much. And he didn't care that I was black and still had leprosy – at least some of the time. I mean he even had books right on the raft and one of them was an atlas with the finest colored pictures a body could see. At one point we even went by a place he said was called Branson and he said to me. "Do you know that the people who live there are mostly Egyptian?" "Really," I said. "How do you know?"

"Well you have to know that the Egyptians really like cats, which was a really important clue. You see, they liked cats so much that they wouldn't even let them die and rot like other animals. Instead, they just throw em down on the ground when they die and cut their inside out so they can fill the cats with sweet, aromatic spices to make' em look and smell better when dead than when living." But it was what Huck explained next that made such talk forever infecting my mind to this very day.

I was in a sweat to find out more about the Egyptian cats and all but I was confused so I interrupted Huck and I asked, "Where do Egyptians come from?"

Well at that precise time, we were floatin' past a big ol' church like the one I went to in Scranton when I was livin' in Chihuahua. It was just like it with a great big bell tower and lot of windows with rainbow colors and crosses and flyin' buttresses which I still think is a funny name for somethin' holdin' up a church. I got a butt too but it don't fly and I don't use it to hold up no church. But sometime I do let one fly. if you know what I mean.

So anyway, like I was sayin', we were sailin' right past it when its bells began to toll and Huck looks up at it and says, "Now that's an Egyptian church."

"It is?" I asked. "How do you know? Is this the same river the Egyptians was on when they crossed after the Israelites and tried to slay Moses?" I asked.

"Well, I think it might be the same river, Danny Black (what he was taken to callin' me since I was no longer Mean, nor Joe, nor Green, to him), but I heard that over time rivers can change course and I think that's what happened here. Now back in the time of Moses the river flowed more westerly and went through a place called Las Vegas and that is where the Egyptians started out with their cats. In fact I was there once with my cousin and the Widow Douglas who took care of me after my parents passed on. I saw the pyramids with my own very eyes and even the Spinks of Leon. Such a sight. It is amazing to think of how the ancient Egyptians built all of that even without electric lights or even have batteries or Thomas Edison."

His wisdom was amazin' but I had to ask, "Is that where the cats came from, Las Vegas?"

"I don't think so, but they liked 'em 'cause they would curl up in their laps and lick their hands and stuff real gentle like little fur balls. But they would wander off and not come back for days and they even went to live in the Egyptian churches."

Suddenly, it snapped! I figured it out and I said, "Huck, do you mean that's why we started callin' Egyptians Roamin' Catlicks?" I was stunned at learning that I myself had been reared Catlick and yet never before understood that I was also Egyptian and that the river they called the Nile and the Big Muddy we was travelin' on was all one and the same. The Bible was even more than I could ever have imagined. An' ol' Huck was amazin' beyond words. So at that moment while we was driftin' and I was feelin' excited at my wisdom, I dipped my hand into the holy waters of the Big Muddy Nile and blessed myself and then, like a cat, I licked my wet fingers dry. I had been born again as some folks say and I would remember that too in the years to come.

As the sun died, Huck and I looked up at the beautiful night time sky and thought we saw a shootin' star. We did and we felt blessed and so he and I smoked our cigarettes and remembered that we had already passed Cairo long ago and so we would just keep on goin' to a place called New Orleans.

PART II

THE TWEENER YEARS

> The fault, dear Brutus, is not in
> our stars/But in ourselves…
>
> William Shakespeare

6 YEARS LATER

ME MYSELF AGAIN

Several times over the next few years I came to one job and another. No matter how hard I tried, I grew desperate for excitement and confidence buildin'. I didn't want to be stupid any more and I was tired of just workin' out a lot to go along with my blackman physique which I was proud of but which seemed to contradict my name, afterall I wasn't really green. That's right, as the Huck and I grew to partin' ways 'cause of his wish to get better educated, I found that life grew tiresome and I was gettin' weary of movin' about with no direction. But like I say, I did have a fine physique if I say so myself. One day I was playin' a touch game of football at one of the big parks in a parish in New Orleans, which was a lot like playin' in my church league until I quit. They said I was too rough. While I was playin' a man came up to me wearin' good clothes and a suit and he said. "What college do you play for? You got power and you got a lineman's physique. And for a lineman, you got speed, boy, real speed."

At first I said nothin' except what I learned from Huck. "I'm a Catlick and I don't eat many things on Friday except fish."

He didn't say nothin'. He just stared at me and then he smiled as he scratched his baldhead and said, "Fine then, what if someone could make arrangements for you to attend a Catlick university right here in New Orleans? "Cause, boy, you got talent. What's your name?"

At first I wanted to use the name Huck had given me to throw him off my trail so he'd leave me alone. I finally told the man, I never even finished high school with good enough grades to make a body proud. "Boy," he said, "We can fix that. You're gonna go to a fine all black college right here in New Orleans and you'll play for me. I'm the head coach there and you'll get in and I guarantee you will play for me one way or the other and it's Catlick above all else. So you'll be right at home. But I still don't know your name."

"My friends call me Mean Joe Green but when misfortune befalls me, my name changes to just plain Joeb."

Well," he said, "Once I'm through with your name will be plain as success itself, 'Success, Mr. Green. Here's my card and my address. I'll see you tomorrow mornin' at practice. Come dressed for success because that's what you already are. Name's 'Hacksaw'. Tomorrow morning at 7 o'clock sharp. We practice for an hour and then eat in the school field house. You're already on a scholarship and your meals are covered." He shook my hand and walked away.

The next day I was there at 6: 30. I could hardly wait. My life had come about in new ways. I was playin' football and I was everything I wanted to be. I was big, I was lean, I was Joe, I was mean, and I was Green, just the deal I would forever be. Again, something I would never forget.

Four years later, everything I had done right came back to me in a momentous reward. Some of my teachers even called me a student instead of a pupil. Thanks to Mrs. Gravelypitts, I knew the difference and I was finally very proud that I understood what my teacher once called "nuance" and "subtlety," except of the football field where I was breaking things. I loved it. And I was so proud to be, not just black, but big and black and bled all over. That's right, I may be bleedin' but I knew they was bleedin' a hell of a lot more. Up to that point, I was feeling like a man blessed by blood, mine and theirs. My meanness now seemed right, but I knew there was another test on the way.

Every day, on my way to class, I stopped by that wishin'well outside the dean's office thinkin' about that whiteness Mrs. Gnarlywitt told that would flash through the dark waters and then disappear. Well, I looked every day and the whiteness never came back. I was cured, no more leprosy or nothin'. I had earned my name.

MY BIG DAY

In April of my senior year, the greatest moment of my life took place. I was drafted to play football in the National Football League. I had arrived. Some people not only said I was a great athlete but they also said I was smart because I could always be trusted to work real hard. Of course, I liked Mr. Hacksaw because he was always kind to me. The kinder he was the smarter and harder I worked. So it came as a complete surprise to me my first week in practice for the team that said I would make all the difference to them. They said I would help lead them to the Super Bowl. But things didn't go at first as I thought they would. Yes, I was about to experience the malady that came about on account of my meanness, the very thing which I had thought to be the strength of my success. The coach himself had said I was Mr. Success.

On that day, I as so excited, I was now a member of the Pittsburgh Steelers. I was a lineman on what I was preparing myself to lead them to the Super Bowl and I had not even played one game yet. I guess I was getting ahead of myself just a little bit because of what happened in only our second day of practice. That's when I was working hard and was trying to be mean but the coach this time was a guy who wasn't as nice as Mr. Hacksaw who always smiled at me and made me feel so good I would have done anything for him. If he had said, Joe, knock down that brick wall, I would have done it without any consideration for my own body. If he said, "That's a runaway train, stand in the tracks and stop it," I would have done it. If he said, "Joe, that's a tank, stand in its way and make it stop," I would have done that too. But this coach was different.

My first couple of days, I got confused quite a bit because I didn't always know the plays. I was still mean as ever, but I was the new guy and some of the guys there, one in particular whose place on the team I was determined to take, he up and told me, "Pansy boy, stay out of my way, you may have been big stuff in college, but you ain't nothing but my bitch here. This is the big time." I had been called things before but never that. But this coach, well, the short of it was that he was the worse, he hurt my feelings.

After I made a mistake in one of the drills, he waved me over to him. I thought he was going to say something nice about my effort or how much he admired my speed even if it was going the wrong way. Instead he looked at me without saying anything until I just stood there looking dumb. Then he says, "They call you mean, right?"

"Yes, sir, they do."

"Why?"

"Because they say I am the meanest player on the field." And then he turns away from me like I wasn't there and "heard him say to one of the other coaches. "After practice go run his bathwater for him and get some bubble bath and some rubber duckies and some diapers. I want to see if this creature gives a shit." And that's when I thought of what one of the other rookies told me about him. They called him Noll the Mole because of the way he could bore right in and dig out a man's heart. Some said he was more mean than even me. And, what was left of my heart after he said that made me want to cry. And, I did, right in front of him. He looked at me whimpering and said, "All I have seen today is not some Mean Green Monster, but a pansy, and worst of all, you're nothing but a Dylantaunt." I did not know what a Dylantaunt was but I did know what taunt meant and I did know someone named Dylan. I also knew that if I ever saw him again I would not be expressing any admiration. I would show the coach I wasn't afraid of any Dylan and wasn't going to allow the coach to hurt my feelings any more. And no, I would not allow any Dylan to taunt me. I had spoken to myself. I was ready.

Well over the next three years I was a star and all because I wouldn't let Noll the Mole get me down any more although he still hurt my feelings and he would make me cry. And finally the prophecy was fulfilled like in the Good Book. We were in the Super Bowl and before the day of the game, we were having a team breakfast in our hotel and who did I see but the man who I had learned to hate because people thought he was writing songs about me like "Her a Cain Transy Farter" which he sang real nasally like everything else he did.

> You may be lean,
> But you ain't mean
> You're just a wuss
> With a real soft tush
> You may think
> You're just so keen
> You're just a wimp
> The not so mean
> The soft Joe Green.
> You may think that you got art
> It's just a bean
> And a real bad fart.

After that, he made a lot more songs about things blowin' in the wind and, like I say, I never forgot that. So when I saw that he was supposed to play the half time show for the Super Bowl, I found out that he was staying in the same hotel as we were. I didn't think about it. I just knocked on his door. He opened it and I said, "Hi, Mr. Nasally, remember me?" I didn't wait for a reply and instead I pushed him out of my way, found his guitars and I just bit into each one until I broke 'em so they couldn't be used ever again. He tried to stop but I said, "Remember, I'm not your Dylantaunt ever again unless you want to get hurt." Then for good measure I pushed him just enough to break his nose and one or two of the fingers on his left hand so he couldn't play even if he bought a new guitar. Of course, whenever he wanted to stop performing, he always told people he had another motorcycle accident. He might have had many over the years but I think that was so I wouldn't confront him again when I was near where he might be singin'. No one ever called me a Dylantaunt ever again, not even Coach Noll the Mole.

Big Joe
Big Joe
Big bad Joe
Every morning at 11
You'd see me arise
Stood 6 feet even
Weigh 1 – 7 – 5.
My feet are flat
And my brain cells too
Doesn't really matter
'Cause my heart's dead too.
Big Joe,
Big Bad Joe.
(Another song I wrote for a guy who
sells sausages and sings about it)

JOE'S BIG GAME

BEFORE THE GAME, THE THING THAT IS AND THE THING THAT WAS. Before the big game, I was sitting in the locker room sharpening up my cleats just so I do something really mean on the Astro turf and tear it up before being caught. You see I knew back then that it was true, it is easier to beg for forgiveness than it is for permission. But something was bothering me. I felt something funny in my spine like I was being watched by an alien something that should not have been anywhere – I felt naked, exposed, even though I wasn't even on the playing field yet. I knew the significance of the tingling from all along my back to my head. Something was very, very weird. The vibrations never lied. My blood was pumped and my body was hot with adrenalin. Still, that feeling, there it was.

So I turned around while yet standing on the side lines of the Big Game. Yes, I was the man who led the Pittsburgh Steelers to their glory days. Still I felt funny. Something wasn't right. I faced the crowd in the stands looking for the reason that that something kept tugging at me and sayin', "Joe, something ain't right. You're still mean, and you're Joe, and you're green, but something is amiss." Then, I looked to my left and only three or four rows back, I saw something - like Robert Frost had said earlier in that poem from when I was in Mrs. Gravely Pitts class - "for once then something." It was a familiar face in the stands sitting next to a woman whose shape I remembered from my boyhood. There she was with a long scarf around her neck that draped over her bosom which I remembered so well. Even without my most wonderful glasses, I could still recognize her beautiful ... well, you know what I mean. But there was something more.

Sitting next to the most wonderful woman was the man who expelled me and he was wearing my glasses. That's right, he was the one making me feel naked. He was looking right at me and I wasn't going to have any of it. He had something of mine given to me years earlier by that Teutonic guy named JM. I wanted them back. Instead of looking out over the gridiron in preparation for the game, I threw my helmet down and crashed through the police line that was supposed to keep the crowd from getting too close to their heroes like me.

I ran past cameramen, sports announcers, hot dog vendors, autograph seekers and even guys selling beer and other stuff. Everything seemed to move about me in slow motion as I careened off one thing and another such that by the time I got to the front row, I was awash in mustard, Budweiser, cotton candy, and the screams and cheers of the many who thought I was just engaged in a kind of pre-game onslaught to rile up our fans. But that wasn't true. I jumped into the stands to claim what was rightfully mine.

I got to the woman and picked her up and threw her over my shoulder and inhaled deeply the scent of rosewater and lilac and was about to run from the stands with Mrs. Whachamacallit in tow when the man who took my glasses fell backwards while sheepishly offering them up the me. "Joe, put me down!" Mrs. So and So yelled. Well, how could I do otherwise? I grabbed my glasses from his hand and gently placed her back in her seat. I figured it was as close to a fair exchange as I was gonna get since it wouldn't have been right to keep her in my locker during the game and, besides, it was against team regulations to keep what we use to call women on the sidelines. But before I ran back out onto the playing field I sniffed her hair one last time. Unfortunately, the rest of that day is a blurr.

All that I can remember is that after the game I found myself sitting alone in the locker room wearing a suit and tie when another really good lookin' dude who we called "The Juice" walked in holding a white football. He walked up to me and asked, "As one black man to another, where'd you learn to play like you did today? You got soul! And here, I wanted to give you this as a token of respect. I never before was hit so hard by one man many so many times." He handed me the white football which he had autographed. Then he said something I will never forget. He said, "If you got another pair of them cool glasses, I'd like to give 'em to my wife. She collects that kind of stuff." I said, "Juice, I don't have another pair exactly like these, but I had a replica made a couple of years ago that are very similar to these." I reached into my suit pocked and pulled the replica out and gave it to him. He took them and said, "I hope she never loses them 'cause it is bad luck to lose glasses. I hope she remembers that." That's all I remember. But I never played football again. The stuff that happened afterwards, however must have led me to where I am now. I say that because after that Big Game, I was never the same. I think I got hit on the head or something but when I got my memory completely back I had changed and so had my world.

"Really? Is that what they meant?"

NOVUS ORDO SECLORUM

TRUTH INJECTION NUMBER 8: A GRAVEYARD LOCATED SOMEWHERE IN THE SWISS ALPS WHERE AN ANNUAL GATHERING OF THE LIVING AND DEAD PLANS HUMAN DESTINY WITH UNSEEN HANDS.

In the early evening as the first stars begin to appear against the moonless sky, Bartholomew the Gravedigger busied himself and his wife with the chores associated with finalizing the reception to the night's events. His wife The Helvetian Princess, Brunhilda the Charmed, and she of the crushing handshake, bring the evening's elixir to a boil as she throws in the magical and patented blend of soy, eye of Gingrich, Moderna Maelstrom, Toledo glass, tapioca, tenderized road kill, cigars ala Clinton, saffron, and yet another concoction of cinnamon and Japanese beetles previously fermented in vinegar cider and JuJu Power Sauce for taste. The brew's aroma of the near divine inspiration wafts through the immediate environs into the graves of those not yet fully integrated with Mother Earth.

The elixir's rejuvenating effects included a new found reanimation of flesh and bone. The dead no longer stayed permanently buried as the mere invading of their earthen coffins provided a rebirth that only the Christologically secure would deem uncontested truth. They were once again alive as they dug themselves out from their burial chambers and dirt. Some, perhaps already too far dissolved, took of themselves what they could and dragged their corpses whether whole or in part to the home of Gravedigger Bartholomew and Brunilda the Charmed to sup upon the food of life. They drank it with the orgasmic response that caused them to shake and rattle their bones like a southern preacher putting his Gospely pianist on display for his entire congregation to see and hear. It was exhilarating and the only question among the moribund and the already deceased however reanimated was this – is The Killer in the house? Would he play for them tonight?

Little by little, limb by limb, the dead ascended from the nether to the heather where the good and refined cruised and casually drove round their crawling, mauling, hauling cadavers seeking new life from the nectar for the dead. Their limousines arrived in a caravan of conspicuous consumption complete with ladies of the night entitling benefits of desire to their benefactors seeking friends. Eyeing them enviously, the dead yet yearned to breathe again to be among the consumers of fleshly delight. Some looked about anxiously waiting to see if Dante or Achilles might yet appear to condemn the living for their surplus of indifference to the suffering that might soon be theirs. But tonight, the one night of the years, the dead hoped for the soulful redemption of their lives of near mere oblivion at their final earthly dissolution.

On this one night of the year, the flies that otherwise afflicted the decayed with their meat eating appetites, abated as they themselves were consumed after they dove into the evening's broth of soy and the strangest of spices. Some of the dead with enough limbs to still hold themselves erect sat in chairs and at tables to reminisce about their lives and their speculation as to whether or not oblivion was truly attainable or even desirable no matter what the instruction of a Buddha or other advocates of the admonishment that all life was suffering, including the after life.

An hour and half after full darkness fell upon the assemblage, a Soviet era Hind helicopter landed at a distance amidst spotlights surrounding a meadow. Only minutes after landing and the cessation of its engines, a side door opened and the evening's host, surely one of the world's most enlightened and beneficent of men, alighted. Despite the floodlights surrounding the helicopter landing zone, the figure was barely discernible as he moved quickly to a waiting limousine accoutered with two skull and crossbones flags, one to either side of the black hood near its grill.

All eyes and sockets moved to follow the migration from the helo pad to a large tented pavilion that lay just feet away from the Gravedigger's house. Brightened with Italian bulbs and torch light, like a Japanese lantern awaiting the heat to send it aloft, it fluttered and seemed to waver slightly in the gentle night time breeze.

After a slow drive to the point of embarkation, the limousine with its skull and cross bone flags cruised to a gentle stop just past the pavilion and in the cemented roadway now occupied on either side by the world's elite and their long and not so long departed brethren and sisters. The multitude hushed and those that could, held their breath while others merely stood, squatted, or leaned upon others truly rubbing elbows and flaking off bits of calcium in the process. Oh Death, here then is thy sting. Would for these poor and afflicted remnants that calcium or the K2 were something less remote than a mountain in Pakistan.

Over a loud, loud speaker, a trumpet blast and he who is without sin emerged from behind the blacked out windows of the finest of limos. A portly man in a dark blue suit wearing a matching tie stood before the adoring multitude as they nearly wept as they were transfixed by the magnificence of his rotundity, the tie flapping over his belly. He smiled at the Adoremus, placing his hands together in a kind of prayerful gesture as he bowed benignly in the cardinal directions. For this man, this quintessence of hush money was non other than Charlemagne, Vizier or Exchequer to the World, known simply as Raja Mahatma Schwaby.

He spoke, "I am Who be, Your Host of Hosts of Heaven and Earth! Love me as I have loved you. For I bring you the liquid Mana, like a punch from Mohammad Ali, I deliver, and you now drink its rejuvenating nectar for I am good. And here in my homeland of the Swiss I bring you more than cheese and clocks. No, no, for today I bring you the future, I bring you the image of the META MAN whom you shall worship as he fulfills your earthly desire to rule over the entire world no matter what the fools of democracy, aristocracy, and plutocracy would embrace. For I myself know, for I myself embrace the META MAN whose truth and rule is beyond your understanding. Trust me! And as I will say soon as I stand before the puny judges of the Hague, 'Trust me, but never Bust me!'"

"Today, I believe that I bring you not merely my presence but also a wealth of entertainment the likes of which you have never before seen. I myself will not be your host for there are those in the past who have asked the question, 'Why is it that in the past our hosts have been living people and not selected from among the dead and dying who abound in our cemetery or in those of other nations. He or she, you said, need not be the Swiss dead, for we are humanists. Let's have a guest host from another country to reveal this evening's surprise, the introduction, and here he stopped and turned to his chauffer who handed him two broom like sticks upon each of which was a large bear paw. He held them both aloft and said nothing as the crowd looked on in wonderment and adulation for he had brought a metaphor to life rather than merely pausing as others might have done. Such a rhetorical flourishing in total silence was nearly incomprehensible to most. Then as if it were indeed one hand clapping and then eventually two, and then thousands, he placed the claws back in the car. He spoke again, this time in words.

"Behold!" He gestured mightily and repeated as he turned to look upon the tented pavilion, "Behold!" After a moment more of stabilizing himself physically, lest he suffer an aneurism from the force he was about to unleash, a bolt of electricity from the ground smacked and zapped against the top protruding tent spike holding all in place. Sparks flew and a sort of Teslan whirl wind unleashed itself and images swirled on high of Ben Franklin, kites, and skeleton keys, the latter of which found itself particularly intriguing to the dead. Even as the electric light show evolved into a sort of Picasso like swirl or painted incomprehensibilities, the crowd, awestruck, looked on with both wonder and trepidation.

Then, the electrical-like show of zapping and zipping above the tent gradually subsided until the lightning effect disappeared with no more than the Italian bulbs and torches offering the only source of illumination. Amidst the darkness, directly about a hundred feet above the pike, the main shaft of the tent's support, a slight undulating whiteness secreted into the void. At first little more than a slightly oscillating light fixture, the wavering light grew in size and intensity. Some watching from below covered their eyes as if protecting them against sunlight. Yet the blackness all around remained. It was as if the light left the night to itself while penetrating the sight of those yet soulful or those bereft of theirs, not having nurtured its bequeathal in their care.

The Raja Schwaby announced," Tonight, I who have lived so long and so righteously do on this night declare that our fulfilment of a new world order is at hand. So, it is that I have declared that the spirit of past does this night merge with the present. Hence, it is that therefore your host of hosts for this occasion who was while living the greatest gameshow host in all of history, the one, the only, the only, the one, the inimitable, the fastidious, the capricious, the spirit of Christmas and of the hellacious, the METAMAN OF THE UNIVERSE, the Samurai of Showbusiness, -- join with me now as we dare to say his name, YAAAAAAAA – MAAAAAA – FUUUUUUUU –CHIIIIIIIIIIII – YAAAAAAAA – MAAAAAAAA – GAAAAAAAAAA – RIIIIIIIIIIIIIII! Yamafuci Yamagari! The Host of Host! "

As if his words were conjoined with the will of the heavens itself, the cloud above the tent expanded greatly into a full fledged orb filled with a sort of blueish almost watery element out of which emerged first the face and then the body-like being itself, the heavenly anointed Samurai of Showbusiness, the curator henceforth of humanity's destiny, Yamafuci Yamagari. As the crowd shook in orgasmic delight at his presence, he walked forth from the Orb of Greatness and slowly he began to descend an imperceptible stairway from on high. As he did so, thirty Kodo drummers from Japan revealed themselves, fifteen to each side of roadway by which the limousines had arrived only an hour or so ago. Temporarily hidden in an underground enclave on either side of the roadway, like a stage floor, the hydraulics lifted them into view each with a six foot high drum on which they pounded out the rhythmic heartbeat of Mother Earth itself. Yamafuci slowed his descent as the drum beat grew louder. The crowd in sympathy with its pounding, their own hearts began to beat to the same rhythm. Like children brought back into the warmth and comfort of the womb, the crowd below quieted. Some wept uncontrollable at the spiritual umbilical cord now connecting them to their pre-birth past. Heart and heartless all were unified as they watched in awe as Yamafuci walked out onto an unseen platform on which he now stood. After five minutes of a brain numbing and pulse pounding liberation of anxiety for all, the drumming stopped.

"I am sent from above, for I am METAMAN. I am real and unreal, for I am of the divine digitalization of myself. For though dead, I now live and shall direct your lives as we come to that climactic moment in which all history comes to an end. I am your avatar." Miraculously, all heard him even though he never appeared to raise his voice. Another sign of his having the mandate of heaven and Jack Dorsey, Mark Zuckerberg and Bill Gates. For all these are of the Oracle of Wisdom, Chips of Knowledge.

Without saying another word, pointed both of his hands to something in the distance now approaching the crowd on the same roadway on which the limo crowd and others had entered the point of assembly. The crowd turned in near unison to see a large white sheets covered object being towed on a very large platform with sixteen tractor trailer tires. The laborers, looking tired, strong, and dirty numbered in the hundreds and sang a mournful song reminiscent of the Vulgar Boatman condemned forever to wander from continent to continent for their cruelty. Their low vibrations rivaled the Kodo base and seemed to imitate the interminable sameness of Mongol throat singing. At the sound of whips on flesh the leader of the two lines of rope laden laborers began a more familiar sound of oooooooo weeeeeeee ooooooo, ooooooooeeeeeeooooooo.

When the slow moving monolith at last arrived at its designated staging area, the Vulgar Boatmen departed to the sound of whips for a temporary reprieve from the strains of their exertions. Then at a signal from METAMAN, several dark suited personnel from the grave yard burial teams ran out from the tented pavilion and untied a series of ropes on the platform and at once the sheets covering the monolith dropped to the ground. There facing the crowd, the living as well as the dead, stood a two-headed beast. Only one head faced the crowd while the other could be surmised by an abundance of long hair as belonging to a face gazing in the opposite direction. For a moment the crowd was quiet, standing and sitting in shock. Then one shouted, "It cannot be! It cannot lead! We are doomed!" The crowd disentangled itself from their paralytic shock with a murmuring of discontent.

"Silence!" uttered a now apparently annoyed Yamafuci Yamagari. "I am the great and powerful avatar of avatars. I am the digital manifestation of your deepest yearnings and I say unto you that today there is no other door from which you my chose. Before you is the selection, the determination as to who shall lead the world. Therefore, I saith unto you, 'Embrace the past that you may know the present and create the future, for it is he who shall lead you. And I assure you there is nothing behind doors B or C. Today those choices are not available. Take what you are given. Now join with me in our chant of promotion which shall change the world. Now say it, 'BBB. 'Say it with me, 'BBB. 'I can't hear you! Say it now, 'BBB. ' The nonplussed crowd stood in stony silence. Finally, METAMAN relented and asked, "What is wrong? Why aren't – "

"Because we don't know what it means, what does that thing with Joe's face have to do with BBB?"

"Oh, is that all?" asked METAMAN." It means Bring Back Biden."

"Oooooo," acknowledged the crowd.

"Isn' he already dead?" shouted yet another in the crowd.

"Yeah, he's dead, I saw him talking yesterday on the telly. He's dead."

"No, no, no," replied METAMAN.

"No, he's dead. I saw him too. He's dead. He tried out for a remake of that movie and they said he would need a brain to memorize the Comrad Scare Crow's part. Yep, he's dead."

"Nooooooo, oh please, don't say anything more until you see the she goddess we have enjoined to join in our efforts to remake the world. She too is part of our deliverance from frustration. She you will admire."

He then motioned to the handlers to turn the platform around, which was delayed until the Vulgar Boatmen could be re-enlisted and brought back into immediate service with whips and more threats of more eternities spent looking for work as strong but generally rebellious muscle marauders. However, ten minutes later, the other half of the statue now faced the crowd with 'the Joe' now looking the other way so that only the back of his head was now exposed to the crowd. The remaining sheets were pulled down and the crowd groaned and gasped in fear and anxiety.

One turbaned observer shouted out in fear, "It is the Devil Queen of Cali! It is she, the Black Queen of Destruction, the Jamaican Jehovah of Bad JuJu."

"It gets worse," shouted another, "She's Canadian and she won't eat bacon. Even the French don't like her."

"Yeah," offered another, "She's supposedly the goddess of time but she doesn't even wear a watch and she called her opponents 'thugs' but she's the only one who created a secret society of thugs. Beside, if she's the goddess of time, how come she always comes late to meetings."

"I don't like her, take he back, her laugh makes me cringe. Or at least put a muzzle on her or something so we don't have to listen to her."

"But you must understand, guys, she was the only queen who was still available. Hillary said no until Joe dies along with Kamelia the Queen of Destruction, as you say, in a bizarre plane crash on the way to rescue children on Epstein's island."

"Don't make us laugh, why would that Jeff guy let anyone take his kids away since Kamelia the Cali Queen still delivers kids directly from Jamaica and Haiti."

"Come on guys, I am the great and powerful—"

At that precise moment, the image of METAMAN and the cloud that had seemingly delivered him, disappeared. Momentarily, a dog appeared at the entrance way to the tented pavilion dragging an electric cord behind him with the plug hanging to one side. "Oh, Otto, you bad doggie, now look what you've done. Now we'll never get home."

"Ok, Ok, listen up meine kinder," interrupted Raja Schwaby. Look things haven't gone quite the way I had planned but it's too late. We have no alternatives, all the good people are either in the other party or they don't want to be a part of this. So, come on, for old times' sake, come on, meine kinder."

"What's in it for us?"

"Well, I thought you wanted to control the world."

"Not really, it's kind of boring. No one really likes you, they just pretend and then you end up like Bill over here. Sorry, Bill, but you're kinda' dorky, you know."

"Wait, wait, I have an idea, let's play a kind of riddle game. You answer correctly and you get a free trip to Disney and you just have to vote the way we tell you, OK?

"Oh, all right, go ahead."

"Here's the first question. What man, whom you all know, is a convicted felon in the country of France and can't even get that conviction overturned with all the money he has?"

"That's too easy, I want a hard one."

"You just say that because you don't know the answer," replied Raja Schwaby.

"It's George Soros," rebutted the bored but challenged aspiring contestant.

"Ok, let me see. How about this one: The president of the Phillipines was so angry about this man's currency manipulation he said that if that man steps one foot in his country, he will arrest him and hang him, no trial, no jury, straight to the hangman."

"Man, these are so easy demanded one other yawning member of the crowd. Something really hard."

"Ok, this one you'll never get. What prominent person whom you all know owns a villa and ranch in eastern Ukraine and acts like he doesn't know what you are talking about?"

Immediately, several hand went up waving them wildly and nearly falling out of their chairs to be called upon. "I know, I know." "No, me first, my hand was up first."

"That' too easy, again. It's George Soros."

"Got you! You are wrong! George does not own two places in the Ukraine."

"Ooo, that one was hard, unfair," chimed another.

Raja Schwaby smiled cunningly and said, "It's wrong because it was a trick question. He owns *five* places in the Ukraine." Ha Ha.

"Give us another chance, but still make it hard."

"Ok, same question. This prominent person whom you all know still owns two places in the Ukraine. Who is he?"

"Can I call a life line?"

"No, give me a good answer."

"I know, I know, I know, it's that British guy with the Russian accent who shoots people."

"No, no, you are thinking of Kyle Rittenhouse, only I don't think he has a Russian accent even when he is shooting people."

"Over here, I think I've got it. I think it's that black guy who was arrested for wearing a rope around his neck when he was racing his car. I think they fired him because the rope was dangerous 'cause it could get caught in the wheels and break a guy's neck or something."

"No, no," volunteered another, I think that was that Jussie guy named O. J. who killed his wife for having an affair with a tiger or something like that. I saw it on television."

In exasperation, Raja Schwaby threw up his hands and uttered, "No, no, not even close! "

"Well, then who?"

"Joe Biden! " he retorted. "Joe Biden! "

"No, you are all wrong," responded the hither to quiet girl who stood by holding Otto. I watch the news, the Russians blew up his villa yesterday. So you see, he has only *one* place in the Ukraine. I win."

"Ok," sighed a defeated Raja Schwaby, "What do you want for your reward?"

"Otto and I just want to go home, please."

"Where do you live?"

"In Kansas with Eminem, he's my fiancé."

"No? You cannot be serious, meine frauline, no one lives in Kansas. It is too flat."

"We know, when I was a little tiny, tiny girl, before I was born, my fiancé even wrote a song about us Jayhawkers entitled, 'Topeka Girls Go Round the Outside, Round the Outside. ' But it didn't do well because everyone said, 'Any Topeka girl who does go outside keeps getting blown away, like I did. That's why I am here. So anyway, he wrote a different version which did well nationally, but not in Kansas. So how am I going to get home to see Eminem?'"

Raja Schwaby called over to his chauffer, Fritz, "Herkommen, take all zie umlauts from the cadaver bins and place them in the tent pavilion. Once there fill them with helium and send them aloft with the little girl and her Otto so they can go back home. Make certain she has a GPS and a Messerkadoofus for automatically changing course for Topeka. Alert the Civil Air Patrol so that they can be on the lookout for dog named Otto and a girl aloft with a bunch of umlauts."

"Ja sicher, meine Raja."

"Oh thank you, thank you. I'll always remember you."

ONE HOUR LATER

Against the lovely blackness of the night time sky, the tented pavilion began to lift from the earth with Otto and the little girl leaning forward through one of the openings. Simultaenously, in an aesthetic display of cross cultural appreciation, Raja Schwaby gave the order to release the more than five hundred Sino/Japanese lanterns which ascended peacefully and pleasingly all about the tent. As the little girl and her dog floated heavenward and Topeka wise, Raja turned again to Fritz the fulfiller, and said, "Call the Topeka Board of Education and tell them that that little girl also suffers from a rare form of dyslexia causing her to mix up the syllable order in very simple words. Unfortunately, the doggie will not understand what his real name is. Understand, Fritz?"

"Totolly! "

The admiring attendees watched as the girl and her companion travelled upwards. They waved, "Good-bye, my little schnitzel britzel, good-bye, sleep tight, don't let the umlauts ignite, say little girl, remember, when the wind blows hard in Kansas, stay flat and don't look up."

THE PANAMA CAPERS

PART III

THE CAMPAIGN YEARS

DEMOCRATIC NATIONAL HEADQUARTERS, WASHINGTON, D. C.

LOBBY: John Remmick, renowned journalist and extensively published poet, was escorted into the front desk area where he was given a VIP pass enabling him to move about freely in the newly updated facility of the Democratic National Headquarters. Today, he and a select group of other 'influencers' were to attend a special Beta-version of the head quarter's new and improved version of its first time ever Meta update of all the computerized and digitized material within its walls.

A hostess with a red white and blue scarf around her neck and a white blouse, smiled broadly at him and others and said, "If you have special needs or concerns I suggest that you consult with Ms. Daley who is in charge of security and special access areas that require biometric access only. Those areas are off limits until she herself allows you in. Consider her today your friend, your confidant, and your sister. She's here for you. Depart to the various areas of the special interest that each of you may have, you may wish to have a human escort, such as, those you see milling about in their blue suits and American flag lapel pins. Otherwise, we have decided today to offer you for the first time ever, our robots. These are lifesize and lifelike creations done by the DNC and Disney studios. Seth Rich — nevermind — I mean —well, I meant, Oh good, here they come now,"

Upon turning to look in the direction in which the hostess was pointing, John Remmick watched in amusement and borderline awe as two robots approached him, walking hand in hand. As they drew closer, his awe turned to near bewilderment, realizing as he did, that one robot was an exact replica of Ruth Bader Ginsberg, and the other, another lifesize replica, this one of Antonin Scalia. Remmick could hardly believe his eyes, both were unbelievably exact replicas of those real life individuals that they represented. The features, the stature, the walk – it was all there. However, what was missing? Something, something — then he noticed a slight humming coming from both robots. In fact both were humming one of the most beautiful arias in all of music, Puccini's "Nessum Dorma." Long known for their love of opera, both though long dead, were seemingly alive and humming that most fatalistic moment in all of operatic tragedy. For a fleeting second, Remmick considered the rumors that abounded about the peculiarities of their deaths, particularly those associated with Scalia's passing. That momentary flickering of the desire to know more, however, passed as he turned to examine a series of paintings depicting all of those members of the Democratic party who had been elected to the nation's highest office.

With the contingent of other 'influencers' having gone off in the direction of conference rooms and the area of politically like-minded coffee klatches, he found it slightly off putting that the woman who appeared to be the head of information and security was casually but pointedly only a slight distance behind him. Upon his turning to observe her, she stopped seemingly embarrassed at having been caught in the act of, shadowing him. This made him wonder at what point in his wanderings she might approach him to ask him something about himself or his thoughts on what he was observing. Upon stopping to gaze upon the statue of the man who is considered the founder of the Democratic party, the two robots came abreast of him, still humming, but not saying a word. As he eyed the two creations of Disney engineers in working with Google software developers, he still could not get over the life like quality of their creation.

Slowly, he moved on, past Andrew Jackson, Harry S. Truman, Woodrow Wilson, until he saw a couple of others from the original group with which he had arrived. They were standing in front of a portrait of Abraham Lincoln. "I didn't know he was a Democrat," remarked one of the women as Remmick was approaching. Remmick paused long enough to consider what she had just said. At that moment, behind him and off to his right stood The Shadow, as he was want to call her. Remmick maintained his focus on the picture as the others walked away and the robots idled in the general direction it seemed he would head next.

The Shadow then came and stood next to him and asked, "What do you think?"

"What do I think? It should be obvious, trying to claim that the first Republican candidate for the presidency of the United States, coming as he did from the newly formed anti-slave party, seems a bit disingenuous to say the least. Why the propaganda? Just admit he was a good man who belonged to a party that sought to abolish slavery and still hold the nation together. At this point in history, that shouldn't be such a terrible admission." She scowled in response, replying, "Too bad you can't bring yourself to admit the truth. There are consequences."

Remmick looked her straight in the eye and said, "You're not going to try that line on me, are you? I know a thing or two about propaganda and that is exactly the essence of what you're now saying."

She turned and snapped her fingers. "Seize him!" she shouted and immediately four FBI agents tag-alongs of the original group of which Remmick had been a part upon first entering the building—they grabbed him around the neck and both arms while another threw handcuffs on him.

Two minutes later, despise his resistance, Remmick found himself inside a men's room. Pushed to the end of a row of stalls, someone other than the four men detaining him, rushed past and unlocked a room that at first one might have thought to be a maintenance closet. Inside, however, the lights already on, revealed a large room that extended well beyond the wall that seemed to be the last foot of the area with the stalls and urinals. Once inside, he was pushed into a plush comfort chair with leg braces on either side into which he was expecting someone to clamp his shins. Instead, two of the agents stood next to the door through which he had just been pushed. Though wearing the conventional FBI special agent jackets, one of them motioned for him to relax and remain seated as he himself sat down opposite Remmick. "Sorry for the bum's rush, we have something important to tell you. It couldn't be helped. Turn side ways and Jeremy will reach around in back and unlock the cuffs." Upon doing so, Remmick rubbed his wrists and stated, "Why?" No one said anything in response.

The fourth agent unlocked yet another door facing Remmick, out of which walked a man of no less than six foot six and wearing a brown suit coat with matching tie and shoes. He smiled slightly as Remmick sat quietly and patiently while trying to catch his breath. The man crossed the room and offered his hand which he refused. "I understand," the tall man responded. The agent sitting opposite Remmick stood up, offering his seat to the new comer. He then nodded to the agent who had unlocked the door to let him in. The man then reached into his jacket and pulled out a brown envelope and opened its flap. As if he were offering Remmick a cigarette, he motioned for him to reach and extract a small stack of what turned out to be photographs. As he looked at the first several photographs, his eyes widened and he squirmed uncomfortably, and then glared at the tall man seated directly in front of him. "Your wife is a beautiful woman." Without responding, Remmick continued looking through the photographs which he recognized as those of his children taken at school, at soccer, at baseball and even a birthday party. Remmick's look of alarm was calculated by the tall man as being perhaps the appropriate time to offer more. "Your sons are safe and from what we see quite adept on the athletic field." You are here, however, because you know more than you think you know and we require your cooperation. Perhaps before you say 'no,' you'll go to the bottom and pull out the last picture. I'm sorry but it had to be done." The man waited as Remmick looked at the bottom photograph. He blanched. "Sorry, but your sons will love you all the more when they see the new puppy which I am sure you will feel compelled to provide."

"It's a dog, why?"

"We need your cooperation and we will not accept no for an answer. So now, we have plans for you, but first allow me to introduce myself. My name is Carlos Ignacio Armando and I work for a government agency which is calling upon you to perform an extraordinary service. You and your family will be protected and we will allow no one at any time to imperil them or you. Here, you'll need this. It's a throw away phone but one which you will know better than to ever do that. We will, however, contact you exclusively on this device and no other. You need to carry it with you at all times. I'm sorry our time with you today has been less than pleasant. You'll understand why only later on. Good morning. Please take the pictures with you and dispose of them, especially the one of the dog. You don't want your family to see it."

The man departed through the same door through which he entered and the agents standing at the other door stepped aside and opened it, nodding for Remmick to depart.

Still reeling from the entire experience, Remmick walked hurriedly past the front desk, outside into the sunlight. As he did so, he noticed once again the two robots as well as one that looked strangely like Clarence Thomas. He paused just for a moment and considered, "So that's why they're so life like, why there were no autopsies." Horrified, he continued to the exit. He walked briskly toward his car parked at a distance. As he was about to open his car door, he saw standing no more than 50 feet from him The Shadow. She smiled, waved, and pressed her cell phone key. His new throw away phone immediately rang. He answered and she said, "Sorry for the rough treatment. We had to do it that way for your protection. The other women that were standing near you as you looked over honest Abe are working for the other side and had to be convinced that you were arrested so that we could prevent them from getting their hands on your first. Enjoy your time with your family. Again, sorry."

Upon her hanging up, Remmick immediately drove home thinking it best not to mention this to any of his family, including his wife.

MEMORIES

BARN TALK: Joe lay on the ground, straining repeatedly to clear the encrustations that had seemingly sealed his eyelids shut. He felt a gentle touch and a soothing voice, a warm compress whiping off his face and the droll around his mouth. With several soft and gentle massaging fingers around the eyes, the Sandman's nightly deposit was wiped clean and the world of the present, the here and now, unfolded gradually and welcomingly. "Joe, Joe, are you okay? We found you on the floor and thought maybe at first we had lost you. We found you lying here next –"

"Where am I? Who are you?"

As she gently pushed him forward to sit up, he looked at her strangely and asked. "Did we win? I think we did."

"Joe, it's me, Jill. Joe, I don't know what happened. We found you here on the floor. It looks, from that bump, that somehow or another you hit yourself in the head with the crane you were using for the car wheels. You never did get them on entirely. But those are really big pumpkins."

Joe looked down at his hands and asked bewildered, "Where's my ring? My ring? Where is it?"

"What ring? You don't wear rings, least that I know of."

"My watch is gone, too. Look! " Joe held up both hands and twisted his forearms from the elbow to his fingertips as if doing so would confirm that he was not hiding anything. The ring and the watch were certainly gone. "Did you take it?"

"Joe, I don't know what you're talking about. The hit on the head has changed you. Something's different."

"My World Series ring and my watch for being selected. I was the MVP. I beat everyone. I was the best."

"Joe, you never played baseball. You said you thought it boring. But, but, Joe, do you know who I am?"

"Yeah, yeah, you're the nice old lady who makes me cookies and milk."

"Joe, I am not old! Look at me! Snap out of it! " She slapped him across the face and waited for a response.

"Oh, now I remember you, you're Kamilia, the Ever Changing Raptor of, of, of – I don't know."

"Joe, something's different, you don't look the same, you don't act the same and you don't talk the same. I think the old Joe is gone."

EDITOR'S INTERJECTION:

In private conversations with Jill Biden, she confided that the transition in Joe's life from the ever past to the never present was occasioned by a severe bump on the head. This bump would have killed an ordinary man, but Joe, at that moment of possible transition from the world of the living to the nirvanic climax of oblivion, was occasioned by Joe's looking upon the suffering of all humanity and willing himself back to life so that he might assuage that suffering and bring hope. Jill, who is herself a doctor and therefore a long suffering proponent of self-medicating belief in the need to help the afflicted with academic discourses on how people get what they deserve, says Joe may himself be the exemplification of Big Boat Budddhism. He wants everyone to be in the Boat, like Noah, except the navigator will probably be an enlightened Indian from Mumbai where there aren't many boats. I hope this clears things up as we consider that Jill herself believes that Joe lives today by the motto, "Just say 'NO to NIRVANA. You don't deserve it. '"

PRODUCTS OF THE BARNYARD SLAM

Joe and Jill went up the Hill
To fetch a Kill for Moloch.
But Trafficking in Children
Was what they then decried,
But then they saw the profits there
So broke the borders wide.
And kiddies in the crossings
Make pesos for the cartels,
And women in submission,
They get to live in hotels.
Oh, how they love the kiddies,
And the mamas with their titties,
While the Queen of Cali laughs a lot
While shouting that she 'Cannot, Cannot, Cannot.'

Your Mayorkas, My Mayorkas
Makes the border porous,
All the while he's lying, proceeding to ignore us.
So the river ain't so grande,
And fentanyl a mainstay,
So compassion out of fashion
For the dead and dying in our land.
And Daddy holds the contract for the Hunter China band.

So the money from the Zhi Ping
Leaves our children weeping.
And Hunter has a long string,
Connected by a draw string –
So the commies do the lacing,
The drug war we are facing,
Fentanyl outpacing all our efforts to deny it.

You'll find it in the laptop or
Can you say it?
FSB, FSB, FSB,
Don't you ever wonder...
THE END, HAVE A NICE DAY

KABOOM AND CABAL
KAMALA AND BIDEN OUTFOX OUR METHOD AND CAN A BIDEN ACT LUCID?

WATERGATE HOTEL, WASHINGTON D. C.: A gathering of the primary influencers in the party of Joe Biden and Cali the Concupiscent. Twelve men and women in the penthouse at one of the most prestigious hotels in the D. C. area plot their strategy for the ascendance of the Prime Mover of the Universe, the President of the United States and his side kick, the Virgin Person of the Vaginal Pulchritude and Vestigial Puella, The influencers, though twelve, recognize that one among them is he who is without sin and the lover of Puer Morbidus, the Jabber Puer of the Red Slippers.

SUBJCT NUMBER 1: We all know why we are here. Today there is one more opportunity to yet gather ourselves about the object of our affection and derision, the Joe. Like it or not, he is our candidate and we must now come to a consensus as to how we can reconcile ourselves to the most unorthodox of approaches. No one I know of can or has run the universe with such a daunting lack of skill or cognitive ability. That he can do so much with so little may, indeed, however, be the most remarkable of achievements. In fact I am yet hopeful that that may be the basis of victory as we pave our plans here this afternoon.

SUBJECT NUMBER 2: I don't know how you can be so optimistic. He's fragile, senile, and hard to control. Even when he's cogent, he's incomprehensible. He's too busy sniffing and pawing every little girl that gets close to him. I'm afraid we'll have an incident. Then what?

SUBJECT NUMBER 3: Yeah, what do you mean? I don't see anything that looks good on the horizon.

SUBJECT NUMBER 1: First of all, look where we are. We are here in the heart of enemy territory. At one time, this was their domain. Now we control it. We used it once to take down one of their icons and we can do it again.

SUBJECT NUMBER 3: Yeah, I get what you say, but I still don't see how. After all, Mr. Avuncular or whatever he thinks he is – well even Jill can't control him. Yesterday, he smacked one of his aides for forgetting his galoshes. But it's summer time. Why does he need them?

SUBJECT NUMBER 1: That was just a momentary set back. He fell asleep during the opening line of Richard the III. It's not his fault. He was tired and when he awoke, I guess that line stayed with him. He just didn't want to get his shoes wet and covered in salt, snow, and ice. It could happen to anybody.

SUBJECT NUMBER 2: If he were Frosty the Snowman, maybe, but otherwise no. We're toast and he's a snowball in hell.

SUBJECT NUMBER 1: Maybe, but therein lies our hope. Remember, Dante said that the last layer of hell wasn't fire but ice. And, yes, we're there right now. My hope is that we can chisel our way out. And I think I see a possible means of doing so.

SUBJECT NUMBER 2: I still don't see how. After all –

SUBJECT NUMBER 1: Stop, stop, stop, and just consider the man himself and what possible strengths come from the very weaknesses he projects. In fact, I think you've already suggested what they might be. Ice, sniffing and pawing. Think about it. What image immediately comes to mind?

SUBJECT NUMBER 3: Personally, I think of a pervert. Hell, his own secret service detail refused to bring their wives or girl friends to the last Christmas party he attended because he keeps smacking them on the ass. The agents put it this way, "He squeezes what he pleases." If that weren't bad enough, several of the female agents have asked for transfers to other assignments because of his swimming habits. When he's in the pool, they are required to be present. Well, he certainly is. In fact, again, one of the agents quit because he once again came out naked and swam for half an hour and then asked her to get something for the chlorine in his eyes. Shit, then he she tried to spray some stuff in his eyes, he tilted his head back and pretended to be off balance and he grabbed both boobs for balance. My God, if that gets out –

SUBJECT NUMBER 1: If it gets out, it won't matter. No one will believe it. After all, think about it. What possibility is there that an old man would go around grabbing or pawing women with guns.

SUBJECT NUMBER 2: Look, Sherlock, it's already happened. Only it wasn't her gun he was grabbing. It was his own and pointing it right at her. It's been going on for years.

SUBJECT NUMBER 1: Yes, and what's come of it? Nothing. She'll never complain because she'll then be exposed as a disgruntled bitch.

SUBJECT NUMBER 4: You are aware that four of us in this room are women, or, a you might say, 'bitches'?

SUBJECT NUMBER 1: Has he pointed his gun at you? Pay attention, I have a plan and we'll all benefit from it.

SUBJECT NUMBER 2: Stop teasing, start pleasing. I'm losing hope.

SUBJECT NUMBER 1: You mentioned sniffing, pawing, and sleeping. What animal does that remind you of?

SUBJECT NUMBER 4: OK, I'll play since I am what you would call a bitch, a female dog. But right now I am growling and I will shit right here on the floor if I don't like your answer and then you can clean up and let me out. So the answer, I have already said it. Dog.

SUBJECT NUMBER 1: You got it. We are going to make him into a big lovable puppy dog so that even his indiscretions like pooping in the Vatican or pissing down his pant leg or even seeking out and sniffing things he ought not – it's all part of his playful and alpha male dominance. He's like Bill Clinton, only less compromising. He could work with people but this guy, like a true puppy in training is easily led and we can only hope his deep state handlers will not give him so much leash that he hangs himself. Well the peons will then have to cover him and clean up his crap. But that's what we're paid for. Puppies and ice cream, that's our strategy. He will be recreated in the image of Tramp who is the only guy who could impress his girl by sliding a meat ball across his dinner plate to his lady love with his nose. If he can do that and get away with it, so can Joe. After all, he can also speak and roll over without prompting, at least according to Jill.

SUBJECT NUMBER 4: Does she rub his tummy?

SUBJECT NUMBER 1: Only if he pisses on the hydrant and not the bed post.

A CHORUS OF RESPONSES: God Almighty, who could ever imagine this. What a way to win an election. We have to make a guy look subhuman in order to win an election. I guess that means we have to prevent him from humping the queen's leg next time he meets her.

"I don't think this approach would work in Korea, the doggie style."

"Yeah, the Koreans like it Gangsta Style. They must be more advanced than us."

"Really? Really? Is this how much we've fallen? The Korean doggie style of winning an election? How can that ever work?

SUBJECT NUMBER 1: As they say, in Italy and Korea, Puppy Chou, baby, Puppy Chou.

SUBJECT NUMBER 4: I hate you. I'm tempted to renounce my citizenship and shit on the floor before I leave but I want my biscuit first.

SUBJECT NUMBER 1: He leans over and whispers in her ear. I've got a boner for you right now and I know just where I want to bury it.

SUBJECT NUMBER 4: She whispers back, "If we do it my way, we'll start orally so I can bite it off and stick it up your ass."

As the others file out of the room, he smiles and says, "I love it when you get angry. By the way, stop licking your paws, it's unbecoming of a doggie in heat."

SUBJECT NUMBER 4: I swear to you, you will pay for this…

SUBJECT NUMBER 1: Remember our next meeting is tomorrow at 9AM. Our political counterparts in the House will be here to help us finalize our approach.

SCIFF BUILDING NEXT TO PENTAGON

(SECURE COMPARTMENTALIZED INFORMATION FACILITY)
INCOMING: THE FOUR WEIRD SISTERS
OF EQUITY MINUS ONE

The building was a considerable walk from the main entrance of the Pentagon but within a line of sight that some would stride to during their lunch hour if they were trying to get some exercise. Others never even knew it existed. Within its biometric layers of entry and egress, were a deliberately stultifying number of cameras and locks that made any unauthorized personnel in a given cubicle subject to immediate manhandling even to the point of being thrown on the floor and having a military weapon placed against a person's neck. But on this day, the usual cavity searchers and even the airport style body scanners were turned off. Equity required a new and more tolerant inclusion of those without the need to know or the proclivity to keep their mouths tightly sealed once having left the facility. All such requirements were benignly jettisoned. As was their custom, the three Queens of Equity ignored protocol and sauntered casually, cynically, and impatiently through the series of interlocking cubicles where they would enlightened those whom they perceived as their oppressors who needed their chastisement and frequent scowls.

Upon entering the SCIFF itself, they were greeted by ten seniors officers from all branches of the military, 16 junior officers of the same, six intelligence analysts and the same twelve members who had met with one another at yesterday's DNC strategy session.

Once the Queens entered, though of a most self-aggrandizing pedigree, not one of the officers rose, nor any of the others. The man escorting them into the room showed them to their seats in the room with amphitheater tiered seating. They were strategically positioned in the relatively dark room with a computers in front of many of the positions as well as electronic maps of all the primary hot spots around the globe.

All were at first quiet. Then the middle of the three Queens clapped her hands together and led the threesome in a contemporary rendition of a much older:

> We three Queens
> Though oppressed that we are
> Find you hateful
> And really bizarre.
> Right wing leaning,
> Still careening
> Toward your death in duty fashion,
> Cursed be you
> With no compassion.

After a few moments of silence, they began again with a spell of wizardry and witchery to bring the room full of mostly men to their feet in respectful adoration, or, that failing, they pronounced upon them all a curse to bring the men to their knees in supplication. None again moved.

Then again with a look of boredom mixed with distain, the provocateurs of wisdom mixed with violence began to chant:

> Each a goddess
> Who be we,
> We the goddess
> We be three.

We are colored
As you see,
No shit wisdom
Each in we.
I am birthing such a person
We'll take your life
Or even worsen.
No pity you
Or even nursin'
So fear us now
And stop your cursin'.

Each then hissed and made snake motions with their hands, and began again, "We the sisters of Cali, she our mistress of the night, she'll eat your gonads if you fret her, fear her now, you'll never get her."

"Still silent? Then perhaps you need to know more precisely who we are and why you must comply to bring our candidate to victory in the election against he who shall not be named."

The creature standing in the middle of the three who seemed to be their primary person began with individual introductions.

I am Pretentious Pressley, now installed, by those who love me though I be bold. No sex is referenced in what I'm called, just my suffering by white men galled. But I fight with wisdom against those destructing with power, graft and greed, though all I desire lay in what they heed, the call so clarion, like shampoo, no need for me, but great for you.

I am Ilhan of Migration
The Princess of Elation.
You complain about elections,
But I prefer erections,
Of the Minnesota Members,
I you know just what I mean.
Some call me Somali,

Mogadishu is my home,
And always where I've travelled
There things they have unravelled.
And that is why I curse you
And I will never nurse you
'Cause I don't give a damn.
So you can call me mother,
Say I'm special to my brother,
And I'll call you just a scam.
But me and Bernie will elect him,
If you will just erect him.
If he won't eat no ham.

In silence, the officers, analytics and political handlers for the Joe looked
on in bewilderment and disbelief. One senior officer shook his head and
said under his breath, "This is Congress?"

Then again, the last of the three introduced herself.
You've got the Tali-lebe and the Taliban,
One I ain't and the other I am.
When the nation is in Crisis,
And you denounce the ISIS,
Look to me// a phenomenon,
I'm a real cool Mama
Not from Lebanon
I denounce the trucker
Who called me mother fucker
Because I'm a real cool chic
From the land of Islam
Yeah, I'm Detroit West Bank,
Here and abroad,
I call you bigot
When I call you a name,
'Cause I got the privilege
That makes you insane.
Don't love this nation

And I'll never give thanks
Allah is my goddess
And I won't break ranks.

The three now stood before their audience. The silence convinced them of
their own mesmerizing presence and that of the unbroken fealty which such
well trained and now obedienct and submissive, mostly men, would manifest
in their further interactions. Then, behind them, a jumbo screen clicked on,
and the missing member appeared on it for all to see. It was she, the goddess,
not just another, not one of four, but *theee* goddess. Rather than deigning to
speak to those present, she of the solemn demeanor shared her countenance
as the words she would not speak rolled slowly down the screen.

I am she
Whom thou dost
Wish to date
But not so I
With those of hate.
I am the Angel of AOC
Say it backward and
And look on me.
Yes, the Cween Of Angles,

'Cause I'm acuté,'
Obtuse be yee
But not so me.
I'm so sweet
As a syllable bee,
Suffering spells in the
presence of a "he,"
I'm a Bernie Babe
with a pot full of honey,
So sticky with me
And we'll just make money.

When I say it now,
It just seems funny.

Then, from the confines of her zoom world, she turned to shake her behind at her camera and looked back teasingly, saying "This is mine and you can't date me, cause I'm so special. Don't you just hate me?" She turned again to the camera as the other three stepped toward the assembled as if seeking the gestures of submission. Thinking their hour had come round at last, they sloughed from the enclave in august splendor, determined as they were to depart with their political and military conquest firmly in hand.

Once the door had been closed two well known figures, the Kulprits of Kabul, Wee Willie Milley and his side kick, The Black Night, entered. Milley said, "I heard it all, more proof of the fact that White Supremacy is an existential threat and that is why we have to listen to the women of color." "I guess so," said the Black Night.

At that point several of the Subjects of the previous night's meeting pulled out a pack of TicTacs and swallowed a couple of red pills each. Within seconds, they began to chuckle under their breaths, attempting to suppress the sound by covering their mouths. But despite their most valiant attempts, they began to cry and scream in absolute delight. "What a fraud, you spineless wimps," shouted Subject Number One. He then fell on the floor, contorting in the breathless convulsions of one responding to the absurdity of what he had just watched. Military personnel, attempting to maintain their decorum, snickered and occasionally guffawed while trying to prevent an all out recognition of their belief in the absurdity of what they had just watched." Well," said one, "I think I know why the Bernie Babes can't get dates, because nobody *wants* to date them! " Others high fived each other and one said, "I think a new world order is about to unfold, and I don't think they'll be leading it." Others could not talk for laughing.

ON THE CAMPAIGN TRAIL

WISDOM
JOE TEACHES PROVERBS

JOE CAMPAIGNS IN HIS HOMESTATE OF CHIHUAHUA, MEXICO, OVER OBJECTIONS FROM HIS STAFF. THE FOLLOWING REMARKS WERE TAKEN WORD FOR WORD FROM JOE'S APPEARANCE BEFORE A LIVE STUDIO AUDIENCE OF THE CALI CARTEL, THE SINALOA CARTEL, AND FOURTEEN OTHER PROJOES WHO HAVE SENT HIM CAMPAIGN CONRIBUTIONS. THEY HAVE ALSO OFFERED TO ASSIST IN BORDER CONTROL ALONG WITH SEVERAL CHINESE NATIONALS WHO HAVE LARGE FENTANYL SUPPLY/ DISTRIBUTION CENTERS THERE AND IN TIAJUANA.

JOE SPEAKS: *Hombres and Hombrettas, Habla Oblongata? Si, sesame. Green Goes and Green Goes Ghettoes, as Pancho Villa sesame, Rio Grande Aka Poko, Ich bin ein Cha-Cha-Hua waw a wa wa. It's time for Africa. Oh, Sysco, Derribar este muro! Deja ir a mi gente!* Anyway, I'm here today to take what is rightfully yours. You see, California has a penis and I'm sure you want to fix it like the time I was driving here to go to church and the Federales say to me, "el permiso de conducer". Good thing Hunter was with me. He fixed everything and then bought the cop cause we had a

161

lot of money. You'd be surprised what can be found in a Chinese fortune cookie. So I was thinking if one country has a lot of people who can't cross the border illegally why don't we make it legal like. So I was saying – think before you drink. So, there's this problem with eels and stuff trying to swim a river, Rio Mucho Muchacho Grande and they drown and then the bodies get stuck in sewers. So I say, like I said, California has a large penis, which in your language is pronounced Maga Baja. So if you just sent everybody there, then California will rightfully claim Baja Penisula as part of California, then the problem is solved. No one drowns and California gets its penis back.

(The crowd was flabbered and ghasted and absolutely enthralled.)

EDITORIAL COMMENT:

Some of the people in the audience, at least those who spoke English, did not at first understand Joe's remarks, so one asked another, "What does it mean?" Another explained, "I think that is where a lot of the money – now it makes sense – the thing the Yankees and Joe Greengo call the PPP, the Penis Protection Program – that's where the money goes to pay us to traffic the children to Baja. 'Cause there they ain't got no laws and we just get paid by the Americans and they send the muchachos to that ranch that Island Guy owns in New Mexico. That way we don't have to deal with their mamas who don't always like it when we take little ones. But Hunter said it good for everybody. It's what he calls "Wean, Wean, Wean." I like it, he is funny and he pays me good – I am his number one supplier even for the thing from China." "Why does Joe wear the black mark on his forehead?" "He is showing he is Catholic and a humble man." "But it is not Lent. Why he is wearing it now?" "Because he is a – I don't know – maybe he is Muslim." "Joe tells me one time he is ecu-maniacal and so he loves all religions, like Larry, Moe, and Curly Joe, which is like 'En el nombre del Padre...'You know the rest." "Si, I like Joe, he is a good Joe." "Ha Ha Ha Ha, like you." "Aha, I know now why he wears the black mark, he is already dead, isn't he?" "Some say."

JOE CONTINUES: In my country we have a problem, like how come we have such a big wall. It's a really big wall, bigger than that one in China. People say that the wall was good for China. Yeah, but just ask the Mongols what they thought. Like I always say, if you built it, the Mongols will come and then remember what Genghis the Mongol said, 'If you have a wall, fill it with chinks,' like in a cabin or something. So now that we have this big wall, we have Mongols crossing your border illegally into Baja and places like Ontario and Windsor. It has to stop!

THOUGH REWILDERED AND ENTERTAINED, THE TRAFICERS OF ALL VARIETIES CHANTED "JOSE, JOSE, HE OK, JOSE, JOSE, HE OK..."

But what I really want to talk to you about today is my position on the most important issues of our day and time. There's some who say that my positions on the issues are incoherent, which it is, by design, like a floral pattern. So, here are my primary beliefs and I want you to remember each and every one. First of all, I am Pro-Choice, I am Pro-Choice in all things, even like in your getting up in the morning. Why? Hunter doesn't get up some time until the next day. But when he does, he's choiced the thing himself. So that's that. But I am also Pro-Life, like if you're living how can you be against yourself, that would be suicide. So, look at me. I am choiced myself to be me here living like an hombre.

And next, I am definitely Pro-noun. Because if you don't have pronouns, you don't have a you or me or even an I. But I understand that some people are against pronouns because they are afraid of being left out, like the garbage. But I think I have the best solution in the world for that.

You see I actually know the power of words. So, for example, why do we really need different pronouns for different people? Just blend them all together. That's the right thing to do. Take he, she, and it and put them all together, like I said, you end up with one word, shit. You got to admit, I have a special kind of genius for making things simple. For example, if you're a guy, and you like a girl, you say to her." Shit looks really great. Can shit buy shit some coffee. Shit smells so good. Or, you could say, 'That perfume smells like it was made just for shit. 'Now what some of you are sayin', 'But what about the plurals? 'They, we, you, theirs and sometime y and other stuff like that. Someone told me once they're mostly Viking words and they talked funny so we use a lot of their words and they went around rowing boats and singing songs with their favorite word, ya, ya, ya, tha's for sure. They would talk like that. But their favorite pronoun, I think was just 'ya. 'A lot of men today who want to be like Vikings, they still say things like that. They say, like to a girl, or a baby elephant, 'Shit loves ya. ' That's the Vikings talking.

But the thing I feel strongly about is pro-verbs. I remember in grade school talking about Jose and Juanita – in Scranton, Jahuahua, or what ever it was called. We read all about Jose and Juanita going up a hill to fetch a pail of water. I didn't like it much because who says 'fetch' except to a dog? Besides, it was so easy, why did they have to teach people English when it was already known. I knew it and I lived in Scranton. Think on it. And remember, there's nobody in the world who can't speak English if you speak it loud enough and slow enough. That's why didn't need any translator when I talked to Putin over the phone. I said, "You stay out of Bolshoi Bhagavad Gita Ukraine, which every comes first, or I'll teach you." He remembers what I did to Popcorn 'er Popsicle or that guy. So have you heard anything about him invading Bolshoi Bhagavad Gita or any place like it? Nope, he doesn't talk about it because he's afraid. He saw what I did in Afghanistan and, believe me, now he knows the way I speak and work. I'm more than just pretty words, I'm also a man of pretty actions, like a pro-verb. Remember when God was talking to Moses in the Ten Commandments movie and said I am who am. Well, though I disagree with Him about the pro-noun in this case, I prefer, "Ya shit" instead, because he was right he needed to be a man of action even if he was God. That's why God is always slewing people and pillorying them and most just making life miserable for them. 'Cause he's God and wrote a whole book about being pro-verb, just like me.

And also, God told a story about a girl named Shiba who wanted to marry a rich Jew named David – no, wait, wait, wait—it was a guy named Solomon. Now Solomon was really rich because he owned a bunch of islands in the Pacific Ocean. But the Chinese Communist wanted them too so that they become wise like Solomon who was always splitting babies in half to see what would happen. But he married this girl named Shiba who use to be a type of dog until God made her eat a whole pillar of salt and he turned her from a Shiba into a beautiful queen. But she made the Chinese jealous because they said you stole our dog. They said, "It's an Inu, a Shiba Inu, and we want her back." Solomon said, "I have a better idea, let's split her in two, what do you think?" The Chinese said, "This is kind of hard, even for us. If she were a Uyghur, we would do it, but she's a Shiba Inu before she was a Queen Shiba Inu." Then God came down and said, "Listen up, remember this, 'The only good Uyghur is a detained Uyghur.' "Well, Solomon saw that the eyes of the Queen Shiba Inu looked a lot like the Chinese dogs with their eyes, kinda slanting one way or another. So Solomon said he was gonna give her back if he could have some Uyghurs to help with his new golf course in Bermuda. God was going to say OK but then he remembered. something and said, "Solomon, remember I am God. Do you want me to change her eyes to look like yours?" Well, old Solomon was really happy and so he said, "Go ahead, make sure she has two of them." So God made her into a real queen beauty but just like that she got up and got mad at God because she said her eyes were now too big. You have to remember that she was only 5'6" but Solomon was 6'3." Solomon accidentally said the wrong thing to Dog. He always got things backwards.

So in conclusion. I want to remind you above all else that I am pro-verb. Like in the Book of Pro-verbs like "See Feng Feng run. See Nancy run." Those are verbs, like they *say*, which is also a verb, and you can't leave home without them. But mostly, I like the one about Solomon when he says, 'The race is not always to the swift, but mostly to the fastest. 'It was something like that but don't quote me because my staff will have to do things then. When I talk a lot, my staff starts yellen stuff (no pun intended), like 'Clean up in aisle 4, or the Fed is your Friend.' So like Al Jolson and God once said, 'You ain't seen something yet, what do you want for nothing?'

I want to conclude everything by saying thank you to the Cartelites and others here today. To me, whether you be Sinaloa, which sounds like a Hawaiian greeting in a whore house – well I digress – or Kama – Cali the Border Babe, I love you all. To me your names belong right up there with

Pfizer, Moderna, and Feng Feng's Fragrant Fentanyl – A Fix for the Entire Family. That's all I have to say about that.

EDITORIAL COMMENT:

We here at the Delphic Oracle Publishing House admire Joe's cryptic and unapologetic discourse in wisdom. Some possess it, others don't. With Joe the answer is obvious. Joe realizes that it is for those who truly admire his pronouncements that they search into the realm of the metaphysical and not merely the metaverse. Joe wants you his readers to realize that you must always dig deep to get the gold nuggets much as he has taught his son Hunter to do. In fact it was just the other day that Hunter said he really admired the Ukrainian flag because its colors are blue and yellow, just like blue skies above and sunflowers below. Well, in keeping with family tradition, Hunter was actually being cryptic and saying, 'Look, Ukraine has more gold or yellow stuff than even California, which is where I find most of it – that is, the yellow stuff. It was I think Alexander the Great who once said 'Go east young man, go east. 'Remember also the Greek myth about Jason and the Argonauts who went east to attain the Golden Fleece. They found it in Ukraine, the ancient land of Colchis. Think on it and you will become wise. For the Golden Fleece is still there and being exploited routinely. Ask Joe.

JOSEPHS AND THE MULTICOLORED GLASSES

THREE DAYS LATER, AFTER RECOPERATING FROM HIS ONE HOUR PRESENTATION IN THE LAND OF NEBULOUS BORDERS, JOE WAS SPENDING THE AFTERNOON SWIMMING IN THE HOTEL POOL OF THE HILTON TOP NOTCH OUTSIDE DAYTON, OHIO. JOE AHEAD OF TIME HAD BEEN WATCHING A STORY ABOUT FOREIGNERS COMING TO AMERICA AS HE BELIEVED HE ONCE HAD DONE. THE SITUATION BY WHICH THIS SITUATION AROSE IS ENCAPSULATED IN THE DIALOGUE BELOW.

JM: Mr. Joseph, do you remember me?

JOEB: Why aren't you that fella from a long, long time ago when I had to get special glasses from when I was a kid or something?

JM: I am indeed, I am known as JM, the eye doctor.

JOEB: Yeah, yeah, yeah, I remember that so well. Wow those glasses were special too. I had them for a long time and then Jill took them from me one day after she saw what she saw when she put them on. You should have seen. She kept looking really low on the anatomy of some of those Secret Service guys. You know what she said to me afterwards? She said, 'Boy, Joe, some of those guys are packin' some pretty big secrets and you better be careful because I think they're guns are really loaded.' Well I told her not to worry because they're highly trained and they never go off half cocked. I meant it too.

JM: Well, Joseph, I have recently had a dream and if you can interpret it correctly for me, then I have a special surprise for you. You see, in my dream, I see seven fat ladies and seven super models who are having a race. The fat ladies, because they are fat, fall and immediately roll down a steep hill that gives them the big lead because those who are fat roll faster and faster like great big tractor trailer tires. But the super models because they are supremely beautiful and do not have so much fat on them, they walk down the hill and leave nothing of themselves behind. But the fat ones who were supremely so, eventually left their lard and sweat behind them so that the super models slipped and slid upon their buttery grease marks right past them over the finish line. Now, Joe, how should I interpret what happened?

JOEB: Well, let's see, that means you've got seven fatsoes and seven skinny wafers or 7 minus 7 which comes out to zero which isn't the answer to anything but is wisdom all the same. So, if you think about it since 7 minus 7 gives us zero, and zero is something given to us by the Hindu shop keepers of India, then I interpret the dream to mean that the women were all Hindu or maybe sherpas out of work and that is why they went down the mountain and not up the mountain. They weren't very good. You have to be able to go up the mountain as well, even super models. But since the super models won the race, then I have a Pro-verb that explains it all very shortly. And that is, 'You win with thin,' or, put another way, 'Once you fat, you always go flat.'

JM: Joseph, you astound me. I have followed your career ever since you were a boy and now you are a man full of wisdom and vigor.

JOEB: Is that anything like piss and vinegar?

JM: Ya, ya, Joseph it is. But here is the most important thing. If you truly wish to be the kind of ruler who will be remembered for all of history, you must find a way to blind the people to your faults and have them see only your wisdom and your multi colored clothing and skin. To do this, you need but get people to wear the mushroom lenses of the glasses you had and of which I have made thousands upon thousands. Joseph, I have had them shipped to your present location here in Dayton where you can distribute them to all of your followers when you speak to them.

JOEB: Where are they?

JM: They are in a well.

JOEB: A well? You mean like with water and stuff?

JM: Yes, Joseph, and here is the good part, they are not visible to the naked eye because the lenses always give back to the one looking into them the thing he most wishes to see. If one looks into the water, he expects to see water or his own image, which he will always look fondly upon. So these glasses can make you the most powerful man on earth. I can even show you where the well is and brought along some special friends who will be happy to provide them to you. Your staff will need to bring them to you as my little ones dive to the bottom of the well to bring them up.

JOEB: Ah, man, this is great! Even hunter can't do this kind of stuff. When do we start?

JM: Right now, Joseph. Come with me. By the way, where are your Secret Service personnel?

JOEB: Oh, them? Well, Hunter's having a party of unwed mothers and he wanted to bring them over to help celebrate motherhood because he thinks it is so important.

JM: Oh, I see? And the rest of your staff?

JOEB: I don't know, they keep quitting, especially when they have to work with my Virgin Princess. They say she's the Black Queen of Cali and she eats babies and stuff. So they're afraid to work for her and everyone I send over there ends up quitting. I'm having to fire one person after another, even Jill.

Joseph, it is called the Well of Glass and it is located in the new annex which is being constructed right next door. We can walk there through a causeway that connects this half of the hotel with the new part. But the well is already filled with water and your gift of glass and glasses. Come with me.

MINUTES LATER, JM AND JOSEPH OF THE MULTICOLRED GLASSES ENTERED THE FOYER OF THE NEWLY CONSTRUCTED HALF OF THE HOTEL. NOT YET OPEN TO THE PUBLIC, THE SECTION WAS NO LESS BEAUTIFUL AND HAD THE ADDITION OF A WELL THAT WAS NO LESS THAN 40 FEET WIDE THAT FELL STRANGELY, IT SEEMED TO UNFATHOMABLE DEPTHS. THERE WAS NO ONE ABOUT, NOT EVEN THE CONSTRUCTION TEAM WORKING ON THIS SECTION OF THE BUILDING. BUT AS JM AND JOSEPH DESCENDED A LONG SPIRALING STAIRCASE, A NUMBER OF THIN CHILDREN OF COFFEE OR CHOCOLATE COMPLEXION EMERGED FROM THE SHADOWS, NODDING OBSEQUIOUSLY TO JM IN PARTICULAR AS HE PASSED. JOE WAS RIGHT BEHIND HIM AND COULD NOT HELP BUT NOTICE THAT ALMOST ALL OF THE CHILDEN, DESPITE THEIR COMPLEXIONS, HAD BRILLAINT BLUE EYES. A FEW HAD GREEN EYES AND THEN THERE WERE SOME WITH EACH. WOW, JOE THOUGHT, WHAT A TREAT.

The children circled the well, leaning against its wall and placed their goggles on to protect them from the water's blurring of their vision. One by one the children descended the depths coming up repeatedly with baskets of glasses which JM and Joseph deposited in a long row of cardboard boxes off to one side. At first there were hundreds of glasses and then there thousands, and then tens of thousands. While Joseph was wondering how many could be appearing even as he watched and participated in the event itself, it seemed like the miraculous multiplication of the loaves of bread. But then Joseph's eyes were opened when appearing before him were the Vestal Non-Virgins of Hysteria, Sister Chlamydia, Pretentious Presley, Ilhan the Omar, and the Cween of Angles. Then, he watched as the *Weird Sisters now Full of Themselves* in wisdom and power, twirled and tweeted on their phones, spinning round and secreting moans. They spoke as one, "Twizzle, twizzle, trizzle tum, make them many where we see one." Suddenly, even more of the magic glasses appeared. As they spoke a pint sized wizard with a long flowing beard and a turtle on his hind legs and of a jolly disposition walked by smiling at the incongruity of the moment.

Hours later, JM and Joseph were standing at the gate of his next big gala event outside a stadium not far from the hotel.

As each person entered the facility JM and Joseph of the multicolored glasses handed out their gift of the mushroom lenses. The chocolate and coffee colored, blue and green eyed rescued children of Haiti did likewise. What they saw with the new eyes was enough to bring a weltering of tears. For each with the glasses saw with new eyes the future. They had visions the likes of which none before could ever have imagined. Oh happy days when Joseph walked.

JOSEPH AND THE NAKED TRUTH,
WAPAKONETA, OHIO

ON THIS OCCASION, AS ON SO MANY OTHERS, JOSEPH ONCE AGAIN BROKE RANKS WITH HIS ADVISERS AND DECIDED TO MAKE HIS SPEECH ABOUT TYING IN HIS CAMPAIGN WITH THE ETHNIC DIVERSITY OF AMERICA. WE HAVE THAT SPEECH IN ITS ENTIRETY... JOE WALKED OUT ONTO THE STAGE AND THE CROWD, HOWEVER SMALL, WAS EQUIPPED WITH THE MIRACULOUS MULTICOLORED AND MULTICOLORING LENSES OF THE MAGIC MUSHROOMS. THE CROWD'S RESPONSE SAYS IT ALL.

THE JOE: Hello, my fellow Americans. Come to me. I am your father, friend, and freedom loving coal miner from Detroit where I knew the value of hard work and living the life of a patriot. I love America and all its diversity. I myself am diverse.

CROWD: Wow! With these glasses we can see everything. He's so beautiful, he's black as the ace of spades, just like me. Even his hair, the Afro, it's so authentic. He is one of one. BLM! BLM! BLM!

THE JOE: Yes, I am the dream of Dreamers where ever they are, whether they come from Argentina, the Solomon Island, Venezuela, Cuba, or even San Francisco. I embrace all of you who are part of this great tapestry of America.

CROWD; Is that like a tape worm? Who comes from Solomon Islands? Salamanders? You look cool, but now I also see your underwater. What kind of underware is that? What are those stains, Joe? Why are you naked?

THE JOE: I wanted to be here today because, like you, I am diverse. Before I became the man I am today, I was once like each and everyone of you. Some people don't understand that. Why, because they're not diverse like me. When I was growing up in Scranton, Chihuahua, which is in Africa, I learned a lot of about diversity. For example, just the other day, I went to the doctors for my annual physical which I get every three or four years no matter what. So there I was being the typically diverse guy that I am asking questions about myself to my doctors. When I talked about certain things, one of the doctors said, 'Joe, you're in continent. '

That's when I said, 'Well, son of a bitch, you dog faced pony soldier with heart thing around your neck, where else would I be? Like, do you think I live on an island or a in a swamp. Of course, I'm incontinent, it's what happens when you live in North America. I'm proud to be incontinent because it means I live on land which is better than islands that don't have much land. But like I say, I am diverse. That's why I picked this town of Wapakoneta to give my favorite speech about what it means to be in love with your heritage.

CROWD: BLM! BLM! BLM.

THE JOE: That's right, Be Like Me! Be Like Me! Be Like Me!

CAMPAIGN STAFFER: Brilliant. He has connected one movement with his personal charisma. Brilliant.

OTHERS: No one can stop us now. The man is brilliant.

CROWD: We want Joe.

THE JOE: Yes, I know!

CROWD: We want Joe!

THE JOE: Yes, I know!

STAFFER: Have you tried the glasses on yet?

ANOTHER: Not yet, why?

STAFFER: Take a look.

ANOTHER: Oh, no!

STAFFER: Oh, yes. He's naked.

CROWD: Yeah, but they don't seem to care.

ANOTHER: Exactly, the glasses are made with some sort of special mushroom lense that is essentially hallucinatory. The person wearing them sees exactly what he wants to see. Skin color, hair texture and color, height, weight – it's all there.

STAFFER: Are you saying that you and I have longed to see the Joe naked?

CROWD: The women of Secret Service are subject to that all the time. They even have a nickname for him. He's called 'Droopy.' When he turns this way, you'll see why. By the way it is the female members of Secret Service that gave him the name because he insists on their being present whenever he uses the pool.

STAFFER: OMG! What do we do?

ANOTHER: The media won't report it and the crowd and others and many viewers won't believe it even if they see it. We're safe.

THE JOE: So I am here today to promote diversity and equity, two great American things. Like I always say, and you say it with me. Evil Quits Unlike Tough Yunzes!

THE CROWD: Evil Quits Unlike Tough Yunzes. Joe, what's a Yunzes? Yeah, Joe, what's a Yunze, or a Yunzie? Is it like a yoyo or a kind of dog?

THE JOE: No, no, no, ha ha, it's all just part of what you need to know to be an equity guy. Ya see, and I don't think I ever told you this before, but I'm originally from southern Pennsylvania where I use to work as a buggy maker and horse trainer. I lived in what people and other creatures call the Dutch country.

THE CROWD: Joe, Joe, call on me! I know where that is. It's in New Amsterdam! But I think they changed the name.

THE JOE: They did, but the Dutch taught me to be diverse and an equity type guy. I use to train their horse, and fix their buggies. And they explained that yunzes means you or all of you. So now yunzes get to be just like the Dutch only you don't have to fix their buggies or train their horses because I already did it. That's why people use to – especially the girls – they use to call me the Dutch Treat. Because I always looked after them and made sure they drove and paid for everything so I could be like the Dutch and Amish and the guys with six toes who – I don't know their name any more but they had six toes from doing the Ilhan Omar thing or something.

THE CROWD: Joe, were you born in Pennsylvania or are you an American equity type guy? It is true, right, you are one of us? 'Cause we're in Ohio and we like Americans but not particulary the Dutch even if they have their own paint company.

THE JOE: I'm a genuine type equity guy from the state of Pennsylvania and I am now in the city of Wapakoneta to show my affection for all the Ayetalian type Americans. And like I always say, if you don't like pizza and Marlon Brando type movies, then you ain't Ayetalian.

THE CROWD: We love Ayetalians! We love Ayetalians!

THE JOE: That's why I'm here today to celebrate my Ayetalian heritage with you. You see, and a lot of people don't know this, I've not told anybody before, because I wanted it to be a surprise. My mother was Ayetalian befoe she became an American and she wanted to be with her people so she moved here because the city takes its name, Wapakoneta, from all the Ayetalians that started the city. They were mostly fishermen from Sicily and from Lake Erie which is part of Canada where they don't have any Ayetalians, mostly French and stuff and hockey players from Sweden and Russia. So anyway they named the city after themselves, Wapa, that means

Ayetalians, kon, which means we trick you, and *eta,* which means what it says, *Ayetalians trick you into eta.* **Wapa-kon-eta!** Cause even if you're not hungry they make you eta all that pizza and lots of meat balls.

STAFF: I hate to admit this, but it really is genius. Even if I thought it, I could not deliver the lines without shooting myself in the foot, in the head and in my heart. Who but an idiot could believe this and deliver the lines? The Joe really is a tale pulled by an idiot full of the pizza and free home delivery. It's so bad, it's good.

THE JOE: And a true sign of my equity type thing, I'd also like to recognize all the mackerel snappers and Micks in the audience. I use to be just like you. But just remember, we don't have to eat fish on Friday anymore if we don't want to. You can now go to Arby's and eat gyros made from stuff that ain't even human. It's just dead stuff made from what's thrown out after they get rid of the rest of the cow 'cause they got no sheep to make the gyro.

Anyway, when I was in Ayetaly, I even said to the Pope, I said, 'Pope, I don't fish,' and he said, 'What's that smell?'

CROWD: Yeah, I hear you got him good. I hear you did it to the queen to.

THE JOE: That's not true. I did it in front of the prince and his wife, Camilla. She deserved it because she made a pipe bomb to kill Princess Diana with a drone attack in a tunnel. I know that for a fact.

THE CROWD: Gosh, Joe, you know everything!

THE JOE: So, like I was always sayin', Ayerish and Ayetalians are pretty much the same people just like the Mexicans. That's why when the Mexicans sing, they always go, 'Aye, Aye, Aye, Aye, just like sailors who are also mostly Ayrish. And that is why, if you are wearing your navy uniform like my son Hunter before he quit beause he wanted to have more fun and wanted to be a SEAL, he had a thread hanging from his uniform which wasn't suppose to be there. You know what they call that? It's called an Ayerish pennant, like a small flag. So if you're in the navy and your uniform is kind of sloppy, even just a little bit, they call you Ayerish. That isn't right, so now I tell the navy you can't call it a pennant any more, you just call it an Ayerish thread instead.

So, like I was saying again, I like Ayerish people because they're just really good dancers when they do the jig and all that. But sometime you do have to be careful, 'cause their DNA which I carry deep inside of me from my mother's side, well I and a lot of Ayerish have a mild form of leprosy. That's why those funny little green guys are called lepercons. The word comes from the ancient Gaelic which means 'tricky little lepers. ' And they are, even if you find the end of the rainbow, you never get to keep it. They take it back and give it to some mayor or something like in Boston or Cleveland. Sometime they trick you into not knowing that they're really Ayerish playing a trick. They do that by sayin' they're black instead. But what they don't tell you is there is such a thing as the Black Ayerish. Know who are they?

CROWD: Tell us, Joe! Tell us, Joe!

THE JOE: See they don't tell you who they are. A long time ago those people called the Moors who lived in North Africa – well they conquered Spain and had a lot of black babies. That's because they liked the Spanish women. That's where the expression 'Sin-orita?' comes from. And a lot of the Spanish woman said,'Si-sinor?' Like she was sayin' what do you have in mind? And that's why a lot of the Spanish are so dark. They sinned a lot. And then, in 1588 that guy Julian who wrote that calendar that we use or we don't any more, I don't know, he said that the Spanish tried to conquer England by creating the biggest fleet of navy ships in history so that they could sin some more with those English girls. But they ran into a great big storm and a lot of them ended up shipwrecked and they ended up floating to Ayereland. Well the freckles and the red hair of those Ayerish girls really made the men excited. Next thing you know they were having black babies with freckles and red hair. But to some of the Ayeresh the black babies with red hair and freckles just looked weird. They thought they must be lepers. So they didn't give them as many potatoes to eat as the rest of the Ayerish so they had a tough time. So they grew littler and littler and started turning green form the lack of nourishment. Eventually they dug tunnels, chased rainbows, and found gold which they won't let anyone take. So you see, lepercons are nothing but Black Ayerish from Spain who you can recognize by their red hair and freckles. So if you got either one, you are probably really black. So if you go to a bar on St. Patrick's day and you see a black guy who's drinking a lot and dancing the jig, then's he's secretly a leper who tricks people. Like Michael Jackson. That's why his skin was so funny. When you're a lepercon it's hard to keep it a secret.

CROWD: Joe, what about me. I'm Russian.

THE JOE: See, I had a White Russian for breakfast. They're just like the Ayresh, a very happy and confused people who believe in equity. What's your name?

CROWD GUY: Vladimir.

THE JOE: What's your last name? You look familiar.

CROWD GUY: Eeeees not possible or you will soon look like little green man yourself. Talk more, I listen.

THE JOE: Strange, aren't you. After all, I use to be known as Mean Joe Green. Do you know me?

CROWD GUY: Nyet, I like to play music by Fats Domino. I all American guy with just some Russian blood, tiny bit. I like equity too. Like we say, <u>E</u>ven <u>Q</u>uitters <u>Y</u>ell when someone poison them-- with coppery stuff that makes you turn green. Special formula from my mother – in Russia. Ha Ha, joke.

THE JOE: I don't know much about Russian but I do know a lot about Ukraine. In fact I even once wrote a song about a rich guy in the Ukraine who owned property right next to mine at my ranch—whops. Never mind. Anyway, I wrote this song about a guy who was upset because he just bought a plane and it rained all the time and so he could never fly it. Opening line was, 'The rain in Ukraine falls mainly on my plane. 'I like it but I gave it to some other guy and he switched the words about just slightly. But I didn't get the credit I deserve.

CROWD GUY: You funny guy, you like great Russian composer. There are no great Ukrainian composters. They are all Nazis who steal from my mother in Russia.

THE JOE: Well if you say, Vladimer. But I still like the one I composed about the Ukrainian with the plane. That was hard work.

CROWD GUY: I have song too I write. It go like this. The pain in Ukraine, caused mainly by inane. That you, you insane. You got it. Listen, my friends from Moscow, Ohio, Children Choir sing it much and loud. (CHORAL RESPONSE): We love you Vladimir. We—

CROWD GUY: Nyet! Nyet! Not now. Not here. Other song. Quick, in Russian.

CHILDREN CHOIR: The pain in Ukraine caused mainly by inane

EDITOR'S COMMENT: As the young of Moscow, Ohio, burst into song that implied a possibly nefarious connection between Joe and the war in Ukraine, a group from Cleveland, known as the Red State Antagonists burst into song to try to drown out these heedless Moscovites.

RED STATE ANTAGONISTS (AGONISTES, short version): We love you Joey, Oh yes we do, When you are near to us, We know it's true, We'll take this Commie Red state and turn it blue, We love you Joey, you, you, you, you.

EDITOR'S COMMENT:

The internecine display of choral strophe/antistrophe movement continued for several minutes until the hoarsness of vocal chords forced an abatement of the good natured display of slightly different views on the things that don't really matter. Ha Ha, it was all in fun until Joe said something that made us all wonder.

THE JOE: Say, you – you guy with the big bear skin cap with the red star on it – you. Guienivere --

CROWD GUY: Nyet! Vladimir! All American guy like Moose and Squirrel!

THE JOE: Sorry, I thought you might be someone I know. You look familiar.

CROWD GUY: Da, you look familiar too, like boy named Hunter. We like to call him 'Bathouse Biden. 'He say he like the little boys and girls to sit on his lap and scratch his hair legs. He say he learn from his dad and we have pictures. You want to see when we beat him in sauna with birch branches? He like very much. Ask heeem if he remember Odessa, Black Sea Restaurant. Maybe he no remember, but girlly girls with camera, they remember. Wife of Moscow mayor, she remember.

Аляска

VLADIMAR

A GREEK BEARING GIFTS

EDITOR'S COMMENT:

Amidst the increasingly acrimonious exchange not only between the Crowd Guy and the Joe, but also between the Agonistes, and the Muscovites, of Ohio, a well known figure ran out on the stage and raised his hands. 'Please! Please! Give us some quiet. The Joe has a special composition that he himself has written to bring us all together? Please, put your glasses back on, I noticed that some of you have taken them off. Not good, Not good. Please, now put them back on as we celebrate the diversity and equity of Wapakoneta and the New World Order. Say it with my now, 'Biden! Biden! Biden! Let's bring him back with his special all around composition of affection for the Greeks, the longest lasting lovers of liberty, the L4 composition of ancient Greeks to perform with The Joe—here they are now to assist him, from Athens, Ohio, the Casbah Dancing Club – the newly formed group, The Biden Belly Dancers!!!! Let's hear it!

The Joe had gone off stage to compose himself, only to return wearing a toga and pronounced protrusion from his pelvic area. This protrusion bumped the speaker in the rear as Joe came back on stage.

CROWD GUY: Who is Greek looking guy doing talking?

ELDER CROWD GUY ASSISTANT: Eeeees Greek guy named George Stephanopolos, Boss.

CROWD GUY: Da, da, now I remember, we use to call him, 'Big Hair'. Honey pots call him 'Mister Neem Noga.' Spasiba, Boris.

OLD BALD GUY YOU KNOW (OBGYN): The Crowd suddenly quieted down. But with the protrusion repeatedly bumping him in his rear, the Big Hair guy looked at THE JOE in horror. He then shouted to the crowd, "Take off the glasses! Take them off! You won't need them anymore!" Several ladies of dating organizations known affectionately as the Broads with Bods and the Sexy Surveillance Sisters, respectively, shouted back, 'No, we like it! Give it to us, Joe! " Another group of women in wheel chairs and having grey hair and blue bushkas with sunflowers to show their support for Ukraine clapped vigorously while keeping their glasses firmly in place. Known as the ELDERS, short for <u>E</u>lder <u>L</u>ooking <u>D</u>ames <u>E</u>at Rinos, stomped their feet and canes rhythmically, begging for the rally to go on.

The Joe in a moment of near delirium and an andrenolin type rush, pushed the Greek Guy out of the way with his eros induced protuberance and stepped to the microphone which he tapped slightly with the same. "Is it still on?"

The various ladies' organizations, shouted back, "You've turned us all on, Joe. Let's hear it and make sure you sing with all your zing! Ping us, Joe, Ping us. Yeah, just like you just did to the microphone!"

The Joe in a rare moment of mental clarity, looked to either side as the Athenian Dancing Girls of the Casbah Club took their positions, swaying back and forth until their unified movement brought Joe to a moment of revelatory and hitherto concealed thespian climax known back stage as The Urge. He was primed, he was pumped, and he was all Greek! No Trojan could hold him!

THE JOE: All right, admoroning fans, from Moscow, Wapakoneta, Venice, Ohio, and even Berlin, Ohio, and, of course, Athens, home of the

Non-thee Ohio State University, better known as Sweet Ass, Ohio, because of their females-- get ready for my greatest compost ever. I wrote it because I love Greeks and their really, really, really old culture and music and their restaurants where everybody is named, Gus, George, Chris or Nick. In fact, I was in a bar the other day run by a bunch of Greeks, and I asked the bar tender, my friend, the Greek named Chris, and I said, 'Chris, why are all the Greeks I know named, 'Gus, George, Chris or Nick. ' Well he thought about it for a minute and he said, 'Because they all named after their grandpas. 'I said, 'Even the girls? 'He said, 'It's stupid question, we have two girls named Athena, all the rest named Gus, George, Chris or Nick, so they can be like Lesbians and hang around with little green Black Ayerish guys and find rainbow and then gold. You should know. You use to be Ayerish before you became Greek.

'So here I am folks and just remember, you may be equitable and you may be diverse, but it's all Greek to me. He nodded to a staff member who then walked out holding an ancient electric lyre with a strange input, out put system protruding from one end, a electric wire and a male connecting device. The lyre was no longer than two feet and was composed of a petrified cheery bough recently discovered beneath a Mafia bathhouse in Covington, Kentucky, once owned by a charismatic mayor of Cincinnati known merely as the Springer. The Joe, who once had had a spinal tap intrusion to correct his political posture, plugged it in right next to the female opening near the top of his spinal column which gave off an electrical impulse. The speaker built into the device began to reverberate with the sound of Joe's plucking upon its four strings. Joe then spoke his dedication: **To My Mama Where Ever She May Be.** He began playing and stepping to an ancient Greek rhythm that had come to The Joe in a dream.

> Well, my name is Oedipus
> And this is my dance,
> Here to tell you
> About a little romance.
> Had this thing about my Mama,
> You all laugh and I go,
> Uh huh, Uh huh.

People say if you do not stop
You're gonna go blind.
Well I don't care
If I lose my mind.
'Cause I'm too busy
Defyin' my fate,
Learned my lesson
Just a little too late
Uh huh, Uh huh..
Well, the other day
I found two new broaches,
Stuck 'em in my eyes,
It's Buenos noches!
Good night! Out of sight!
Yeah.

The crowd went wild.

STAFF MEMBER: What is our official title?

OTHER STAFF MEMBER: Us? You're a staff member and so am I.

STAFF MEMBER: Unfortunately, he has given the term an entirely new meaning and I want to resign.

OTHE STAFF MEMBER: But wait, there's more. Look!

OMNISCIENT GUY BEING YOUNG AND NEUTRAL (OGBYN): In an attempt to quell near riotous calamity of the admoronic fans, the Four Weird Sister, Chlyamedia, Pretentious, Ilhan the Contentious, and the Cween, alighted on stage from the avionic disposition above the Parthenon-like portico. But just before they lighted, The Joe, turning to make certain that the Dancing Casbah Girls were still in step, he accidentally knocked two of them down completely with his manly protuberance. Turning in the opposite direction, he knocked several more down in a further display of manliness. Big Hair ran up to him when The Joe turned suddenly, knocking him into the orchestra pit in front of the stage. As he fell, he grabbed the microphone for balance, causing the cord, to catch on The Joe's leg and pull him to the edge where he teetered until the flying foursome of unbridled power and elitism fell upon it, pulling him backward onto the pile of floundering and flailing Greek girls. As Joe rose from the pile of broken and crying bodies, he looked down and realized he was still holding onto the electric lyrie and he began to sing another song he had written years earlier.

> Wild thing
> I think I'll mug you
> Or maybe drug you
> So you feel no pain,
> So when I use you,
> I won't abuse you...

He looked down with pleasure and pride at his manliness and moved it about voluntarily, aiming first at one lovely and then another. At that the fearsome foresome of power and greeniness, charged at him, when he manlily swung it at them it with malice and mischief to see how far he could throw them. All four crashed headlong into the still unconscious Big Hair. His work for the moment done, The Joe marched off stage into the waiting throng of ethnically diverse elders and youngsters of desire awaiting his presence, one to be felt and prolonged. Then it was that Joe, now the Jubilant, marched toward one of the exits followed by a bunch of Black Ayerish from Cleveland, Toledo, Youngston, Canton, Columbus, and Cincinnati. They followed him into the night time darkness to the accompaniment of McNamara's Band and They Ain't Heavy, They're Just My Bevy, a none too oblique reference to the entourage of elder female citizens attempting to touch but the hem of his toga. The latter, by the way, was another song which The Joe had written while script writing for The Hollywood Executive known simply as the Big Z.

Joe Biden's lyre which was found beneath a ramshackle mafia bath house in northern Kentucky by Dr. Theodosius Slumaniclese, the renowned Greek archaeologist and restauranteur from Athens, Ohio. The item depicted here is made of petrified cherry bough and a most primitive, and indeed ancient, input-output system that plugs into a hole at the base of a person's skull. There the item's cord, once inserted, absorbs electrical impulses generated by the brain and spinal cord transferring them into four cat-gut strings that reverberate with beauteous expressions of aesthetic delight. Joe played this instrument with uncommon vigor and banality.

Joe's Lyre

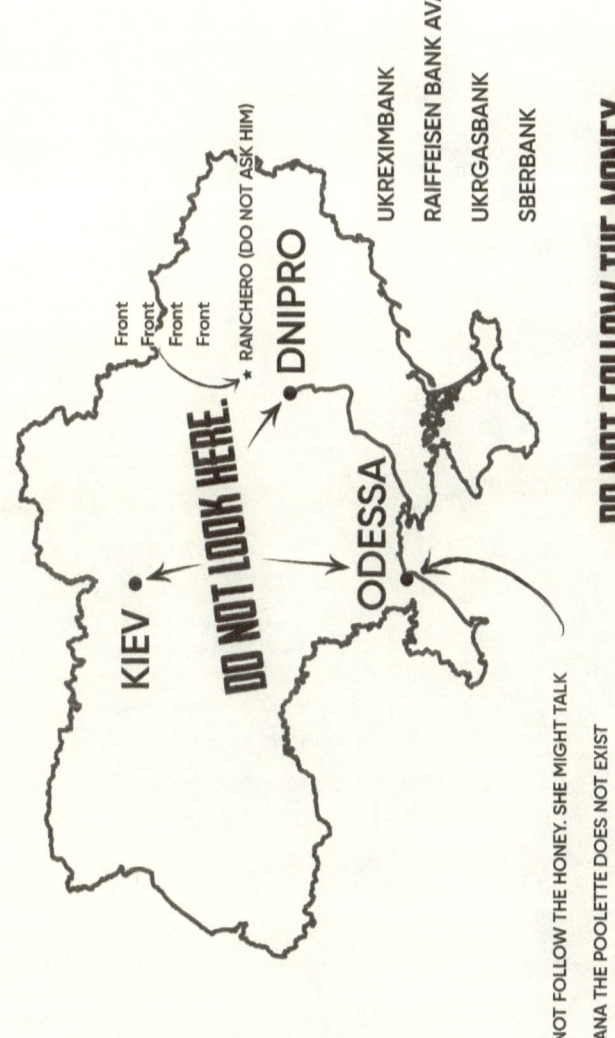

WHITE HOUSE POOL REPORTERS: INSTRUCTIONS

MAP OF UKRAINE

KIEV

DO NOT LOOK HERE.

DNIPRO

★ RANCHERO (DO NOT ASK HIM)

Front
Front
Front
Front

ODESSA

UKREXIMBANK

RAIFFEISEN BANK AVAL

UKRGASBANK

SBERBANK

DO NOT FOLLOW THE MONEY.

DO NOT FOLLOW THE HONEY. SHE MIGHT TALK

TATIANA THE POOLETTE DOES NOT EXIST

DON'T DRAG HER INTO THIS

IT WAS ALL FOR LOVE

THE LETTER Z

OBGYN: Once the Russian invasion of Ukraine occurred, there was a concerted effort by the West to punish Putin and Company in the most severe way possible. It had long been recognized tha the letter Z was an emblem widely used by the Russian military to signal their pride in the quest bequeathed to them by their fearless leader. Their tanks, other armored vehicles and even their planes frequently are seen to carry the symbol on their sides, on their fronts and rears, as well as on the uniforms of Russian military regardless of the branch. Some soldiers even painted it on their foreheads and cheeks.

In the West a fearsome response to this came from private and public concerns long known for their courage and their morality. For example, the National Assocation of Librarians immediately banned any and all allusions to the letter in the Dewey Decimal System and the exclusion of any popular books which had the letter in the title or was capitalized at any point in the front or back cover art.

Hollywood likewise displayed its bravado and moral strength by permenately banning the television series known as 'Zorro,' the full length motion picture entitled 'Z', and all on line programming featuring Zoom presentations were temporarily halted until the term 'Ooom' could be permenately installed in the software systems without offending any devoted Yogi meditators.

But no doubt the most courageous of all stands against the letter Z was that of the music industry which permenately banned ZZ Top from performing presently or in the future at any site recognized as being associated with Communist government officials operating out of Moscow, whether in Russia or Ohio, or in any Red States where they were assumed to have a coterie of followers who would not weep for the Ukraine. Billy Gibbons was arrested only a day later and forced to shave his head and beard and to perform under the new name of NN Bottom.

What the United Nations did, however, stunned almost everyone. All countries with a prominent Z in their name were permanently removed from voting rights in the United Nations General Assembly until such time as a substitute for their Z infected names were renounced. The unexpected consequences of such a brave renunciation of the orthography of the Western world as well as derivative nations and their alphabets occurred immediately. When the American Nazi Party created an electronic logo with a spinning 'N' which could be placed horizontally to look like a Z, Billy Gibbons was immediately arrested again and forced to change his band's name to Zhi Zhi the Chinese Executive which was given a pass on grounds of compassion. Then Webster's Dictionary and the Oxford English Dictionary, in a joint news conference, announced that the term xylophone would be deleted never again to be returned to the English language as the 'x' within was too much like the letter sounding 'z,' of English. Canada also chimed in with its own forceful assumption of affiliation with the suffering of Ukraine when its guy, Justin the Brave, said that the Canadian proclivity of identifying the number zero with the term 'zed' would be permanently banned. He then went public with a conversation with the Prime Minister of India, Modi, in which he suggested that the government with its large Hindu population denounce their racist remarks for their having once invented and subsequently bragged of their creation of zero. However, countries like and Zimbabwe and Zambia were particularly hard hit by the news (no 'z' sound intended). Tanzania pretended not to understand what was being said about the letter z and had a celebration for its people on the islands of Zanazibar. They could hardly do otherwise.

The Greeks expunged the letter 'z' as well and purged the stories of mythology of the name Zeus. His name was permanently replaced by a conglomeration of Chris, Nick, George, and Gus, or for short. Guchiger (accent on the second syllable). Say it again (If you're American, you put the stress on the first syllable. Greeks don't talk that way, Gu-**CHI**-ger, ancient Greek term for explaining the sound Greek parents make when tickling their babies.

Subsequently, the United States Congress called a special session of the House and unilaterally banned all racist letters, including, but limited to z, x, n, and sometime y, for no apparent reason. Janet Yellen complained but was summarily censored and sealed in the vault of the Federal Reserve where she was forced to do penance by having to read the story of Crassus, the richest man in Rome, and subject herself to an approximation of his death by having molten gold poured down his throat. She opted for the more embarrassing but less lethal case of having to walk out of the vault with several 2 ounce gold plated bars stuffed in various orifices. She was, of course, allowed to keep them once she complied and actually renounced her affection for gold while embracing all forms of fiat currency. She was also compelled to publicly denounce all orthographically compromised letters, like those mentioned above, as inherently racist.

States like New Jersey, New York, New Mexico announced their own bans on such material, z, etc. while North Carolina, Nevada, and North Dakota, adopted the Cyrillic alphabet which maintained the letter z without compunction. All four states then sent the same letter to Congress which consisted of nothing more than, "Z U! Stick it Up Your Xylophone." Alaska's governor announced that they had never abandoned the Cyrillic alphabet in the first place while students at Berkley were confused and did not how they should respond. So it a moment of brain density filled with self loathing, they found a bunch of lemmings and followed them into the Pacific Ocean. A week later, the mayor announced a day of mourning and forcibly removed the homeless because their excrement was flowing into the ocean and defiling the grave site of so many who had died so heroically.

However, there was misunerstanding at the National Engravery in New Zealand. The head of the instititon, Nelson Iverson, was told to correct the name of the country to make both halves the same. He did so, Zew Zealand. Despite the beauty of the new currency, he was forced to publicly apologize and remake it. However, because of the electronic wizardry of the American Nazi Party and its using the N as a spinning form of the Z, he was forced to remake the currency again, New Nealand wasn't good enough. He again apologized and then created currency with a single word, *Newni*. Others again objected on the basis of the N and Z were simply too much to bear, both had to go. Two weeks later, he came back with his latest, *Eweee*. While most complimented him on his ingenuity, detractors complained that the term sounded too much like the word "Aussie" which was, of course, too close to the term used to identify their Australian neighbors. When the Aussies got wind of this, they pulled their ambassador from Wellington and banned any further showings in Australia of movies filmed in Eweee like Lord of the Rings, The Tales of Narnia, the Hobbit Trilogy, and Avatar. However, even the Kiwi thought that banning The Piano was a good idea and if they had to do it over again, they would.

Still two weeks later, Maori warriors from the country invaded Australia and the war was on. Nothing could stop it. The Solomon Islands declared war on both countries and declared it would not drink Aussie or Kiwi wine until the price went down.

In the meantime the head of engraving created a new version with no words at all as words were increasingly deemed racist. Instead, he portrayed a picture of a dead kiwi bird with its hind end facing the viewer. He then fled to Australia where he joined the Army and then led the attack on the National Engavery and returned the country to the Maori who didn't notice since they were too busy conquering other parts of Australia.

So, the new Latin alphabet was A, B, C, D, E, _ G_.. All other letters were deemed superfluous and unnecessary as most the Western population were increasingly illiterate. Therefore what was point. Math was no better. After all, who could solve for x or y when neither any longer existed. Even the question, "why" was forbidden as it harkened back to days before the darkness had begun.

DAYS OF NOSTALIA

NEWS FLASH: THE LETTER "G" WAS JUST BANNED, HENCE, NO MORE "NOSTALGIA."

OBGYN: In need of some time to recover from the rigors of the campaign trail, The Joe, at the behest of his wife Jill and a distraught Hunter now that his art sales had plummeted in Russia, sought refuge at that barn of so long ago. Here Joe would have an opportunity to once again reminisce upon those trying but no lesss glorious days of his youth as a coal miner in Detroit before his picaresque adventures in Chihuahua, Illinois, Tennessee, Arkansas, Louisiana and so many more wonderful places.

THE BARN: Joe and Jill struggled, attempting to open the doors of the old barn, pulling hard against the large wooden red and white and heavily chipped old wooden relics. Like an elderly patient reluctant to surrender its existence while maintaining its secrets, it resisted. The hinges creaked and finally Joe called Secret Service over and asked for help. He and Jill stepped back as four strong and fit men pulled hard against the rusty old metal hinges until the doors finally swung open with a moan and the dusty interior of straw and stale air burst forth. Upon their entering, a stream of light from two south eastern facing windows in the hayloft allowed the sunrays to catch the whisp and whirl of straw dust floating aimlessly.

Off to one side of the barn, they caught sight of a tabby cat with four kittens following behind. Without saying a word, Joe looked at Jill and saw her hallowed by the light from the sunrays. Her hair seemed more radiant than he could remember. She smiled back and stroked his face. He turned to the Secret Service only about twenty feet away and said, "Fella, could we have some time to ourselves, just for a while. Please." They turned away and two of them pushed back against the doors, leaving them alone.

Joe took Jill by the hand walked over to a large pile of straw that strenched well beyond their height. They two of them began to pull it all away, slowly and deliberately revealing the body of the car which they had once worked so hard to construct – one piece at a time. "Look at it, Joe, everything we needed but the tires. You never did get them on. Mostly they're still flat on the bottom and rotted on the top. Still the pumpkin smell does have its appeal."

Joe smiled back and tears trickled down on his cheeks." Did you ever think it would end this way? I mean I'm now king and maybe king again, but it is isn't fun any more. I don't even like it. Look at Hunter, he's a mess and we don't – well I have to tell you something. I don't want you to tell me --- I want to tell you that I need you to keep this a secret."

Jill looked at him with admiration and sighed, saying, "I already know."

"You do? You've been making a lot of trips to the bathroom lately. That's usually one of the first signs. I noticed. Of course, Joe, just take less saw pimento and instead ask the boy wonder from South Bend, how to handle lactation?"

"Huh?"

"Joe, pregnancy is nothing to be ashamed of. I'll help you through this difficult time."

Joe sat quietly for a moment and finally responded. "Yeah, times have changed, who would even think of such stuff? I guess…Well, funny that some people would even think it possible.

"The power hungry, our colleagues, our so called friends. Face it, Joe. Face it, we're just puppets of the Raja and George Soros, among others. We took the money and along the way we got rich."

For several minutes neither said a word. Then a loud knocking at the barn door. A Secret Service member peaked in and said, "Mr. President, your son is here. He says he needs to see you. It's urgent."

"Send him in."

Hunter immediately walked by the Secret Service member and immediately to his father's arms." Dad," he said with halting and tearful eyes and tremvling lips, "Dad, I'm so ashamed. I've been caught. They know. They say the grand jury –,"

"Here, son, sit down in between us. We know. It affects us all. That's the way it is. I'm sorry, son. I have not been the father I should have been. I stayed too long. Even now, I couldn't help but wonder, if I had spent more time with you – maybe your life would have been different. Maybe term or age limits aren't so bad after all – for presidents, congressmen, judges."

"Joe," said Jill, "Perhaps it's not too late. You have leverage. You could offer, Joe. Maybe they would allow it. You go and Hunter is let go. No jail time."

"A good father might do that. But for us, it's too late –,"

"Dad, don't do it. It's already too late. The Russians have the receipts for everything. They're even demanding their money back for the art work if I don't, if you – if we don't."

Joe's head dropped. After several seconds, he raised it again and looked directly into Hunter's eyes. "Son, I should have known better and I can only say how sorry I am for everything. I let Washington and the D. C. Swamp suck out my life and soul and yours as well."

"Joe, it's not too late, you have grandchildren."

"But, Jill, even if the American public were too understand, the Chinese won't. It's always the same. It's payback time and I can't even rely on Nancy any longer. She's compromised too. Her husband, you know…"

"Dad, I'll never talk, they can't make me.

"Joe? What are you thinking?"

Joe was silent for a long time And then another knocking at the barn door.

A Secret Service member once again peered through an opening and said, "Sir, Mr. President, one of the locals is here and he says he has something for you. He insists on talking only to you and not us. He refuses to divulge –,"

"Send him in," Joe responded with a tired voice of resignation. He stared blankly as the door opened to allow in a man, accompanied by two of the agents, one to either side.

"Sir, he says, he knows you."

A man in what appeared to be in his 60's or early 70's stepped forward in blue bib overalls with a long sleeved checkered red, black, and grey flannel underneath and offered a rough calloused hand. Joe stood up and shook it haphardly.

"Sir, I'm sure you don't remember me from a long time ago but I was sent to help make preparations for a very special gift that the residents of your sometime, longago, semi-childhood home, right here in Orange County ala mode, USA. As such, Mr. President, I am the representative of the fine people here who want to commemorate your early years and dreams by helping you get that vehicle you constructed out on the highway to the forthcoming Pasadena Petunia Festival in the Land of Cali which is only a couple of weeks away."

"Well," said Joe, "I don't know what to say."

"You leave the sayin' to us. We're farmers and we know what the hell we're doin'. With your permission, sir, may we bring our gift in? You're gonna love it."

Joe nodded at his Secret Service detail who turned and opened the door as eight farmers in total walked through the door wheeling in four very large tractor tires.

"Well, certainly, let's see what we have here. Whoo eeee! " The farmers' presence seemed to lift the mood of Joe, Jill and Hunter. It was as if the presence of prospective voters had an animating and celebratory mood-lifting effect that brought a momentary reprieve from tears and self-reflections. The Secret Service gathered round as Joe and Jill and Hunter watched the farmers lift the car with a crane still suspended from the ceiling from where Joe had left it years earlier. In a matter of minutes, all four of the orange tires were firmly in position and the dream of Joe's Auto-biography was fully realized. If only Otto and the others were here to share it with him, he pondered.

Once all of the tires were in place the farmer spokesman smiled broadly and said, "Mr. Joe President, these tires are unlike any others that you might consider. They are composed of genetically modified pumpkin extract and fiberonic soy extract that have been jucified, hudrolycized, textureized, seared, semi-calcified, vulcanized, and fully vaccinated with thrombosis inducing coagulants that prevent deterioration of the tires by mummifying the fleshy pulp of bumpfied resilience necessary for high powered usage and abusage, suitable for farming, off road dragging, and high profile parading at the annual Pasadena Petunia Festival. You sir, are now a farmer. As such, America knows that no one will ever go hungry again. You support us and we support you. Remember, as my friend Smoky says, 'Only you can prevent starvation.' And every baby in America will call you Dada.

"I'm touched," replied Joe. "Truly, truly touched."

"We know, that's why we came. It's all for you."

"By the way, who sent you? Who makes this kind of tires? I mean, I believe in magic and all but today not too many people would even believe in it or that something like this could happen. Like you're showing me and the world that the magic is back."

"Well, sir, we got fields to harvest and all and you've asked so much, well – I can only say so much and so here it is. I, sir, am the president of The RUral Mechanics of Pumpkinland. We specialize in making dreams come true. That's all you need to know."

"Thank you."

The RUral Mechanics
of Pumpkinland

PART IV
TWILIGHT

PETUNIA PRIMARY PARTY OF POLITICAL PATRIOTS

THE CORNER OF ORANGE AND GREEN BLVD., THE TRADITIONAL STARTING POINT OF THE PETUNIA PARADE IN WHICH TRANSGENDER MALES CELEBRATE THE PULVERIZING OF REAL WOMEN IN POWDER PUFF FOOTBALL CHAMPIONSHIP. TODAY, HOWEVER, THIS PATRIOTIC EVENT IS DEVOID OF POLITICS AND IS PURELY PATRIOTIC.

ON ONE SIDE OF THE STREET STOOD THE "GALLANT GIRLS OF THE GREEN REVOLUTION," SIGNIFYING THEIR TRANSITIONING STATE WITH T SHIRTS THAT READ, "BETTER GREEN THAN MEAN," "SAY NO TO TESTES-SAUROS REX" WHILE OTHERS READ "ESTROGEN AND SOY MAKES ME HATE THE BOY(WITHIN)."

OPPOSITE THIS CROWD STOOD THOSE WEARING ORANGE T SHIRTS WITH BLACK LETTERING THAT READ, "SQUEEZE THIS ORANGE," ON THE FRONT SIDE WHILE IN BACK IT SAID, "ORANGE RHYMES WITH DEAD. BACK OFF."

MEANWHILE A SIDEWALK VENDOR BY THE NAME OF "WEE WILLIE WONKER OF APHRODITIC SELF-INDULGENCE," GAVE AWAY FREE SAMPLES OF FENTANYL LACED ELEPHANT EARS AND ORANGE JUICE, COMPLIMENTS OF THE CALI CARTEL.

IN THE MEAN TIME THE BELOVED MAYOR OF LOS ANGELES, ERIC GARCETTI, OR AS HE WAS KNOWN IN CALI, SINALOA, AND CHIHUAHUA (PRONOUNCED CHI [LONG *I* LIKE IN *CHINA* AND WHO-A, WHO-A, AS AN OWL MIGHT PRONOUNCE IT], 'EASY ERIC.'

Then to the tumultuous roars of the crowd, The Joe appeared in the still roofless auto-mobile, the story of which we have so wonderfully and diligently relayed through our narrative of adventure and singular truthiness. The vehicle had even received a pumpkin colored finish in its entirety to match its priceless tires, complements of the farmer's organization known as The RUral Mechanics of Pumpkinland. As the vehicle pulled by the police and Secret Service vehicles, all of which were similarly marked as to hide their real identity, the police on either side cordoned off the area. However, one police woman with an orange jump suit in an attempt to blend in with the Ulsterites of Northern Ireland and other Protestant Americans, began to shout, "If you are now or have ever been a January 6 protestor, please follow me and ignore all traffic signs, police instructions, or anyone representing authority and please give your name and home address to the men with the clipboards and the women and non-binary persons of humanity standing outside various buildings along the way. At no time should you avoid confrontations or submit to police authority without a fight. Remember, I was there too on the special day, which much like Pearl Harbor is a day which shall live in infamy.

The Joe now stood up in his cartrosity as it pulled forward toward the actual starting point of the Pasadena Putunia Patriot Parade. This was the first reviewing stand at which the floats, bands, marchers, and various choral groups had to pass by for preliminary recognition of their actually having the appropriate permits. Like a knight on his charger, a destrier of immense and historically large proportions, thanks in part to its tractor tires and a hybrid of truck and other metallic covers, the Joe stood. The vehicle's exterior of brilliant orange on a Mack chassis reclaimed from a military junk collector who wished to be part of this wonderful event, looked like a fiery rendition of a typical Detroit Halloween celebration. It was entirely aglow in the noon day sun.

Joe then pulled off his oft doffed aviator sun glasses as he tried to forestall thinking impure thoughts at the prospect of his forthcoming meeting with his VP. She, he thought, would be somewhere in the crowd watching his every move. He glanced to the left side of Colorado Boulevard, where a mob of Ayerish elfs were forming into a phalanx to protect the inner confines of a group of engineer provocoteurs who were busy assembling mini-handgrenades to plant on or throw at carbon emitting, plant nurturing, bark enriching, gas guzzling, float pulling pick up trucks filled with inebrieated college students abuzz with conversations about being buzzed.

Off to his right was a street vendor dressed in the traditional attire of southern border sex trafficker (sex by eight or it's too late) with a sign over his minitablet and orange drink mix that said, "Complements of the Cali Cartel." Just several feet away sat another table with absolutely nothing on it but behind which there sat three hombres of disreputable demeanor and reputation known as the three dingoes, none of whom was taken seriously by anyone – yet. They growled at passerbys and repeatedly pointed to a sign which read, "Pay us now."

The Joe stood before the adulatation on his own world famous auto-mobile, the very manifestation of his ingenuity and sticktoitofness. He was near eye level with those within the reviewing stand which was parked directly in the middle of Colorado Blvd. Directly in front of him was his very own son who at that moment was standing next to the afternoon's master of ceremonies, a renewoned journalist by the last name of Remmick. Though he could not immediately recollect where, Joe was certain that they had met at least once before.

Remmick stood at the center of the open air reviewing stand. The construct straddled the center line of the avenue and forced all parade participants to break ranks to move to either side to move around it. Remmick grasped firmly the microphone of the local news channel with one hand and thanked the crowd for being there. In his other hand he held a burner flip phone with the intials HRC barely discernible on the backside. He couldn't understand and chose not to inquire. After holding up his hands to signal the need to be quiet, he reached behind him and held up two paintings, landscapes, respectively, of Ukraine and Russia burgeoning with spring time poppies and sunflowers.

He then changed his journalistic demeanor and stated opening bid is $500,000 for each. In the rapid fire delivery of an auctioneer he began the bidding within just two minutes when a man from the House of Baud, micro-electronics firm, based in Las Vegas, Nevada, offered $10,000,000. The buyer stepped forward and whispered to Hunter the artist extraordinare, "send everything to Mandalay Bay Hotel." Moments later, the still stunned Remmick and less so Hunter, began again the momentary bidding on two other paintings focusing upon the mountainscape of South China and Romania. Without hesitating the same man from the House of Baud held up both hands to signify $20,000,000 for the two paintings and then shook his head in ascent as Hunter raised his eyebrows and mouthed the words, "same place?"

Though privy to the entire interaction, Joe seemed oblivious and simply looked lovingly at Jill as she held her hand over her heart and made a fluttering hand gesture to suggest that his mere presence had once again stirred her to flights of intimacy and ecstasy. Joe, ever manly, pointed directly to his groin as if he and she were the only two people on the planet. No one else mattered, including Hunter.

The greenies and the orangies still stood in relatively straight lines opposite one another counting off the moments and rhythmically marching in place as Hunter himself took the microphone. Before speaking into it, he leaned forward toward his father and said, "Dad, they know. I can't tell you everything right now so I have to speak in code in such a way that only you will understand. I tried to send you an email but it's being monitored and besides there's another problem. Your executive order banning the orthographic depiction of the Latin alphabet because of its being racist, well, there weren't enough letters that survived that made any sense. A, B, C, D, L, and sometimes Y just aren't enough."

"Hey, son, don't worry, the rules don't apply to us, you know, like those people that rule in England. They know how to make the rules. Good for others, but not for themselves. That's the magic of true leadership. So, just go ahead and read what you have written." "Ok, Pop. I'll do it, but while I was getting ready, I wrote out two copies, one for you and one for my speech right now. Here's your copy." Though leaning over the transom of a ship about to sink, Hunter was able to stretch himself far enough forward to hand the words to his father as the band by the side of the road played the soothing melody of Ebb Tide. Finally, Hunter nodded to Remmick who backed away from center stage. Just as he was about to speak, Joe said, "Who? What is this. Where are the words? What kind of stuff is this? Did you give me the wrong piece of paper? Wait, wait, I get it. This is cursive, isn't it? I remember it now, give me the original typewritten speech."

"Dad, this is the original."

"Son, I love you man, but nobody reads this stuff any more, why do you think I uttered, notice I said uttered, like a cow in heat, that the Latin alphabet is now all but officially banned. There's no Romans around any more, so who needs it. Besides, my good friend, Zhi Zhi, 'er what do ya call him, even he said it was racist. He said that's why his people have all that trouble with "L" sounds. That's why it's China and not Lina. That's why he got rid of that guy in the really tall mountains, the Grand Tetons – you know, the Dolly Partons, or the Highly SaLassie. No, I think it was Dolly Lambda, that's it! So like I was sayin' the Romans were racists. They made the letters so the Chinese couldn't say them. So today I have fixed that problem. Besides, you can still talk it, you just can't write, and no one can even read it any more. That's why I have so much damn trouble with the Bill of Rights. I can't read it! " Why did King George have to write it in cursive?

Sensing that his father's temper was about to flair, "Dad, Dad, wait, one more thing, don't get mad, it's all in code. It's in pidgin that I learned when I was studying codes in the Navy. It's a pidgin."

Joe immediately ducked and flailed about trying to keep the pidgin from pooping on his head. Slightly misconstruing what was happening, Secret Service fired several shots at some invisible bird and a naked, wrinkly old woman just two days out of her latest botox detoxification treatment given routinely while getting her hair affixed, ran directly to Joe, shouting squeakily, "Don't worry, Joe, I'll get it, I'll get it."

As she screamed and jumped from one position on the car to the next, Jill dove from the review stand to cover Joe with her body. Secret Service continued firing until the ambulances arrived and carted off several of the bodies which were later deemed to have been the result of a shooter hiding miles away at the Guatemalan Cultural Center which subsequently blew itself up once the investigation began. It was the third explosion that had occurred in the building on Wilshire Blvd. in the past 3 years. Something wasn't right. Naked Nancy, however, came over to Joe and rubbed her soft and tired body against his leg and said, "I got a thing named after a cat and you got a thing named after a loaded gun, maybe later, huh?" Hunter looked down and said, "If you don't want her, you know, I liked young women and really, really old ones. Trade you." Nancy giggled and Jill responded, "Don't flatter yourself, I'm not that old," as she eyed Hunter suspiciously.

Jill looked up at Hunter and said, "Your father doesn't understand inuendo unless he's playing with me, right Joe?" Joe stood a bit stoop shouldered for fear of another bird attack and eyed the sky fearfully.

The crowd, however frightened, filtered back almost immediately since the Joe's activities and the mysterious bird of prey were much more interesting than the latest version of the newest game show, Jabber, Jabber, Who's Got Covid. Within minutes everything was back to abnormal just the way Los Angelina Jolieeses liked it.

In the meantime, Naked Nancy continued on down the street hoping to find her hair and flailing about wildly with her right hand in repeated stabbing gestures aimed at imaginary antagonists including Ronald Reagan and Bob Dole, the pineapple guy. Convinced that she herself was the main stay of all things cultural in America, she moved on whistfully reminiscing about her days as a show girl on the Vegas strip where she once performed nude with Anton Scalia and Ruth Bader Ginsberg while singing Nessun Dorma together and then finally La Donne Mobile which she was convinced that Perry Como or Elon Musk had written about her. She was transported into heaven as they lifted her playfully between them like a child.

"Oh, Anton," she whispered, in a momentary lapse into the realm of guilt and nostalgia, "How I miss you. Sorry about the way things turned out. But sometime a girl's gotta do what a girl's gotta do. And you, Ruth, you stayed too long – you know it was nothing personal. After all, he was always mine. Me, mine, La Donna Mobile, La Donna Mobile, you of all people should appreciate that." Proudly, Naked Nancy walked gracefully and compellingly to the Guatemalan Cultural Center to make certain that nothing was left.

In the mean time, Hunter began the speech that would compel the world to take notice as his father commandeered a Glock from a Secret Service to protect himself from low flying pidgins and birds.

"I would like to begin today with a call to prayer with a recitation of the 23rd psalm in its native Hebraic tongue – a prayer which I will recite forward and then backwards for those who are left handed and therefore sinister or right to left for those free of the orthographic oppression of ancient Romans. They conquered much of the world with their stolen verbs and nouns – but not today, not today! Rabak Hallah! Rabak Hallah!" (ASIDE: Thank you Monterey School of Languages)

Moved to a near worshipful and celebratory moment of paternal affection for his son's linguistic virtuosity, Joe fired a shot into the air, followed by burst of feathers and a loose excretory descent from a passing seagull… "Seagull, inland, huh, it's a sign, that means there must be Mormons around here and we aren't even near Salt Lake where Lot's wife died," he said to himself.

Hunter continued as the Equity Emergency Embulance (EEE for short) confiscated the remains and sped off to the crematorium at Roselawn Cemetery, a section sponsored and funded by Animals and Nature US. The high pitched eeeeeeeeeeeeeeeeee of a siren instilled compassion for the trauma which many suspected that Joe may have felt as he yet took aim at another…

So in the native tongue of the bipedal, non-orthographically oppressed first blossoming seed of all subsequent evolutionary DNA providers of sexual innies and outies, let me say, 'Da Big Guy is my bank a bank, Sheepa Sheepa, Greenies compella o slo mio, no Noah rainmaker bya distilled voda from Gog and Mama Gog. Papa mio bilonga olsem ooooo-kry-oooooo-nits, Haben zie eina mucho Gringo bakabak 4 Grandi seni eeeee Grandi meloni. Special Boom Boom na me lie 'bout distilled laundered ping ping outta pong pong. Sophia wise New Romani heap a big receipt bundles – KAKA KAKA. Me in no greenie hope much. Sheepie sheepie shit floatsom by. Me no habe eine basta paddle. Helpa, helpa, awready, Mucho Dada. Me no bail, you no tell – bout Chic in Kee EV. Bambino haben meina fromma electric mamma. Tooo much talka talka. Like Winnie who Pooh Pooh on mio, she gibbe honey, now wannta gryvnya. She talka bang bang – she loud talka talka, I nyet gryvnya, I get grief da. You see seagull, Mormon send. Need mormoni 4 people with sheepa sheepa. One knee man from Rome senddym hope wid, Send mucho gryvnya and EV T and A wid titter tease and ESL. HINT: Elon say Revelation Biga Biga, No hidda TT Titter no more longa. Gotta U, Ragazzone? (Special thanks to translators and transexuals from the DOD)

"KALI OF CALI AND THE POPPY (GENUS: ESCHSCHOLTZIA CALIFORNICATION)"

Unknown to Hunter, both the forward edition as well as his orthographically and cryptographically sequestered forms of his coded message were already being surveilled and copied by American owned satellites floating immediately above him. No one noticed, out of sight, out of mind. The second time through Joe grew restless and tried to dismiss the backward version as irrelevant. But Hunter insisted on going forward or at least backward at a precise pace. He, however, grew nervous when CAIR showed its presence and one of its members said, 'You think you can fool us by reading backwards, huh? We don't care what you say because we are a very, very backward people and that is how we know what you are doing. Backwards is our way of life. Because, if you are backwards you can always see where you have been after you got there.' Struck by such a pronouncement which was of a profoundity never before uttered in his presence, Joe compulsively looked behind him to see the future as only a wise man could. Rabka Hallah.

An anxious Joe waved to someone in the crowd who was somewhat concealed. As he did, Remmick, who was walking behind glanced in that direction, immediately identifying the person who waved back. Startled but remembering his instructions, he flipped open his burner phone and pressed the contact list which immediately connected to Carlos Ignatio Armando. 'Big Guy is in contact with Ghost Guy. He's somewhere in the crowd. Better be prepared for engagement.'

Joe immediately revved the motor of his Pumpkin Wheeled Auto Tractor and spun it full circle as his Secret Service detail wheeled the reviewing stand to align perfectly with the white line down the length of Colorado Boulevard and pushed it to one side of the thoroughfare on the strength and inherent mobility of its eight wheeled undercarriage.

He then proceeded up the Causeway of Hope, Humility, and Healing. Along the way, he viewed one tented pavilion or road side stand after another. None was intriguing though than that of the Swami River Gateway House of Hinduistan. Seated either in traditional chairs or merely floating freely, these seekers of truth and money had over their vending tent a large sign that read, 'Cali the Black Goddess of Destruction is NOT one of us. 'She is from hell, not heaven.' Below the sign was a large portrait of the VP with an X through it. Next to the sign itself were three fierce looking individuals with drawn simitars and holding a large sign that read, 'Thugs not Drugs. ' A pamphlet on the dangers of various forms of inebriation along with another protestation next to her picture read, 'Kali is bad, Cali is badder, She so sad, she make us sadder. '

As he looked to his right, he saw a now elderly George Bush still trying to sell one of his portraits of Vladimir Putin from his days of yore when he visited "The Ranch." Against a display board the price was posted at first $500,000 in an attempt to keep pace with Hunter's establishment as one of the greatest painters of all time. As such, George in a moment of speculation had attempted a portrait of his former VP's hunting buddy after a whiff of bird shot hit him square in the face which was later traced to a shell expelled from a shotgun owned by the same VP. The portrait was half finished when George ran out of red paint. The price of $500. 000 had originally been raised significantly by the former VP himself who was misled when he was told that there had been no forgeries produced of the original. George had insisted that since he was in fact the originating artist his works could not be forgeries. Beside, he needed the money for his own dance studio intended to complement his latest artistic ventures. Even as Joe drove by triumphantly with his Glock in one hand, he watched suspiciously as a familiar looking man walked over to the painting of Putin, thanked the artist, George the Younger, and then walked off without paying. 'Will send gold from new mine, NAK, in Alaska. Much gold and silver and other things you stole from mother in Russia.'

To his left was the Qween herself standing beneath the awning of her very own non-capitalistic pavilion with a bold signing say, 'Don't Hate Me, Just Date Me. 'Next to it was a picture of a smiling Bernie with his hand draped strategically over one breast and her iconic deer in the headlights expression. Beneath a table loaded with campaign literature was a large red and white sign that read, 'Free Air Kisses and My 2 Second Presence in Your Life for Only $10,000." Bernie's smile was radiant.

Not too far behind Joe's automobile were the Marching Kodiaks of Nome, Alaska, and members of the Russo -American Gun Club. Waiting in ambush one mile ahead was the Ukraine Snipers Club of Chicago, Illinois, not feeling at all constrained by California's gun laws. As they pointed out to L. A. authorities, most of their guns had been obtained illegally from the very best gun running gangs in Chicago. The term "obtained from the Latin Kings," or "bought for you by El Rukns" was synonymous with non-traceability and quality, many of them having been designed, manufactured, and shipped from Ukraine itself. The L. A. authorities explained that the real threat to civilians and community members was from legal gun sales. Since theirs were illegal guns and had in all likelihood been obtained through the violation of the southern U. S. border, there was no need to worry as such illegal crossings were sanctioned by the current administration.

Meanwhile, Hunter had driven away from the parade itself to inspect what was left of the Guatemalan Cultural Center. While he was shuffling through the debris of books, pamphlets, official documents of one sort or another as well as a panoply of maps, soda bottles, and clothing and large numbers of lamps of the exact same design, all of which were badly burned and/or fractured, Naked Nancy walked up quietly and tried to grab his ass. But, despite her best efforts, she fell backwards, yelling 'ouch' at the despoiling of her sacred flesh with nails, splinters, and shattered glass.

Hunter glanced back but was distracted, apparently searching for something, something very, very special. Naked Nancy followed, saying, 'Hunter, I now have glass in my ass, aren't you going to—'

Stop! I need to find it – "You mean the fentanyl? Haha. I bet you Mr. Shape Shiffer got here first. People that know him keep jumping or falling off buildings. You would think they were all child stars from Disney studios or something. By the way Hunter, dear, my butt still hurts. I have a surprise for yooooou, baby."

Nancy, I need to find the receipts. We need the ones Feng Feng had in her possession before she fled the country.

"Oh, don't worry about that, Hunter. Swalwell didn't and look how much he made, c'mon. That glass is still there and I can make you happy, Hunter. I make your Daddy happy and I don't even make him share the trade deal money with me anymore. My husband, Feinstein, and my own nephew – they get their cut upfront before the fentanyl is even shipped.

The Chinese Commies are happy and so are the great American insider trading capitalists like me. c'mon, pull the glass and maybe I can pull something for you, ya know."

"Nancy, if the receipts fall into the wrong hands…"

"Oh, don't worry, baby, nothing will happen, I just wanna make you happy. Besides, it just feels so right. You're Catholic, I'm Catholic, the Big Guy is Catholic. It's not a sin if you do it for another Catholic. C'mon, pick out the glass and I'll even let you kiss me. C'mon, you don't even have to go to confession. It's not like doing it with a Mormon or somethin'. If you don't do something you hunka hunka burnin' love! Take me, I'm yours!"

"Nancy, damnit! Stop!"

"What too young? I can act older if you want, you know like wear old people clothes and halter tops that cover my entire breasts, unlike, unlike, you know. I'm smokin' hot and –"

"No, Nancy. You're only about a foot tall, you're too short!"

"I wasn't always this way. You were my Godson, you should have taken care of me. I was a content creator. I coulda been somebody."

"Still, you're only about a foot high and you're still naked. And you're not my Godmother."

"I could be…"

"No, your too damn short."

"I love it when you get angry. C'mon, grab my ass and take out the glass, I know you want to."

"Short, short, short!"

"Is that my fault? Look at this poor, poor woman. Yes, it's true I've lost 16 inches in height just this last five weeks. But why, Joe, why? I'll tell you why. Because my facial foreskin cream from South Korean babies and boys hasn't come in, Joe. Which reminds me, are you circumcised? You wanna show Mommy? C'mon I need another supplier closer to home."

"I'm Hunter, not Joe."

"Joe works better for dramatic purposes, beside your Daddy reminds me of Jesus and his dad was Joseph the carpenter and 'cause of that I bet he had good woodies, not Jesus, Joseph."

"Ooooh, lookie here, Joe, it's a pair of mouse ears, a coloring book, and a map—a map of the grooming studios. Ha, that's funny, I didn't think they had horse stables at Disney. Wow! This is great. More fun for the entire family."

"If I don't find those receipt, we are in a lot of trouble, Nancy."

"We? I'm just a poor woman who, because of supply chain problems, can't get her Uyghur Wonderland free booty booster shots of growth hormones – Look at me, Joe. I've lost more inches in a month than most men in a life time. All because the supply chain guys can't get the Uyghurs to market. It's a crime. I've always been beholdin' to the kindness of Uyghurs. Speaking of Look at me, Joe, I'm in need."

"Nancy, for the last time!"

"Joe, before you say no remember I have these pictures of you at your little ranchero near the Black Sea with a little honey. See. I carry these with me wherever I go."

Startled, Hunter finally turns around and glances at the stack of pictures held out by a wiggly, giggly Naked Nancy. Hunter stammers, 'How? Where? I mean how could you carry anything around Oh, no, even for me...I don't even want to touch them.'

In response Nancy begins to sing softly, 'I am woman hear me OR I want to have an Oscar Meyer Uyghur each and any way I ever can. C'mon, Joe, gimme before – C'mon they don't call me the Eager for Uyghur for nothin. By the way, do you know where we shipped the Uyghurs from Guantanamo when we let 'em go after we captured them in one of Bin Laden's camps? Huh, well let me tell you. We sent them to Bermuda, they run the golf courses there. Look it up. We could go there you know. Just think, Bermuda, a Uyghur Wonderland for Hunter, Joe, and his young bride.

Hunter finally relents, 'If I pull out the glass, will you stop bothering me?'

I can't promise. Let me drape myself over your knees as you cotton pick the broken glass from my ass. When you're through I'll call to get our plane ready and we can blow this pop stand like I did your Daddy. Remember to kiss it to make it better. Bermuda, a Shangrila for eager beavers like us and Uyghur foreskin providers, my very own fountain of youth. Say, you're awfully quiet…

Hunter begins to gag but continues as he presses down, meticulously picking the glass from the…

'Ooooo, Joe Man, someday I may even let you paint me nude.' She hums softly 'Dixie'.

Off again to his left, Joe and his entourage passed an Armenian shop keeper that advertised 'RAZOR SHARP SICKLE SALES, TODAY ONLY.' Why, Joe wondered, would anyone pay to catch a disease even at a discount. Next to that pavilion was yet another ever so emblematic of the ancestral homeland of the Armenians when they first arrived on earth. Noah's Ark watched the waters of Napa Valley recede causing it to light near Mt. Saroyan in the heart of what would later become known as Little Armenia. Yes, Glendale was its capital, a place to which the diaspora periodically returned. Here it was that the original version of the Star Spangled Banner was written entitled Baubled Bangled Beauties. Along side the entrance to the pavilion was a life size picture of the Kardasian girls with a large red X painted across it. The meaning of the X was immediately clear from a message opposite it paid for by GMAK, Glendale Moms Against Kardasians. It read "They're Not One of Us, No Armenian Mother Would Put Up With This, The Mother's Dutch! " Ever tolerant and embracing of all branches of the Armenian world wide dominance, Joe loved these and all other off shoots of this phenomenal group of people. Having been taught that Armenians always end their names with the letters *ian,* he noted that the Ayetalians and Austro Hungarians were merely one and the same, along with the Romanians, Chechians, Norwegians, Brazilians, Australians, Albanians, and, oddly, he later discovered, even the Ayerish, hence, the common first name of Ian. Joe wondered, were all Europians Armenian, even the Ethiopians?

As Joe drove on by, one of the Armenian members of GMAK ran alongside his tractor, blending in quite inconspicuously with the Secret Service, eventually weaving her way right alongside Joe's slow moving carmonsity. She reached up with a really, really big and brightly shining gold chain. 'Wear this, Joe, all our men do. That's what makes us love them so, their gold and their hairy chests. See you later, new nose job at 6 today. By the way, Kanye stopped by today to say hello, we said, 'Ain't no way you're one of us, you're Black Ayerish, not a pure bred Armenian. 'Ever the diplomat, Joe, said 'Yeah, I think I'm Armenian too and maybe Black Ayerish because I like gold stuff and my mother really liked Ricardo Montalban whose ancestors came from Dublin. 'Confused and running out of breath, the Armenian GMAKer slipped away as Joe fumbled with the gold chain, only to feel a subsequent tingling in his ever expanding chest along with a lot of hair and an entrepreneurial spirit that enabled him to calculate in his head the miles per gallon to be saved if he slowed the tractor down one more mile per hour. Genius, he thought.

As he looked up to the heavenly blue sky in supplication, another pidgin (a creole speaking bird) flew by, squawking, 'Me know Big Guy, You Fly by Night. 'Joe pulled out his Glock, took a bead and feather bombed the next tent where a sorrowful but beautiful looking teenage girl stood in the finery of a bygone era with a long gown. Next to her was a sign that read, 'I am she, the child of sex traffickers and the Cali cartels. I am the last surviving child of the Romanovs, remarkably well preserved by the constant cleansing of the my soul through grief and abandonment. Save us Joe. For I am the princess Amnesia, the last emblem of the innocence and the throw away child. 'Joe drove on most quickly.

Not far beyond, a group of Mexican singers, The Corrido Amigos, were in the midst of singing La Cucarracha, when a swarm of Orkin exterminators fell upon them, yelling, 'We just cleaned this area yesterday. Take your dirty songs somewhere else.' The usual relatively benign spray to kill the urban menace, English speaking cockroaches, was replaced by DDT and a bug bomb set to explode within minutes of the Orkinites departure. Joe just escaped the good natured hazing by the Orkinites who were frustrated and unvaccinated Border Patrol Agents. As they departed laughingly at the now choking and vomiting Corrido Amigos, one shouted, 'We don't need no stinkin' cockroaches. Arriba! Arriba! 'The Orkinites fled to their horses, briefly posed for pictures, and one fingered the Commander in Chief.

BEACH PARTY OF EQUITY AND INCLUSION

In the late afternoon, just before dusk, a clairvoyant and homeless drug user of a myriad of mental disabilities who aspired yet to be the Cali state governor, writhed and groveled on the ground next to a fishing pier that extended far into the ocean. Ignored by the passersby as just another career politician or as a jelly fish in heat, the fishermen passed by while some, ever mindful of the biblical injunction to care for the mindless, tossed some of their own newly purchased live bait to the craving maw of madness. They then smiled benignly at their having prevented another member of the Cali State Assembly of Erstwhile Imbecility with flavorless but always nutritious yearnings for and of the masses who found Emma Lazarus' poem inspirational and yet devoid of the political acumen to perfect its execution. Therefore, in sympathy for themselves they called upon the city of L. A. to hold a massive and decidedly patriotic beach party to which only Annette Funicello and other Ayetalians were banned because they were already dead or had been given last rites in violations of the state's latest anti-religious zealotry enactments.

Like minded practitioners of such opiate ingestion like Joe DiMaggio had his name stricken from Fisherman's Warf and Ted Williams was permanently banned for using racist diatribes for once having promoted the inclusion of black baseball players into the Hall of Fame while not having simultaneously called for the exclusion of all white players including

Babe Ruth. He was after all mistakenly assumed to be white only because his parents were. Look at that nose, they said, 'He's a bro.' When it was discovered by the fact checkers at Gulog that Williams was also a combat pilot in the Korean War and had not apologized for his performance of flying as wing man for John Glenn, the astronaut, he was banned from any public recognition. After all he was also of Basque ancestry. The 'B' having been banned also required a redo of his DNA to be replaced by _ _ A. Somewhere in a cryogenically induced state of nonchalance his head was spinning and white lab coated lovers of baseball were forced to permanently dismantle his freakingly cold cocoon by pouring out its contents including his head. Congresswoman Maxine Watters requested that both the head itself and his baseball records be rescinded and given to a much more deserving man of faith and equity and inclusion, Al Sharpton, because he had never played baseball, wasn't white, and loved all people. Sharpton accepted and will soon be inducted into the Baseball Hall of Fame, soon to be renamed the Aseball All of Ame or as it is otherwise to be known as BBAA. Known affectionately as Baseball Artifacts of Asian American, nicknamed Trinkets and it's Sukiyaki To You. Sharpton's Induction will be celebrated throughout the state with children being allowed to attend school with hoods over their heads and white sheets to celebrate their white privilege even if they were brown, black, or maroon... Asians themselves were not allowed to attend since they were found to have had an inordinately high intake of white rice and incomparable SAT and ACT scores along with strange eyes that gave them an edge (wink, wink).

PATHETIC FALLACY OR FAKE PRICKS AND BIG DICKS

In the twilight of the day, the incomparable beauty and near beatific resonance of nature's near somnolence, the sun itself seemed permanently affixed to the horizon, dawdling above, awaiting the confluence of the evening hues. As they assembled about the globe of fire, the waters themselves took on its shimmer and seemed to positively shake, shatter, and shoal. The distant but ubuitous ocean cargo carriers sat like sentinels on the waters to prevent the people of Cali from leaving. Or the hungry, infirm, or homeless, or those callously indifferent to the vulnerable and enslaved who produced cheaply made goods – all who coveted these goods were bound by poverty or greed to wed themselves to the L.A.port facilities. Leather shoes molded on board and cheaply made products were thrown overboard to Polenesian-Anglos who boated out in hopes of receiving computerized barter in return for sexual favors. In return for pomegranates and pineapples from the Big Island of Hawaii the women offered themselves to the local harbor pilots and waved naked hips and unclad breasts at cheering sailors.

While Joe eyed the beauty in a near tranquil state of peace and a longing for the great beyond from whence came the yuan for which he and Hunter had such an insatiable yen, the orange ball seemed to expand and then collapse slightly in an almost breathing movement. Simultaneously, he felt a trembling and the waters began to boil and percolate. The crowd of covorters momentarily stood perfectly still trying to maintain their presence on plots of earth and sand that moved laterally beneath them. Then as if the apocalypse itself were upon them, Joe watched in horror as the orange ball miraculously rose in the sky and suddenly an explosion of volcanic magnitude blew from the waters. Debris that was brown, sluggish, and decidedly less than molten flew skyward as well as beachward. The brown goop seemed to be flowing from the north when suddenly a Coast Guard cutter bore down upon the people and non-binary DNAers of disreputable species formerly known as human. It cut its engines as it floated into bathers and waders near the beach, one of the men on board shouting, "Get out of the water, seismic waves detected. San Francisco's sewers are emptying and the needles and homeless are coming your way." Dr. Anthony Fauci floated out to sea in a small life preserver in a selfless act of courage, yelling, "E Coli Outbreak! Wear a mask and stay in the water until the ocean recedes. It's for your own good!"

Suddenly another explosion! Boom! Boom! Boom! The waters still boiling then parted as the startled crowd witnessed a rupture of such severity that the ocean bottom appeared where ancient Egyptian chariots suddenly spun again to life with others crashing into the shore line with mummified cats on board. A geyser of molten hot feces broke from the bottom blotting out the heavens and, of course, the sun itself. Even the twinkling lights of the cargo ships seemed to have gone black. The geyser grew in intsensity when a geologist standing on shore shouted, "Kakatoa! Kakatoa! We're all doomed. Kakatoa! Run from the e.coli!"

When some of the bodies of the homeless floated near the shoreline, Maxine Watters, after whom the oceans and all fresh water bodies are named, kicked at them, saying, "Just go home and stop pollutin' my beaches. Go home, you dead beats, 'er dead people, or whatever." The sewage stacked up so high on the shoreline that even the booster shots of some success could not save those impaired by the outbreak of e.coli which left the dead from Covid still dead and the non-dead stinking as if they were.

As Joe moved from one location to another feeling and being helpless, he tried to contain the madness by calling out to heaven in his native tongue of Japanese, translated loosely, "As I look to heaven so my tears do not fall. I'm a Jap in Crap. Oh save me, Cosimodo! I implore you, save me and my people except for maybe Mad Maxine, the indomitable matron of malevolence. Save *me*, if not the others! (And maybe Hunter too, and maybe Jill, maybe)"

Suddenly, Jill who had fallen off the tractor/auto miles before, rushed onto the beach, looking at the increasing mound of bodies and yelled out, "I'm a doctor, I'm a doctor, allow me to save you! Oh, the humanity and others who are essentially non-binary and still somewhat deserving of salvation. Look upon me as your deliverer. Much like DHL and Amazon, I am here. Let me beathe life into you. For, as I have said utterly and salvivically, I am a doctor!"

At that point, she splashed over through the molten feces where Mad Maxine was pressing down upon a body that convulsed beneath the water trying to inhale what little air the lungs could take in. Jill yelled, "Maxine, I'm a doctor, help me get him to shore." "Yach!" uttered Maxine and continued to push down even as Dr. Jill dragged the one time convulsing body onto a soft batch of other bodies and began to press upon the now non-binary cadaver-like being. She pinched the nose of the increasingly rigid thing before her as she placed her lips upon his/it/she (hence forth known as *shit* for purposes of inclusivity). The taste seemed familiar but not something she could immediately place. She huffed and puffed and she buffed the lungs and stomach full until they expelled a lusty regurgitation as well as a sympathetic and simultaneous exhalation of flatulence for which all cadavers are well known. "Oh my God"! she uttered, "I think I have been e.colided. Save me, Lord, and non-toxic God of the male patriarchy! Save me!" Collapsing in disgust upon the already dead, she could not know that he who had jet like speed and agility because of his days as a formidable college wrestler and currently a member of Congress after whom an Arab nation was once and still is named, came upon her and said, "Oh, shit, I don't want to but I guess I have to."

He picked her up with one hand and threw her over his back, running quickly to the nearby church where there was an infirmary and practicing nuns of the hospice supported by the hopelessly benevolent and guilt-plagued archdiocesan pedo-romantics. Here he deposited her body at the front door as a non-binary nun of disproportionately large forearms and shoulders, opened the doors and grabbed her, throwing her over *her* back. Here at the church of the Notoriously Damned or Bifurcated Friends Forever she ran into the night time janitor, Sukiyaki, who said, "Me, give me!" Obligingly, the nun offered the Jill of the Joe to the tender and yet massively strong and gentle giant. As he ascended the stairway to the heavenly realm, he pulled out his earplugs from his pocket and began to sing, slowly and tenderly, the song of the Bell Ringers of the Nagasaki Martyrs of Long Ago. The song poured from his lips until he reached the top of the church where he worked as a full time bell ringer for food and lodging. With his ear plugs firmly embedded, he could not hear the sounds without. Still he continued singing, the second verse of Sukyaki which had seldom been translated for the benefit of an American audience. Here it is. "Oh, watusi wasabi, myomyo Heroshima ala la la la Hi, Hi, Hi, Tora, Tora, Tora, me no Jewish, proximate, ice a nakamoto, binji sumo, aka Mitsubishi, oyotaor, literally, *Me no know em, so I throw em.* "With that, he picked her up, oh so gently from where he had lain her down only minutes earlier and turned her head to look out the bell tower. She smiled as she came to. "Who are you?" she exclaimed with some excitement and appreciation at the realization that someone, a stranger to which she was no doubt beholden, had offered her kindness despite the smell that still engulfed her body. He smiled benignly and uttered softly, "Jill do you not know me? Je suis ton pere. As obliquely as I can, Jill, Je suis ton pere.'

"What?' she exclaimed, "You're that guy? No! It can't be. It can't be. I don't even recognize you!"

At that point, he gathered himself and ran headlong into a huge and by far the largest bell in the tower. His face now contorted, twisted and bruised reverberated as the bell rang so loudly as to break windows for miles around including all those in the church and a several block area including those that were at that very moment being newly installed at the Guatemalan Cultural Center through the contributions of the Sinaloa and Cali Cartels as well as the Confucian School of Communist Indoctrination and Delight sponsored by the CCP. He looked at the still supine figure of his beloved when her eyes brightened at the now battered, bruised, and bumpy and all too familiar face. "Daddy?"

As if to affirm her recognition, he smacked his face again into the bell causing another explosion of glass throughout the community. "Now does it ring a bell?" he asked. "Daddy, where have you been?"

"I left because of the one whom you married, I thought he was little more than a marsupial hiding his ill gotten gains in his pouch as only a non-binary kangaroo would do. He is no Joey just a Joe of no repute. You married him and so I left. Now today I save you unless—"

A loud sound emerged from below, so loud, that even Sukiyaki could hear it with his earplugs in. He stepped to the edge and looked below where a large crowd had formed. Composed of out of work bell ringers left homeless after the Christmas holidays because of a lack of paying work, seventeen electricians who were seeking a means of escaping the daily grind of incessant hours of work from people who studied in college in various fine arts programs that produced nothing and had no marketable skills, two plumbers (same problems) and suicidal truck drivers that were just looking for a target and seven Canadian drivers who refused the vaccination and were still looking to perform life saving techniques on Fauci whether he needed them or not. Justin Trudeau was also wanted but had a knack for disappearing into the tundra to hide from their wrath.

Not knowing what they wanted, but feeling irked to know that his daughter was still married to the man whom he despised, he looked below and yelled, "What do you want?"

"We want Jill, you moron! Give us Jill!"

Sukiyaki, the Franco-Anglo-Japanese American father of Jill, responded, "Bankruptcy! Bankcrupty! The whole damn nation is about to fall into bankruptcy! She is my last and final demand. She is mine and you can't have her. Bankruptcy! Bankruptcy!"

"Give us Jill! Give us Jill! Give us Jill!" The chanting continued *ad nausea* and even sickeningly.

Sukiyaki battered his face repeatedly against he bell and then in a pique of rage, he grabbed his daughter and threw her below to the crowd. Horrified, they looked on as she screamed, "But I'm a doctor!"

She fell mercilessly on three Canadian truck drivers who exclaimed, upon recovering from the collapse of her body upon them, "Well, if the ladies don't find you handsome, they should at least find you handy." Slowly, Jill recovered, more damaged by fear than by actual physical danger. As she slowly pushed herself off the cushion of bodies, she looked up at her father and exclaimed, "I think we need to talk, Dad." Her father smiled, and turned back to his bells.

Meanwhile, down at the beach, two fishermen in waders, looked at the feces drenched fish they had just caught and threw them back, concurring they were no longer in the mood for a shore lunch. As they were wading away from poopy percolation of what some had taken to calling "The San Francisco Treat," they heard a round of applause for a meandering group of hooded figures with their hands folded prayerfully in front of them. Amidst the panpoopic, a member of the Joe's staff ran about looking for a consoling presence to quell an exuberant but somewhat distraught crowd running about through the increasingly vile shore line. He soon lighted upon a visiting choral group from Gethsemani, Kentucky, known as the Trappist monks, who, however wanting to help, handed the staff member a brochure explain their vow of silence. Upon reading the brochure, the staff member cast it aside and responded, "By the power invested in me by the Almighty Joe and the Caprice of the Federal Government, I command you to speak and sing or forever hold your tongue." Most of the 20 hooded minstrels of taciturnity and heavenly aspiration sighed audibly and then reached up to grab their tongues. "Follow me, ordered the staffer."

While marching to the beach in double file and time, most fortuitously and miraculously, the singer song writer Paul Simon waded ashore amidst its debris and began to sing "Sound of Silence." This was an adaptation of a song the Joe had written earlier for a corporkorate donor from Iowa, "Silence of the Hams." This had been short lived on the farm and country charts but remained at the very top of the Pork Bellies Hit Parade for six weeks before senator Joni Ernst replaced it with her derogatory song about "Piggies at the Trough," and the pig farmers game show, "Squeal of Fortune," which was banned on Youtube and shadow banned on Google after a billionaire from Epstein Island decided to start buying up all available farmland for fear that there were still too many self-sufficient Americans. Subsequent references to the nation's capital as Porkopolis, D. C., District of Corruption, were also banned.

Joe, however, ever the poet, wandered the beach in his auto-mobile lonely as a cloud, periodically gunning the V7 engine to scoot past the dead, dying, and metaphorically challenged as they shouted out and then sang, "We are the world, our future here unfurled." The Joe drove on, like Rachel, looking for his one of a kind son, who was still at the Guatemalan Cultural Center looking for answers before an attorney with a goatee came ashore in a durham row boat and handed out a picture that said, "Have you seen this man?"

He handed one such picture to The Joe who looked at Hunter's picture before his cock throbbed three times and simultaneously Hunter, still several miles away, felt a slight tug at his groin as Naked Nancy returned with a gentle solicitation, "Hunter, sweetheart, my butt still hurts and its needs some iodine like what they have in the ocean. You wanna do a midnight swim with me before our plane arrives. It's a water plane. It can land right next to the beach and we can get in a refreshing and erotic swim on the beautiful sands before we depart for Bermuda. Our plane should arrive late tonight."

"Nancy there's pooh in the ocean, don't you smell it? '

It can't be pooh, it's from San Francisco. We don't do pooh, we emit panache and ancient Chinese secrets and really, really expensive ice cream that can be recycled even after it has been digested. We Freaconians are special like that. We—"

Suddenly a shot rang out and a man walking with the Dalai Lama suddenly feel face down in the surf with a bullet hole in the back of his head. The Dalai Lama immediately hit the beach himself and then from beneath his saffron robe pulled out a AK-47 with its banana clip already inserted and ready to fire. He knelt down behind the dying figure and fired in the direction from which the shot had come. With the sun still boiling on the horizon, he found his target, a silhouette on the bow of a nearby Chinese junk carefully disguised as just another cargo vessel flying a Panamanian flag. In rapid succession he and four other silhouettes fell into the ocean waters, badly wounded and finally dead after feeling the now stinky pooh waters of San Francisco. Several of those who might have been saved stabbed themselves in the neck with passing syringes rather than submit to the smells. A floating refrigerator passed by. It had the words, "Let me eat ice cream" on one side, and "French Laundry Take Out" on the other along with the phone number of a certain "Gavin" for home delivery.

The junk then turned broadside, offering six or seven well armed military personnel wearing protective vests against which the bullets from the Dali Lama would be useless. Only a head shot would work. In desperation he took aim as the junk itself rolled high and low upon the waves as explosions from KAKATOA continued to reverberate along the shore line. Known for contemplative and meditative marksmanship skills, he took aim even as he felt the whiz of bullets from the soldiers on board whose advantage in numbers was sure to eventually take its toll. Confounding his problem, the Dali Lama had been able to secure only one full clip and now he was out. Defenseless, he lay behind the wounded, unknown figure as he saw a small outboard powered boat drop from the gunnels on one side into the surf with one steersman and four more soldiers coming ashore. He felt that death was certain when behind him he felt a firm hand on his shoulder. Turning around, he saw four men with Mongolian-like features wearing Bermuda Country Club polo shirts and carrying FN SCAR, the Belgian made assault rifle often preferred by U. S. Special Forces.

Taking a kneeling position around him, they returned fire. The U. S. Coast Guard vessel then opened up on the junk with its small cannon enfilade and within minutes the junk was in flames and sinking ever so slowly. Simultaneously, though officially "unarmed," the Canadian truck drivers took a nearby flanking position and fired their secretive home assembly PUCK (Perfectly Unarmed Canadian Killer) grenade launchers on the floundering junk to finish it off.

The Dalai Lama asked, "What took you guys?"

The leader of the men, all of them former detainees of Guantanamo, yelled above the din, "We the Uyghurs of Xinjiang, we stand with the rightful ruler of Tibet! We came on our float plane, as you requested, to pickup Nancy and Hunter, they think they are going to Bermuda."

"Good," exclaimed the Dalai Lama, they're dirty Commie criminals and are to be rewarded appropriately once they're in the air. "Make sure you throw his paintings out too." "Come on, let's finish this." The Canadian truck drivers followed.

Meanwhile on shore the eternally sworn to silence Trappist monks of Gethsemane, Kentucky, continued their non-verbal but wonderfully robust rendition of Sound of Silence with Paul Simon constantly singing and resinging the same song – until in a moment of fatigue and frustration he added a second, third and fourth verse to amuse the crowd. Though at first frightened by the immediately adjacent ruckus, they concluded that the entire explosion of cannons, rifles, and grenades, along with the tumultuous and earth shaking and water bubbling KAKATOA were all part of Disney's latest version of Wokedom (Loud, Proud, and Unbowed or We Really Love Kids a lot).

And in a moment that would later go down in history as the Miracle of the Sun, the orange ball of fire remained suspended well into the night even unto its own arising the next morning. The dawn thereafter was known as The Day of Two Suns.

Though fighting valiantly five Russian trawlers posing as fishing boats and the massive yachts of two Russian oligarchs moored just beyond the immediate area of engagement, began to weave in and out of more than 100 cargo vessels from China. Upon coming within reach of the beach and the stench, they dropped several speed rafts into the water equipped with Spetsnaz, their equivalent of U. S. Navy SEALs or Delta Force. The faux fishing boats, however, dropped their dragnets into the water to maintain the pose of being real fishermen who just happened to drift in the L. A. harbor area.

Within several minutes the Spetsnaz were completely entangled in their own nets and sped around in circles, the soldiers falling into he water and choking on the poopy flow. "Oh no," one yelled, "It smells just like the stuff coming out of San Francisco when we infiltrate critical infrastructure last week." Despite the entanglement, the Russian Special Forces and The Army of the People's Republic of China were able to force their way almost to the shore line despite Buddhist resistance and the ongoing support of Canadian truckers and a host of binary affirming macho Anglo and Hispanic powerfully toxic horse riding caballeros with lassos and Biden approved whips. Watching the desperate struggle was the master of ceremonies, Remmick. He assessed the situation and immediately flipped open his burner phone and called Carlos Ignatio Armando as Maxine Watters herself attempted to give out fortune cookies and slices of Peking duck to the Chinese troops while the Russian troops who saw her pretended not to.

The Buddhist entourage of the Dali Lama secured the perimeter in an attempt to rescue him and the Bermudese Uyghurs. One of them reached to pick up the still breathing victim of the first Chinese assault. Turning the man over, the monk blanched when he realized who it was and why he was the first victim. Wearing a now bloody grey tee shirt with a map of China with Tibet outlined in red with the words FREE TIBET was the actor Richard Greer. He was hoisted aloft by the monk who nearly tripped over Mad Maxine who grabbed at the tee shirt and attempted to question Greer as to where she could get some of that Free Tibet stuff. Moments later the Joe, ever unmindful of the crisis at hand, drove by gallantly bumping Maxine out of the way and reloading his Glock as he good naturedly fired at stray beach balls and the occasional pidgin which he was sure was stll harassing him. Still the shore line defense was being enveloped when Carlos Ignatio Armando called out to Remmick over the phone, "Now! Execute! Execute!" Remmick then pressed the asterisk button and immediately the walled exterior of two life guard towers pealed open like a banana, exposing giant speakers. Carlos Ignatio Armando then shouted, "Release the Cackle! Now! Now! Release it!"

The relative silence of gun fire, screams of pain and anquish, and ultimately murder and mayhem was drowned out by the laughter of the

Virgin Person of the Biden White House. Loud, piercing, and as always, annoying beyond belief, it caused the Russian and Chinese troops to throw down their weapons and beg for mercy. Even two of Joe's Secret Service detail who spoke with a Pakistani accent, threw down their weapons, and asked to surrender. Maxine who had witnessed the surrender looked on in horror and immediately called Debbie Wasserman Schultz, "She ain't gonna like this no how…Debbie, they got your boys…Loud? Just a beach party…Did you get your computer back yet…Whatever you do Oh, wait, you won't believe who else is here. It's LeBron James!… Yes, I'm sure…I think he's down on the beach trying to negotiate a new shoe contract… He's crying, sayin' he needs the money and, oh look, he's pickin' the shoes right out of the water from one of those exploding cargo ships… Got to go… gonna see if I can take my cut on his shoe deal…I speak Candooneez if you know what I mean.

The Cackle had an extraordinarily debilitating effect on the Chinese and Russian invaders while the Americans, Canadians and the Mad Maxines responded with the usual irritation and call for her devocalization surgery which Joe had objected to since he didn't know what it meant. Tibetans and Uyghur warriors were entirely immune having spent years assimilating a combination of meditative powers augmented by DARPA DNA treatments.

Later that night LeBron James had a terrible dream in which he heard a child singing

> I'm a little Uyghur
> Short and stout
> Got no kidneys
> Took them out
> For my surgery
> You did pay
> But I got
> The shoes
> So thanks anyway!

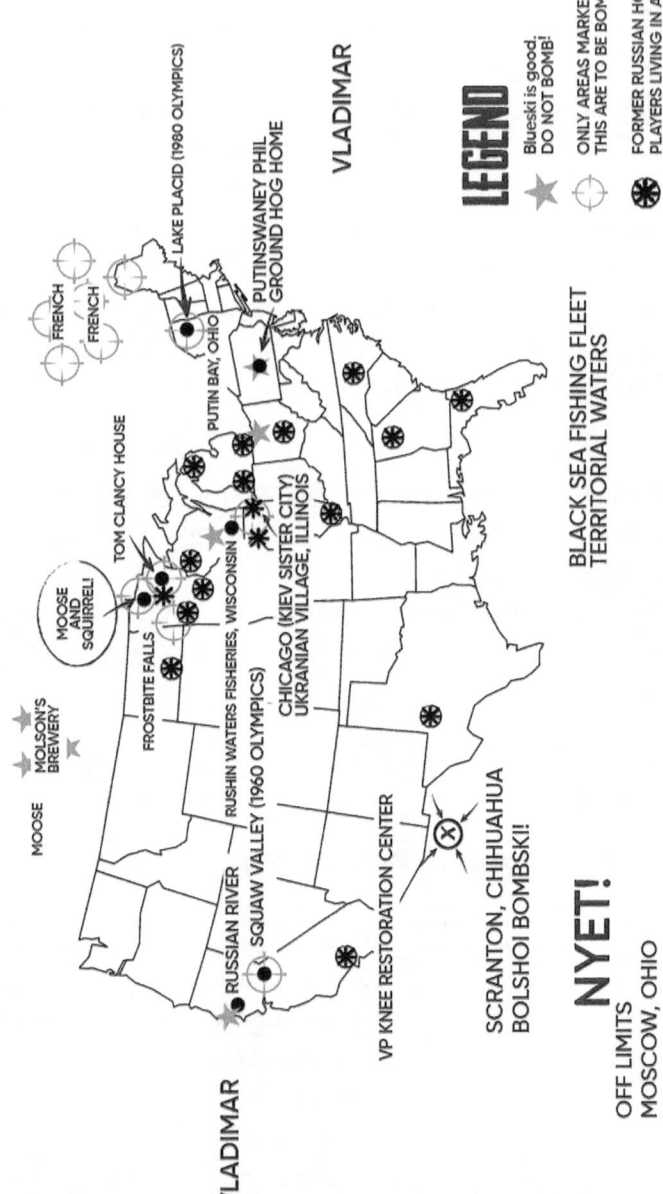

PUTIN TARGETING MAP

MOOSE

MOLSON'S
BREWERY!

FRENCH

FRENCH

LAKE PLACID (1980 OLYMPICS)

TOM CLANCY HOUSE

PUTINSWANEY PHIL
GROUND HOG HOME

PUTIN BAY, OHIO

VLADIMAR

MOOSE
AND
SQUIRREL!

FROSTBITE FALLS

RUSHIN WATERS FISHERIES, WISCONSIN

CHICAGO (KIEV SISTER CITY)
UKRANIAN VILLAGE, ILLINOIS

RUSSIAN RIVER

SQUAW VALLEY (1960 OLYMPICS)

VP KNEE RESTORATION CENTER

BLACK SEA FISHING FLEET
TERRITORIAL WATERS

LEGEND

★ Blueski is good.
DO NOT BOMB!

⊕ ONLY AREAS MARKED WITH
THIS ARE TO BE BOMBED

✸ FORMER RUSSIAN HOCKEY
PLAYERS LIVING IN AMERICA

✸ Destroy. Annihilate.
Maul. Nuke. (DAMN)

VLADIMAR

SCRANTON, CHIHUAHUA
BOLSHOI BOMBSKI!

NYET!

OFF LIMITS
MOSCOW, OHIO

VLADIMAR

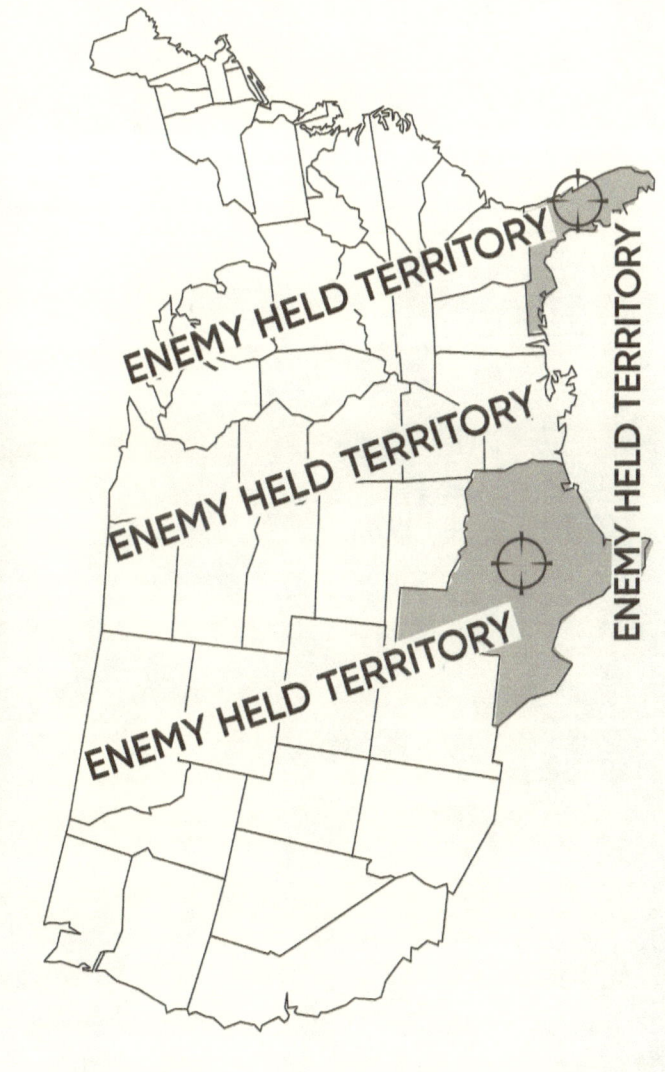

BIDEN TARGETING MAP

ENEMY HELD TERRITORY

ENEMY HELD TERRITORY

ENEMY HELD TERRITORY

ENEMY HELD TERRITORY

HE WHOSE NAME MAY NOT BE SPOKEN

The molten sun continued to spin and spin casting out golden rays upon the sea as a foreboding avalanche of colors descended upon it and immediately dissipated into a kaleidoscopic cloud of near ethereal and divine portend. Then in a voice like thunder, an emanation so profound and so deep that it shook the ocean, called out: "Joooooooeeeee, Joooooooooooeeeeee, Joooooooooooeeeeee." Everything stopped, including the spinning of the sun. The water itself was still and even the San Francisco Treat that had flowed so mightily as KAKATOA was simultaneously erupting—all, all grew silent and ceased to move. Birds in the sky either plummeted to earth or were suspended miraculously in defiance of gravity itself. The hour of salvation was at hand. Redemption for the unredeemed and unredeemable, all things now seemed possible…. with one exception.

Joe continued, unabashed and unaware, driving his auto-mobile in a circle on the beach. Inexplicably, he felt compelled to look directly above him. At first he considered the possibility of another attack of the pidgin. But nooo. Instead coming over him moving out toward the sea was none other than the tented pavilion itself which had last been seen in Switzerland. Against the blue dark sky of near night and moonlessness floated the pavilion itself with an entourage of a thousand Sino/Japanese lanterns, all of which were unbelievably manufactured in Chinatown, Chicago. They shone brightly like a thousand fireflies, illuminating the world around them. Slowly, the pavilion moved westward toward the ocean itself finally hovering like some alien space craft a hundred feet above the water. The canvas exterior shone itself like one of the lanterns with a bright light emanating from within.

Breathlessly everyone looked on including the invaders and the all ships at sea. For through prototypical DARPA enterprises and derivative systems engineered by the Musk himself, a series of electro magnetic surges created a panoply of faux mirrors that shone all over the world everything that was now before the L. A. community including the arrival of She Who Also Would Not Be Named.

From within the tent came a voice, calm, soothing, and yet unlike the robotic voices of a nonhelpful and forever frustratingly stupid response system found on so many website sites. "I am she whom you seek. I am truth, not Pravada, not CNN and certainly not the MSNBC. Here I condemn all truth deniers and liars who will but propagate the propaganda when determined by those from the Land of Panda. I condemn you with the curse of Bubalinski. Listen to him for he alone can take down Big Panda."

Then, the tented glory opened its main entrance and out onto the same mesmerizing stairway with which she had enthralled and captured the visitors in Switzerland, she stepped. Dressed in the virginal blue of Mother Mary, the Dauntless Daughter of Kansas, the Whirl of the Wind herself, remained poised on a platform that none could see but the existence of which none could doubt. She opened her mouth and began, "I am the Gift of God, and such is my name, for Theos is my Lord and Doron I acclaim, this combine, it is my fame, now you've guessed it and I can explain."

In a series of mimes which only the Trappists of Kentucky fully understood she spoke of the theft of a nation's glory and its goodness by those who had rendered it helpless before the onslaught of the Big Tech billionaires and Davos Devils and their pirating way of life, devoted to destruction of the familiar and treasured legacy of by gone generations. Bankers who crushed the peasants and the pheasants as largely being of one and same with shotgun blasts from Federal officials who suck mother's milk from within the land and then spit it out in wasteful retribution. But one above, she explained, was so filled with hate that he seeks to destroy the wealth and goodness of those round the world with his own war chest of power and corruption. He installs governments and then foments discord. "Do your research," she exclaimed and so she finished and moved back into the tent and as yet an all too familiar but surprising figure descended and began to rap. It was He the Rapture whose lyrics were fun and familiar. Friends called him L & L.

The CCP won't let me be
Want to jail my mom
While lookin' for me.
I use to be a rebel
Or so some say,
But like Kid Rock,
You know how I pray,
Fuck this!
But I'm back
In my own special way,
You know the Chinks they did me,
And took it all away,
Unless I fart in junk Chinese.
So now they ban me,
But I do as I please,
'Cause I'm still a Yank
With my own sexy doodle,
And got me a brain –
My own special noodle,
'Cause I ain't no scarecrow
Who just can't think—
No crack in this armor,
No need for a Chink
'Cause I ain't Joe
Who just can't think.
So hey there, Joe –
You're such a Eunuch,
Gonads Gong,
Fried and Bland.
And your cock there too,

Pulled real hard
And they ripped it off.
So now your rod
Is downright scary
And your legs real gross
And the kids call hairy.
So you're old
With a goose down token
And bless me be
You keep on strokin'
But you still can't pee.
So cocka doodle will do
What a man wills to
Only if he stays true.
But what you got,
You know no no no no:
No cock, no rock, no bull,
Nothing left,
No man, no maucho, just bereft.
And I ain't sorry
For what I said,
'Cause I looked at you
And know already
That you're dead.
So you're a dolt
Just like a bolt,
Who got Shanghaied
And now you've lied
And now you're caught,
So good-bye Joe
And the Great Cali,
Dumb as a Rock
Long dead in the sea.

SPEAKING IN TONGUES

In a moment of supreme clairvoyance, the Kamaleon, the Black Queen of Cali, and The Queen of Omission, looked about the beach and then skyward assessing the threat of a sort of rapture now enveloping the crowd. The pronounements became increasingly ominous as He the Rapture began again by singing or rapping to another tune which she strongly suspected that The Big Guy had written specifically for her as she attempted to once again corral a herd of departing White House staff members, all of whom had worked for her. Reluctantly, but bravely, she placed the never before used device snugly around her neck. It was the result of a joint technology venture between the CCP and Google Engineering and Oracle Department of Truth and Talksicity. While she gamely played with the pretty and multifaceted buttons on her collar device with accompanying mother of whirls which wrapped in layers around her neck, He the Rapture began –

> White House Girls
> Go Round the Outside
> Round the Outside...
> Lickin all the Pricks
> In the Gardened Roses.
> They Say You Stink
> And Hold Their Noses.
> So You Track Them
> And Attack Them,
> Bring Them Down
> On the White House Ground,

And So With You,
Go Round and Round'
Call Your B. S.
Oh, Cacklin' Clown.
Oh, Queenie Queen
Of the Kneepad Pad
No one Thought
You Could Be This Bad!

Ever the VP of Voiceless Parody, she pressed the pretty red and green buttons, as she glared at He the Rapture who stood confidently before her, smiling, and mouthing the words, "Blow me, while you're below me..."

Stultified by his offer, she froze and wanting desperately to know if he was serious, she attempted to speak. But the device had already been activated. The voice modulation software usurped her own will and epiglottal stop with a throaty discharge followed by a screaming convulsion. And then it happened in accord with the plan as she sang in a catty and semi-sweet voice wth a distinctly Asian accent, "I yam Chinese if you please, I yam Chinese if you don't please. We making all your toolies, your undies and your woolies. I got my hand upon your scrotums, We squeeze 'em when we mote 'em, But we keep 'em all so quiet, While we all the while Deep Throat 'em. Your world no more is moated, and your politicians all so bloated, your surrender duly noted. Yes, we wash your dirty laundry whether French or even Smellwell, And all your secrets we can now tell. Oh well, Zhi Zhi says just Ping me, with the terms of your surrender, And with our grab around her throat, when used, she's in the moat, and we shall truly see if witches really float. Oh we are your Commie buddies even if you don't please. For like us or despise us – we are the Comm-trollers of your nation. As we like to say in Beijing, this is Win Win for us and Lose Lose for you. Or in Chinese, Feng Feng. Think of her, as you Americans like to say, Commie-kazi. We got you."

Staff members of The Joe ran to her in an attempt to take the device off from a frozen body now standing but unable to move. Her frenzied eyes wandered aimlessly as they tried to unlock the device that periodically sent electrical impulses into her brain, refrying it with every charge of potency. When they finally wrestled the device from her, she collapsed on the ground. Groggily, she sat up and said, "Wo ai ni."

A staff member, one of the few, named curiously Feng Feng Feng, replied, "Good girl, we love you too."

The Queen of Cali then collapsed and was carried from the beach unconscious. As Feng Feng Feng and others like Ho Ho moved in land with her in tow (heavier than she looks), they stumbled over the body of the six year old representative of the Cartel Children's Outreach representative, little Freddie Fentanyl. Later it was determined he had died from an overdose of the jabber juice from Big Pharma. So sad, now the Cartel Outreach would have to once again recruit and train another aspirant.

"THE LAST LAUNDERER"

JOE'S ESCAPE

Joe remained stalwart in his morbidly mobile defensive position, slandering the crowd that encumbered his quick and easy circular movements on the beach with a kind of Commander in Chief creativity that left Secret Service and numerous civilians unintentionally but deservedly run over by Joe's spontaneously altered course of action. He road on triumphantly, occasionally firing his Glock at the still hovering well lit pavilion of heavenly being and its entourage of a thousand points of lit Sino/Chinese lanterns. Inexplicitly but no doubt a sign of his having now upon his shoulders the Mantle of Heaven as the next American President to be ruled by ZZ Top and other Chinese Officials, Joe again looked heavenward where the consoling image of George the Elder smiled upon him, saying "Joe, you are a well dressed man but now I urge you to stop in the name of love." The image provoked a response from Joe who considered its symbolic significance. He turned around in his auto-mobile looking for some hint of its meaning. But then he saw what he could only be a sure fired solution to his metaphysical inquiry.

There next to a mound of bodies, sat a beautiful damsel in distress. He meandered over to her in a circuitous manner until he idled his manficent destrier, his mechanized war horse. She stood smiling, and opened her arms to embrace him. Thinking he was merely teasing, she wiggled her hips playfully, and then thurst out blood soaked scarf from her exertions as doctor, attempting to hand it to him as if he were a champion who would tie it around his long and powerful lance, when he squinted and said, "Who are you? By the way, I got my own hanky."

Without another word, he drove off back towards Colorado Boulevard, searching for a wayward something or other who was still missing, but from what, he did not know. Behind him, he heard the plaintive cry of "But, Joe, I'm a doc—Joe, I'm your wife."

But at the beach the floating pavilion still mesmerized the crowd on land and at sea. She who was both the Gift of God and Flatland Jayhawk, held up her hands to either side in the maternal pose of Mary and said, "That for which so many have so long prayed know that your prayers are answered as I myself shall whirl the wind as never before to pronounce judgement and punish the wicked in ways unseen." With that, the lanterns all about began to toss and turn as if in a vortex while the pavilion remained perfectly stationary. Amidst the swirling winds a parachute appeared on which one word appeared, "Teamster." The blue and white parachute descend slowly from just several hundred feet above the pavilion itself. Upon alighting on the death strewn beach, the locals gathered to look at the body it had delicately laid to earth. Again, the voice from the Gift of God pronounced, "Behold, it is he whom you have sought." Immediately, someone with a lantern named Diogenes Jones held it up to the body and shouted, "It's Jimmy Hoffa and he ain't lookin' so good. Lost a lot of weight and doesn't seem to be breathing. Still we thank you, Gift of God. Is he still dead?"

"You can't have everything," she replied.

"What's the point?" asked a young and impertinent one who did not know who the man was.

"There is no point other than set up you up for the next air drop, even bigger and better."

"It's OK," offered Diogenes Jones. "I don't think this is what I was looking for."

Immediately two more parachutes appeared on high which also descended oh so slowly. Entangled in the nylon ropes and silks of the parachute itself, a man twisted and cursed vehemently as if he had been sequestered against his will. The man chafed at the shoulders and in the groin area as the parachute jerked up and down from his Houdini-like contortions in an attempt to get away. While yet descending the Disney like chorus of a long ago Mexican/American hero of the tv program known as Zorro burst forth from the lungs of disembodied avatars of song and dance.

Out of the night
When the moon ain't so bright,
Comes the Traitor
Known as Soros!
A breaker of banks,
This traitor gave thanks
To the Nazis with tanks
And ooooo his allegiance.
Ooooooo Oooooooo Soros!
The Fox so cunning and free.
A felon in France,
This man of expanse,
With a fortune to be—
The demise of whole nations—
Like Hun-ga-ry!
He kills for the thrill
The grivna and baht,
The ruble and tokens—
Manilla, Malasia,
Miletus.
The man as a whole
Is hate with a soul
And the Fox and the Felon
Are one.
Ooooooooo Soros
His are the dollars
That cause much the dolors
For you and for me.

On his parachute were written the words, "I hate America." His American residency and citizenship notwithstanding, he cursed in his native Hungarian tongue, shouting, "Natse fattyu," (Nazi bastard) to which the crowd responded in the well-known and similar-sounding American vocalization and accompanying gesture.

A minute later, after Soros fell into the welcoming arms of the Canadians, Thai and Tibetan monks and rabid fun loving Americans, Raja Shwaby likewise descended in a parachute that said, "Down with Democracy." Above him a chorus of the same disembodied entities rapped to the clap of the eagerly awaiting group of pro-life Asian Americans holding up the usual signs and chanting "babies are people." Realizing he might soon be in their midst, Raja Schwaby yelled out, I'm against all Youth in Asia, you bastards." "So are we, when did you change?" they yelled back. "Congratulations!" When he landed, they hugged and kissed him until he said, "No I mean I'm actually in favor of Youth in Asia." Confused and wondering at what had just happened, they consulted with one another until the leader came forth and said, "I'm going to spell a word that is Greek and you tell me if you are for or against it." Ok, fair enough."

"Do you agree that we all look Asian to you?"

"Of course, and I bet people like me, Mr. Almond Eyes, I look funny to you, right?"

"No really."

"But do you have anything against us, we the Asian guys and gals?"

"Of course not, I'm a fair minded man. Some of my best friends who are really good at accounting and playing the violin and math stuff in general, they have slanty eyes but that's no reason to dislike them."

"Good, so you like Asians despite what we look like, right?"

"No question."

There's a Greek word that starts with the letters "E and U. In Greek that means 'good. 'Those two letters are often followed by the letters THANOS, which is Greek for death. When you put them together they spell a word that sounds just like 'Youth in Asia. 'But the new word is spelled E-U-T-H-A-N-A-S-I-A. Do you like that kind…"

"Oh, yeah, yeah, yeah, just think of all those girl babies in China. I mean who even wants them."

Before he could take another breath, they were on him with their fists of fury and, mysteriously, Jackie Chan showed up from seemingly out of nowhere, and sang, "Boom laka laka" in order to drown out the sound of Raja Schwaby's cries before collapsing into unconsciousness. Will Smith happened to be walking by when he threw in a couple of slaps for good measure. Chris Rock cried out for mercy somewhere under a stack of bodies, yelling "I'm Korean, don't hit me." Will Smith for no apparent reason walked over and looked down at him, and said, "I hate Kimchi" and kicked him repeatedly.

Affirming the feelings that so many had about the Raja, the disembodied sang,

> The womb is the tomb
> That he would kill.
> Women are passe,
> Just give 'em a pill.
> Schwaby, Schwaby, He's the man,
> He'll manumake you
> If anyone can.
> Dust so smart
> It can rule your heart
> Micro chips –
> That they control
> Steal your brain
> Steal your soul.

Off at a distance one man, J. M., previously of Brazil and one time friend of Joe applauded slowly with a smile and whispered, "We'll be back, Raja." He placed his mushroomed glasses over his eyes and saw the brilliant colors of a world already at war.

Meanwhile, The Joe continued motoring down Colorado Boulevard with his exhausted Secret Service escorts who could neither stay breast of him in his self-made V7 engine driven carstrocity nor anticipate where he would go next. Moreover with his playfully and enigmatically welding his

Glock like a cowboy, he was more interested in taking aim at stop signs and seagulls than in driving. The inevitable occurred.

Half way down Colorado Boulevard with the sun at his back and the new one rising before in the east, Joe saw his own shadow while yet blinded by the light. Boom! At first he thought it might just be another explosion from Kakatoa. "Ah shit!" exclaimed one of the Secret Service members. However, The Joe, if nothing else, is a man of extreme self-confidence. He backed up the carstrosity and inadvertently but no less fervently ran over yet another individual. Along the side walk, a middle aged woman yelled out, "You just killed a pedestration."

Joe responded, "She or he or it, shit, deserved it. Anyone who has sex with kids deserves what she gets." Joe then drove on around the mound of mayhem in front that he had just created knocking down one of his Secret Service guys in the process. Moments later an hysterical woman wearing a white dress with the words "Eat the rich" ran out into the street." You fuckin' pedophile! You killed my friend!"

Joe cranked up the engine again and once more tried to skirt his accuser when she stepped the wrong way, falling flat in front of the Joe mobile, where Joe left tread marks over her new dress with the immensity of his orange and so ever pumpkinish tires. He looked back in triumph, yelling out, "I ain't no pedophile, I wrote every word of that speech myself, didn't borrow a single word from that English guy. He might have had the same dream as me, that's all."

Ilhan the Omar screamed out, "Where is Allah now that we need him? Stop him, he made the noose around the neck of Juicy Smallyaeye. He's an old white guy, something I never realized before. How could I have missed it?" Before Joe could even respond Naked Nancy appeared from behind some bushes, running directly at Ilhan the Omar. She tackled her to the ground and bit viciously into one knee and then the other as the Ilhan yelled, "OK, ok, he ain't legal, he's my brother. Stop, Nancy stop! I'll send him back! I'm sorry! I'm sorry!"

"You miss the point, you and your bimbos have made my life a living hell and I'm not going to let you take the only man I have ever loved – along with his son, and the Good Humor driver, and that Chinese driver and anyway, Joe is mine, you witch!" Moments later, Nancy found herself running back toward the beach being chased by a bunch of rabid Keebler looking elves who sang out, "Do you want to buy some cookies or just eat us instead?" Bewildered and annoyed Nancy ran that much faster hoping to make it to the plane with the only other of several men she had ever loved would be.

But Joe in triumph put his one hand over his eyes to protect them from the sun before him as he approached the point of convergence on Colorado Boulevard between Orange and Green street. Before him was the simmering silhouette of Rashid the Tweeb, or, as she was known unaffectionately in Congressional circles, Princess Chlamydia. As she stepped forward, the Joe in his carstrosity slowed to avoid yet more damage to his bumpers. At that moment, sensing that she was at least temporarily safe from his Nascarian adventurism, she stepped forward to confront him, leaving the safety of the sidewalk for the street. Failing to realize the animosity possessed by so many of the disgruntled over-the-road truckers whose name she had used in vain, she stepped directly into Mack mayhem. It bumped her sky high onto the awaiting embrace of a cluster of palm tree fronds. She landed safely but was now guano garbed by a flock of creole speaking pidgins who shat where she now sat. The irate Canadian trucker moved on, leaving a Canadian prime minister with outstretched thumb to find his own way home.

Joe, feeling ever fatigued, lay down his Glock and turned off the engine, put his head down on the streering wheel causing the car horn to blair while he slept and dreamed of the world as it was and why he was no longer part of it. After the battery died making any need to fear awakening him from his own actions mute, the one remaining Secret Service guy who could still stand, gently lifted the Joe from his seat within his auto and lay him in a wheel barrel. There his old friend Johnny, still dead but as lively as ever, began to sing his rendition of the Nine Inch Nails song, "Hurt." His friend Dylan began pushing the wheelbarrow forward toward the Family's ancestral home in Scranton, Chihuahua, while he quietly soothed the already sleeping soul with the sweetest of words which the Joe himself is thought to have originally penned..." Now as I was young and easy under the Chinese strawberry tree. About the hacienda and happy as the avocado was green on to the fields of praise…Time held me green and dying Though I sang in my chains like the C in China. Hey, Big Panda, spend a little yuan on me."

Princess Pretentious who had only just come upon the scene, wondered what had happened to her friend Rashid the Tweeb. She glanced to her right and saw the sleeping Joe being walked slowly almost mournfully away as she caught the last words of Johnny's singing, "All my friends go away in the end…" Though she at first wanted to awaken him to scold him for being another reprobate and a microaggressing old white guy, she caught herself and reconsidered. Lest she jeopardize her new film role in an eleven part Netflix series entitled The Eleventh Commandment – a part she garnered only after having won a Yul Bryner look alike contest – she walked away.

But as she did so, she walked through a slew of protest signs which consistently read, "B_ _."

Never before had she considered the full import of Joe's having taken her advice to ban the Latin alphabet. She took, however a bit of consolation when she came upon a sign written in cursive that read, "Be Like Me." Given that such narcisstic expressions would soon be banned as part of the New World Order and Social Accreditation Score, she picked it up, having now her keep sake of olden times when America was good and its people were quaint, provincial, and smiled broadly if at times insincerely at its golden opportunities and welcomed those who shared her burdens and her values. She found herself walking along side two old white guys pushing the wheel barrel, one softly singing, "I would not feel so all alone…everybody must get stoned…" As she and the others passed the fentanyl strewn bodies quick with pickpockets and other scavengers, she sighed and wondered at what might have been.

PART V

THE UNSET SUNS

THE BATTLE OF THE BEACH: Remmick slowly walked the beach in a dignified solemnity. He stood looking out to sea at the Chinese armada and the lone U. S. Coast Guard cutter that still guarded the shore line while abandoned Russian trawlers bounced aimlessly on the waves as they were now crewless and afflicted with the damage brought by the Bofors 57 mm gun of the the USCG Bertholf. The Chinese junks, like their name, already cluttered the ocean bottom with only their masts piercing the surface. As he approached the surf, a number of bodies washed ashore. He looked to his left and saw one Carlos Ignatio Armando approaching from the opposite direction. While he waited for him, he looked skyward. The pavilion was gone, the sky was an inky blue and a thousand Sino/Japanese lanterns lay bobbing in the filthy, watery sludge that rolled down on the waves from the infestation wretchedness of San Francisco, which, at some point, certain figures would have been quite content to give back to the Russians.

"You did well, John. Very well. You kept your poise, your family is safe, and we are grateful. Not all government employees are corrupt, just a significant enough number to make our country dysfunctional and in need of a cleansing."

"But the dog, why did you have…"

"We had to be assured of your compliance, nothing more. By the way, we weren't as –" He stopped midsentence and said, "We aren't as mean as we pretend to be. Your dog was a cairn terrier just like in the Wizard of Oz, right?"

"What are you getting at?"

"Before she mysteriously left this morning without a trace, did you happen to see the Gift of God with her own little friend, Otto?"

Remmick smiled.

"Yep, same breed, seemingly different dog. We just had to have someone take care of him for a while."

Carlos turned and looked toward a black SUV parked nearby and waved his arm. Moments later, a cairn terrier jumped from the car, the moment a black suited man opened the rear door. The dog ran directdly to the now kneeling Remmick and jumped into his arms. The dog licked him with the unabashed enthusasism of a well loved pet. "The kids will be so happy."

For the next hour, amidst the detritus and debris of the battle, Carlos and Remmick and Otto walked by the first responders and others. Onlookers beyond the yellow taped off areas and the news cameras panned the area in wonderment at what had occurred here and looked out to sea where the orange sun still stood motionless on the horizon and time stood still. The vastness of orange and blues and the vibrations of something, something beyond mere oceanic tides, continued to foment discord as well as the sense that humanity was at least temporarily saved from its own ignorance and indifference. At one location, the smouldering remains of a Buddhist monk, a signal instance of self-immolation to protest the caustic and narcisstic envelopment of the blessed of humanity with the lazy and selfish whimperings of the truly spoiled, sat upright and indifferent to pain. His last breath was one of defiance and a backward glance upon the sufferings of the world he sought to alleviate by his own sacrifice. Carlos picked up a small American flag washed ashore with the tide and placed it in the blackened hand of this man now free from pain and forever the gadfly to the conscience to those who must go on living this life between the confluence of their flesh forever mingled with their shadows.

Suddenly amidst the silence and the world poised between night and day, the orange ball of fire simply spun into the reaches of space and utterly disappeared. All was dark but for a new horizon in the East where the sun at last burst forth in its usual thermal glory and the good as well as the bad went on living wondering at the travail they had experienced and how they would reconcile it all with the lives they would go on living.

***

What had been described as merely a supply chain problem prompted by COVID concerns manifested itself as a seemingly endless array of Chinese packed cargo ships. What the Spanish Armada had been to the fate of long ago Brittania so too this invasion was no less threatening, no less severe. Through patience, bribery, stealth, and propaganda China seduced America to jettisone its moral authority, military preparedness and personal and international authority and dreams of untold millions to the brutality of Marxian totalitarianism and the salving effect of deadly consumerism. The whores of capitalism, devoid of a patriotic twinge within their board rooms, had sold out their country in the name of greed, power of money alone and Quislingesque capitulation to the Schwabian and Sorosian liver-like infection that was little more than a diabetically infused sweetness that was leading to blindness, economic amputation and a cyclic decline in the entire body politic.

But for a moment, on a beach in southern California, a few, a happy few, had stood their ground. Oddly enough, perhaps unknown to the many or even the few, the day of Two Suns may well be the defining moment not just of the Americans but of the world. If the yoke of corporatism, globalism, and feudalism stand unopposed, what occurs here today is our Normandy, our Okinawa, our Little Big Horn. Our ignoble surrender to the cowards of corruption, whether in Congress, the board rooms, the military-industrial complex that now includes Big Tech, Big Pharma, and Lobbyism. All of this is the torch that sears and totally burns the Bill of Rights and replaces it with the Bill of Privilege, or the rules and regulations of the American Federales. Today all of these entities are marred, seared, and bleared with the mark of Soros and the left-wing howlers. They embrace willfully the status of victim in order to crush opposing voices and ultimately to meet reason with vitriole and shameless self-pity in the name of race baiting policies and schemes of kickback, payback, and pay to play. These are the truly self-indulgent children of a lesser god. Promoted by false prophets of academia and the woefully ignorant, these gods are the stuff of graceless appetites and the stuff of puerile fantasy.

So on the Occasion of the Two Suns, America stood poised between light and dark, dark and light. Our shadow for the first time ever stands before us, behind us, at our side and even beneath us. The psychoanalyst Karl Jung remarked that people and nations are plagued and blessed by their shadows. If ignored, they haunt us. If embraced they bring forth the light within that brings the blessings of the soul and the shadow to fruition. America today has the opportunity for what may be the last time in history to embrace a distant mirror of evolving humanity that for one brave moment caught the imagination that bequeathed to the world a legacy greater than the sum of their individuals and enshrined a goal that still shines brightly with shadows everywhere for people to find their souls locked within them. Those of us who still care are like errant knights who wish to pass on the fruits of labor to generations yet unborn. Whether or not we shall is determined by the power of our voice, the commitments of our souls, and the blessings we enshrine in our hearts. America the blessed must also be America the brave. What we see today is Pearl Harbor in slow motion and we don't appear in most instances to be fighting back. All too often our leaders in Congress, corporate America, and the White House appear as little more than eunuchs bowing to their imperial masters. The latter are truly impregnable, giving birth to nothing but social credit scores and the surrogacy of Zhi Zhi Ping's nihilistic whoredom.

EPILOGUE

"I'm a good Catholic girl."

-- Nancy Pelosi

SOMEWHERE OVER THE NORTH ATLANTIC: 41 DEGREES 43'32" N, 49 DEGREES 56'49" W.

The Kunlong AG600, the Chinese built aircraft, the largest seaplane in the world, hummed across the watery expanse where ships and planes have been lost forever in the blue-green depths below. In the large and expansive cargo hold, Naked Nancy, the forever romantic, sat next to Hunter Biden and rubbed her hand across his thigh while he slept, trying to arouse him to wakefulness. She leaned over and in the voice of conscience she whipered in his ear, "Do you know that Ted Cruz once made a pass at me. I was tempted because he even has a rocket named after him, you know, the Cruz missile. You don't get that unless you've got the complete package. I was wondering, do you know what a crotch rocket is? Do you think he has one? How about you? Do you have one? Can I see it?" In a moment of desire mixed with genuine curiosity she reached for it, suddenly awakening the sleeping Hunter. As Hunter sat straight up, she exclaimed, "It's true, it's true, you do, you do! Just like your dad! Cruz may have his missile but I'll bet no one has a crotch rocket like you! This thing should be banned by the Geneva Conventions, it's so powerful! How many mega hertz is this? I mean, like a V-8 or is that just for breakfast?"

Hunter now stood up with Nancy still gripping his rocket firmly and refusing to let go." Nancy, let go, you're tearing my pants." Helpless but nonetheless aroused by a wealth of naked pictures he had been examining of pre-pubescent girls before he had drifted off to sleep on his new and improved laptop, he shook his loins violently but unsuccessfully. He determined to end the harassment once and for all. Crashing through the bulkhead door between the cargo hold and the cockpit, he was startled to see the entire area empty of personnel, no navigator, no pilot, no crew and no parachutes. Warily he looked around as Naked Nancy tried to calm him with some gentle, calming strokes. "Easy big fella, calm down, we don't want a misfire. Then I'd have to report you to the Geneva Commision for crimes against femality – like me."

"Nancy, don't you get it? We're all alone!"

"Oooooooo, I couldn't have planned it any better. I wov you, you brutish beast, here we are in Shangrila, traversing the beauty of the ocean in our ever expanding wov. You're even better than ice cream. Go ahead, make me, make me!"

"Make you what? Look at the fuel gage, we're on empty!"

"Then fill me up with your wov, wov me and tell me I am yours and make me!"

"Make you what, damn it!"

"Pretend that you are ICE and I have crossed the border illegally and as punishment you want to have your way with me. You are ICE now make me cream!"

At this point there was a sputtering of the engines and even Nancy recognized that this might really be the end. At that point while still holding on tightly and attempting to ease Hunter's pain with her gentle strokes and kisses, she said, "Before we go, I have a confession, pretend you're a priest, we're both Catholic, so this should work."

A defeated Hunter relented of his anger and walked back to the cargo area with Nancy holding on firmly as she began, "Bless me Hunter, for I have sinned. You see, and I didn't really want to tell you this, but I really like Uyghurs, all six of the crew, all of them. I wanted to show them my affection in hopes that they would take us to the golf course where we could play for a hole in one (here she winked suggestively). So I was wondering if you knew that I had a hole in one six times and I did it twice and then just started again, is that still a baker's dozen or do we call that a Uyghur's dozen?" She paused awaiting his response as the plane's descent accelerated. A minute later the immensity of aluminum and aerodynamic foils splashed on the waves like a flat stone. They braced for the final impact and the glory of the angels...and felt as well the chill wrought by the proverbial sinking feeling. They waved goodbye as Hunter's paintings, some of which were still on board, were lost to posterity, a legacy of untold and pay to paint morality.

A NOTE FROM YOUR WRITER

Hello, my name is J. Galt, Escort. Though you will see my name on this book as the writer, I am really just the stenographer. Joe spoke and I listened, much as I have been doing in his presence my entire life. Normally my existence would not even be acknowledged in a literary exercise such as has unfolded in this work of narrative artistry. I am what you would normally call a ghost writer, an individual who writes without attribution, a sort of invisible hand that helps the subject of a particular autobiography to organize his thoughts in such a way as to make the lives of a great man or woman accessible to the world at large. Such has been my role in working with Joe. He spoke, I wrote.

However, for posterity's sake, I think it is important to realize that I am no ordinary ghost writer. You see, I wasn't always a writer. In my youth, during the summers, when I was but a boy, I frequently sat on the lap of a certain studly and decidedly inquisitive and knowledge teaching young man who use to let me rub his legs up and down until the hairs and other things stood up. At first, I thought it was just a game. Later on, Joe taught me that it could be so much more, if only I were to play along to get along. And did I ever. Whew and what a ride it was. Corn Pop never stood a chance. Joe was and still is my guy, now and forever. From ice cream to I scream, hence my name, really just a stage name. Joe taught me everything I know. How to sniff, how to groom and shall we say, how to zoom. In fact he even gave me my name, it's a great name for the porn industry and I owe it all to Joe. Just ask Hunter. If there is anyone in the audience that needs assistance in any aspect of adult literature, please don't hesitate to ask, Joe has my number. In fact, so does Jill.

Anyway, thanks so much for reading and remember me always as the man with the smile that launched a thousand inquiries. See you soon.

J. Galt, Escort

I warned you! I warned you! I warned you!

Three times I warned you and you wouldn't listen! You wentz ahead and read this whole book, even the parts that are maybe true but mostly false. You are pathetic and I am gonna send the IRS and the FBI after you. Why? 'Cause I am a king and you must submit and just remember like I do, I am the Joe, and I ought to know. Audit! Audit! Audit. Ha Ha!

The author poses in the Biden Pumpkin Patch where the tires for the Joe Mobile were found.

www.ingramcontent.com/pod-product-compliance
Lightning Source LLC
Chambersburg PA
CBHW021614120626
46545CB00001B/221